PRODUCTIVITY MANAGEMENT:
PLANNING,
MEASUREMENT AND EVALUATION,
CONTROL AND IMPROVEMENT

PRODUCTIVITY MANAGEMENT: PLANNING, MEASUREMENT AND EVALUATION, CONTROL AND IMPROVEMENT

D. Scott Sink, Ph.D., P.E.

ASSOCIATE PROFESSOR

Department of Industrial Engineering and Operations Research
Director, Virginia Productivity Center
Virginia Polytechnic Institute and State University

JOHN WILEY & SONS

NEW YORK   CHICHESTER   BRISBANE   TORONTO   SINGAPORE

Library of Congress Cataloging in Publication Data:

Sink, D. Scott.
  Productivity management.

  Includes bibliographies and index.
  1. Industrial productivity.   2. Industrial productivity
—Measurement.   I. Title.

HD56.S5 1985   658.3'14   85-3247
ISBN 0-471-89176-2

Printed in the United States of America

10 9 8 7 6 5 4

To my father:
"If you are honest, sincere, persistent, consistent, patient, and work hard you can accomplish anything you put your mind to."

To my mother:
"Anything worth doing is worth doing to the best of your capabilities."

To my wife, Beatrice, for her love and patience.

Thank you.

PREFACE

This book is an attempt to define productivity in a generic fashion. There can and always will be different views of productivity, but there must be a mechanism for integrating those views into a common definition. The ideas, methodologies, and techniques presented in this book have been developed over the past eight years. Much of the material has been tested during the past ten years in a three-day short course for managers and in an industrial-engineering graduate-level course at Oklahoma State University ("Productivity Measurement and Improvement"). In all, over 2000 managers, educators, and students from various organizations, disciplines, and professions have been exposed to this material. As a result of this process, it has become more pragmatic in certain places, and more academically refined in other areas. The reader should thus find here a useful blend of theory, research results, and practical techiques.

This book is about productivity management: planning, measurement, evaluation, control, and improvement. In a larger sense, though, it is really about management. This book is intended to assist managers in all types of organizations and positions to better understand the concept of productivity and to understand how they can begin to better manage it.

The material in this book is presented in a sequence that has been found to be successful in facilitating effective communication of ideas. Part I focusses on productivity from a variety of viewpoints. In Part II, Productivity Basics, the conceptual foundation for the remainder of the book is developed. Fundamental definitions, a generic conceptual model, and the basic productivity management process as it relates to the larger performance management process are discussed.

Part III presents a comprehensive look at major productivity measurement and evaluation strategies and techniques that have been or are being developed and used today. A pragmatic and increasingly popular methodology incorporating the Nominal Group Technique as a basis for developing measures is presented and described in detail. (The use of the Nominal Group Technique and the Delphi Technique as mechanisms for developing productivity measurement and improvement systems was first developed and tested at Ohio State by Morris, Smith, Sink, and Kirkbride from 1975 to 1977. I have continued the development of those concepts and the basic methodology.)

Another measurement technique that is becoming popular is called the Total Factor Productivity Measurement Model, or, more commonly when used at the company level, the Multi-Factor Productivity Measurement Model. A number of U. S. corporations developed applications of this model during the 60s and 70s. The basic components of this model are presented and discussed; in addition, recent enhancements of the model (undertaken at Oklahoma State University) are also presented.

Other major measurement approaches that have been proposed or developed by others in this field or in related fields are also reviewed. Part III is thus an in-depth look at state-of-the-art productivity measurement strategies and techniques in use today. These technqiues are appropriate for the work-group level up to the firm or organization level. The multi-factor model is more appropriate at higher units of analysis (i.e., the division or firm), whereas the methodology incorporating the Nominal Group Technique is more appropriate at lower units of analysis. Industrial engineers, managers, and business students will hopefully find that the information given in Part III fills a void in their repertoire of measurement techniques.

Part IV (Chapters 9–13) examines basic approaches to productivity improvement. A productivity improvement taxonomy is used to provide a structure. Each discipline and profession has its own set of productivity improvement techniques; therefore, the purpose of Part IV is to develop a conceptual framework within which a student or manager can place his or her own strategies for improving productivity. The techniques presented in Part IV focus upon the human element in sociotechnical systems, since it is the major constraining factor. Our focus is on how to generate a greater willingness among managers and employees to search for new ideas. The techniques or approaches presented are motivation, planning, participative problem solving (productivity action teams), and productivity gainsharing. In these areas there has been significant interest and activity, particularly with respect to labor productivity. (The techniques presented assume either a background in management basics or an outside study of suggested references.)

Part V (Chapters 14 and 15) speaks to issues surrounding the design, development, and implementation of productivity management systems. A "grand strategy" is presented and discussed, and a variety of case studies or examples supporting the logic of the strategy are provided. The intent of Part V is to assist students and managers in the process of "putting it all together." It is unlikely that every reader will have the opportunity to design or participate in the design of a productivity management program. However, it is still valuable to think about the manner in which a specific productivity measurement or improvement technique would fit into a larger strategy for productivity management.

The book has been designed and written to be used in a graduate-level course on productivity management; a professional continuing-education short course focussing on productivity management with an emphasis on measurement; or possibly for a senior-level capstone course in industrial engineering or management.

There is a growing suspicion, in many diverse circles, that profitability has been overemphasized at the expense of productivity. This book is about how to get back to basics. The first *test* of sound management is well-managed productivity. This book should help present students and potential managers pass that test.

ACKNOWLEDGMENTS

A number of persons deserve special acknowledgment for their role in the development of this book. Perhaps most important are the invaluable contributions of Drs. William T. Morris and George L. Smith. Both have had a unique and significant impact on the initiation and continued development of many of the ideas and techniques presented herein.

The guidance, support, and encouragement I received from Drs. Joe H. Mize and Kenneth E. Case during my tenure as a faculty member in the School of Industrial Engineering and Management at Oklahoma State University strongly contributed to my continued research in the area of productivity. Without the quality of their leadership skills, this book would not have been completed. In 1976, Dr. Mize founded the Oklahoma Productivity Center. As its director, he initiated a research, development, and extension effort that has provided many learning experiences for me.

The reviewers of this book deserve individual credit for their insights and constructive comments. They are Dr. George L. Smith, Dr. John Imhoff, Dr. Rudy Yobs, and Dr. Thomas Tuttle.

Graduate students play a very special role in the Oklahoma Productivity Center and have contributed in many ways to this book. I appreciate the efforts of Mr. R. Vanchingathan, Mr. J. Swaim, Mr. R. Brower, Mr. W. Viana, Ms. L. Swim, Mrs. C. DeYong, and Ms. S. DeVries.

And, last, I would like to express my appreciation to the professional staff at John Wiley. Bill Stenquist, Editor; Jane Kenneally, Administrative Assistant; Martha Cooley, Senior Copy Editor; Janice Lemmo, the manuscript copy editor; Karin Kincheloe, Designer; and Philip McCaffrey, Production Supervisor, did an outstanding job facilitating my efforts.

Scott Sink
Stillwater, OK
July 1984

CONTENTS

PRODUCTIVITY MANAGEMENT:
PLANNING,
MEASUREMENT AND EVALUATION,
CONTROL AND IMPROVEMENT

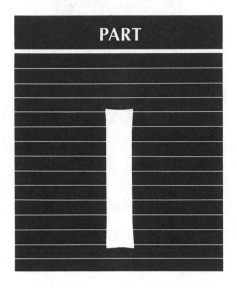

INTRODUCTION

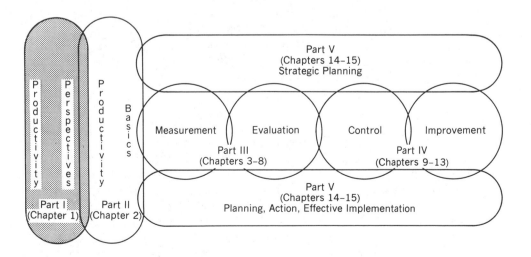

CHAPTER
1

PRODUCTIVITY PERSPECTIVES

HIGHLIGHTS

- What Is Productivity?
- Productivity in Developed Nations
- Productivity in Selected Industries
- Productivity in Professions
- An Overall View of Productivity
- Case Study
- References
- Questions and Applications

OBJECTIVES

- To expose the reader to the concept of productivity.
- To present a variety of perspectives from which one might study productivity.
- To develop an appreciation for the characteristics of the productivity challenge in the United States at the national, industrial, and firm levels.
- To show that productivity management is essential to a firm's survival and success and that successful productivity management takes hard work, effective planning, consistency, persistence, patience, and discipline.

Productivity is a ubiquitous term and concept. One finds it in the titles of numerous books and articles. Many professional conferences are dominated by the theme of improving it. Discipline after discipline is jumping on the bandwagon to promote its perspective on what productivity is and how to increase it. Economists, politicians, industrial engineers, civil engineers, statisticians, quality-control personnel, consultants, labor unions, trade associations, special-purpose centers, manufacturing engineers, dietitians, managers of all types and from all sizes of organizations, and many others have been paying more and more attention to this concept called productivity. The U.S. government even sponsored a White House Conference on Productivity in September 1983 (*Productivity Growth: A Better Life for America*, April 1984) to focus specifically on this critical issue.

What is productivity? Why so much interest over a term that we hardly ever heard about in the fifties and sixties? To answer these questions, one needs to sort through a lot of rhetoric from people with varied backgrounds and tremendously different perspectives on the subject. The purpose of this chapter is to present several prominent perspectives on productivity and some data on why productivity in the United States has suddenly caused this much attention and interest.

What Is Productivity?

Productivity is simply the relationship between the outputs generated from a system and the inputs provided to create those outputs. Inputs in the general form of labor (human resources), capital (physical and financial capital assets), energy, materials, and data are brought into a system. These resources are transformed into outputs (good and services). Productivity is the relationship of the amount produced by a given system during a given period of time, and the quantity of resources consumed to create or produce those outputs over the same period of time. Figure 1.1 depicts this relationship schematically.

Regardless of perspective (political, economic, psychological, engineering, managerial, and so forth), the basic definition for productivity always remains the same. What does change, based on perspective, are the boundaries, size, type, and scope of the system being examined. Most politicians and many economists are interested in a macro-systems perspective. They focus on a nation, a region, a state, or perhaps an industry. This complicates the productivity issue somewhat because there are so many organizations within the boundaries

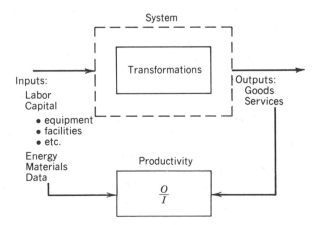

Figure 1.1 General Productivity Concept

of each of these systems, with so many different types of inputs, transformations, and outputs. Nevertheless, much effort is and has been expended to analyze relationships among outputs and inputs for such systems as a nation, a state, and an industry.

The task of measuring productivity is also essentially the same regardless of the system being investigated. The economist or politician interested in measuring the productivity of a nation has to operationalize the same relationship of output to input that the plant manager must to measure the productivity of a plant. Of course, data sources, collection methods and devices, analysis approaches, and so forth, will vary from system to system; however, the same basic relationship is being operationalized.

Although the primary focus of this book is at or below the level of the firm or organization, this chapter presents broader and larger perspectives. Chapter 2 defines more explicitly and intensively basic productivity concepts for organizational, plant, division, function, and work-group-type systems. Chapter 3 operationalizes the concept of productivity measurement and evaluation by presenting techniques appropriate for these systems.

Up until recently, much of what we read in newspapers and magazines about productivity was based on the inflation-indexed gross national product (the constant value, base period = 1967, of goods and services produced by the United States economy). The Bureau of Labor Statistics would collect data from U.S. manufacturers and compile productivity information by dividing the GNP by some measure of labor input (employed persons, labor hours, and so forth). The data were then often presented to the public in the form of a percentage change in productivity over pervious periods (usually quarterly).

Recently, however, other perspectives on productivity have emerged, although they still do not get as much press attention as reports on national productivity changes. These perspectives focus on improving productivity without focusing heavily on measuring it. This book attempts to combine the measurement and improvement aspects. It hopes to promote a better integration between planning, measurement, evaluation, and control and improvement of productivity at the work-group, division, function, plant, and firm levels. Before we tackle this task, though, it may be useful to present some information concerning what has been happening to productivity during the past 15 to 20 years.

Productivity in Developed Nations

There is an abundant supply of detailed information on productivity and related data at the national and industries level. (*Productivity Perspectives*, 1979, 1980, 1981, 1982; *Multiple Input Productivity Indexes*, December 1982; U.S. Bureau of Labor Statistics; numerous productivity-related reports as referenced at end of this chapter). It would be impossible to reproduce all of this material here, but, select data are presented to give the reader a feel for trends in productivity across nations.

The United States has been and continues to be the most productive nation in the world. In almost all sectors of the economy, the level of productivity in the United States exceeds those levels for other developed nations (see Table 1.1). However, the news is not all good. In terms of relative rates of growth, productivity for the United States is poor compared to other developed nations (see Table 1.2). Furthermore, Table 1.3 reveals that for the current decade, the United States is projected to experience competitive but somewhat lower pro-ductivity growth rates than are France, Germany, and Japan. The impact and implications of low productivity growth have been felt by all of us through inflation, declining competitiveness in international or world markets, lowered standards of living, unemployment, and so forth.

The causal relationship among the many variables and factors affecting pro-ductivity is very complex. (Figure 1.2 presents a simplified conceptualization of the impacts of low productivity growth.) However, we can easily state that productivity is the only source of real economic growth and progress. A nation must maintain competitive levels of productivity in key industries to even main-tain its standard of living in what is becoming an increasingly competitive world market. To maintain these competitive levels of performance, industries and organizations in a country must be effective and efficient; maintain high quality; be innovative in product and process; assure quality of work life; maintain competitive productivity levels; and, of course, maintain acceptable levels of profitability. In this book we show how organizations can achieve these goals through improved productivity management.

Table 1.1 Levels and Trends in Real Output Per Employee for Selected Nations by Sector

Levels—1979[a] (U.S. = 100)

COUNTRY	NATL. ECONOMY	AGRI.	MFG.	MINING	CONSTR.	UTILIT.	COMM. & TRANS.	SERVICES
U.S.A.	100	100	100	100	100	100	100	100
Japan	63.7	24.3	93.2	114.0	73.9	94.0	46.8	65.0
Germany[b]	85.0	39.2	85.4	25.2	117.3	134.2	71.8	90.6
France	90.0	53.8	90.0	42.5	89.7	100.6	62.6	88.2
U.K.[b]	58.0	60.2	52.8	58.4	77.0	58.1	55.1	54.9
Belgium	90.8	85.7	96.9	33.1	111.3	172.3	78.8	80.6
Singapore[b]	39.6	36.6	36.9	58.9	59.3	40.0	39.7	38.2
Korea[b]	12.5	6.5	19.3	13.2	16.6	38.9	19.1	14.4

[a]Output expressed in basic data in terms of 1973 exchange rates.
[b]Data for these countries is from 1978.

SOURCE: *Productivity Perspectives* (1982 Edition), American Productivity Center: Houston, Texas. Reprinted with prior permission. (original source: Japan Productivity Center)

Table 1.2 International Comparisons of Growth Rates, 1970–81

Country	MANUFACTURING OUTPUT/HR[a]			MANUFACTURING EXPORTS[b]			UNEMPLOYMENT[c]		
	1970–76	1976–79	1979–81	1970–76	1976–79	1979–81	1970–76	1976–79	1979–81
United States	3.6	1.5	1.0	10.9	6.4	3.1	6.1	6.7	6.8
France	5.2	5.2	1.5	12.1	11.5	−13.5	3.2	4.9	6.9
Germany	5.6	4.7	1.9	14.0	14.9	−5.6	1.6	4.3	4.4
Italy	5.3	3.8	3.9	10.6	9.7	−16.8	3.3	3.7	4.0
Netherlands	7.3	5.0	1.1	12.4	7.0	−12.8	N.A.	4.2	5.4
U.K.	3.4	2.7	3.1	1.7	9.1	−15.8	4.0	5.5	7.6
Japan	7.2	7.9	4.9	12.5	10.2	15.3	1.5	2.1	2.1
Canada	3.8	2.4	−0.7	6.0	4.9	0.2	6.1	7.8	7.5

N.A. = Not Available
[a]Average annual rates of change
[b]Annual growth rates deflated for inflation
[c]Average annual rate

SOURCE: *Productivity Perspectives* (1982 Edition), American Productivity Center: Houston, Texas. Reprinted with prior permission. (original source: Angus Maddison)

Table 1.3 Estimates of Productivity Growth Rates, 1980–90

COUNTRY	1980	1981	1982	1983	1984	1985	1986	1987	1988	1989	1990
U.S.	0.2	1.2	−0.1	1.5	1.7	2.0	2.0	2.0	2.0	2.0	2.0
Canada	−1.5	0	−0.2	−0.2	−0.1	0	0.5	1.0	1.5	2.0	2.0
France	1.4	1.7	3.0	2.5	3.0	3.0	3.0	3.0	3.2	3.0	3.2
Germany	1.6	1.6	2.7	2.8	3.0	3.9	3.2	3.3	3.3	3.3	3.4
Italy	4.1	−0.8	1.2	2.3	2.0	2.0	2.0	2.4	2.3	2.0	2.5
U.K.	1.3	6.1	3.5	2.2	2.2	2.0	2.0	2.0	2.0	2.0	2.0
Belgium	2.4	1.7	2.0	2.0	2.2	2.3	2.4	2.4	2.5	2.5	2.5
Netherlands	0.4	0.6	1.25	1.25	1.8	2.0	2.0	2.5	2.3	2.3	2.5
Japan	3.3	2.1	1.7	2.8	3.2	3.2	3.5	3.3	3.5	4.0	4.0

SOURCE: *Productivity Perspectives* (1982 Edition), American Productivity Center: Houston, Texas.

7

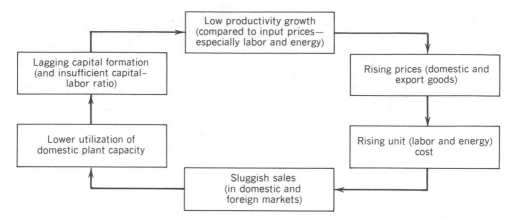

Figure 1.2a   A Model for a Low-Productivity Trap (from *Productivity Perspectives* (1980 Edition) American Productivity Center: Houston, Texas)

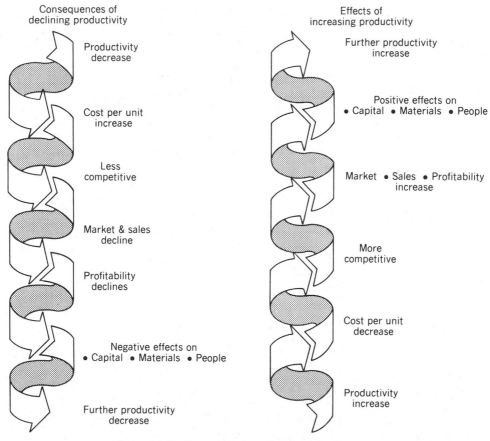

Figure 1.2b   (from American Productivity Center)

Productivity in Selected Industries

As we have seen, the United States has become significantly less competitive in terms of productivity growth rates compared to other nations. The economic success of a nation, however, depends largely on the performance of its major industries and the organizations within those industries. Arguments over the causes of the decline in the productivity growth rate of the United States have taken place for the past 15 or more years among politicians, economists, managers, and professionals from other groups and disciplines. Many economists and politicians favor theories that focus on the role of macroscopic issues, such as tax laws, investment policies, regulatory reform, trade legislation, and industrial policy. They explain low productivity performance in many industries as simply a reflection of poor, inappropriate, or antiquated policies. Clearly, certain aspects of the overall system—political, economic, and otherwise—affect decisions managers make and thereby affect long-term performance. Many policy-related interventions that can positively impact productivity performance can and probably should be made on a national and industry basis. Chief executive officers and other high-level managers can influence economic, regulatory, taxation, and investment policy and should communicate concerns over perceived dysfunctional policies. However, the essence of productivity management is to achieve the highest level of productivity performance possible within the current business environment.

Many managers overemphasize the role of "uncontrollable" factors on poor productivity performance within industries and given companies. Overregulation, Japanese cultural differences, energy costs, lack of industrial policy, and poor labor/government/management communication and cooperation are overused scapegoats for the poor productivity performance of selected American industries. In reality, the United States is experiencing a simultaneous depression and growth period. Certain industries and regions of the country are in the depths of a severe depression. Other industries and areas of the country are experiencing a "boom" economy. The dilemmas of many of our critical industries could be resolved by improved productivity management. Clearly, industrial policy, regulatory policy, and investment policies can and should facilitate productivity growth. But good policies are not a substitute for poor management practices. Good productivity management practices enable management to obtain the best performance within a given operating environment.

Keeping this philosophy in mind, let us examine productivity performance for selected industries.

As you would suspect, productivity levels vary from industry to industry. Figure 1.3 and Table 1.4 examine productivity for selected sectors of the U.S. economy. Note that farming has been a leader in terms of rate of productivity growth. Much of the success this sector has experienced is attributed to technological advancements, larger and better managed farms, and, perhaps most importantly, the national cooperative agricultural extension service, which has facilitated product and process innovation. This service has been so successful

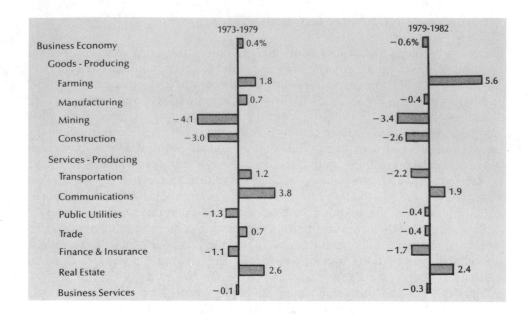

Table 1.4 Share of Output and Hours Attributed to the Sectors of the U.S. Business Economy, 1948 and 1981

Sector	SHARE OF OUTPUT 1948	SHARE OF OUTPUT 1981	SHARE OF HOURS 1948	SHARE OF HOURS 1981	LEVEL OF REAL OUTPUT/HR (IN CONSTANT 1972 $) 1948	LEVEL OF REAL OUTPUT/HR (IN CONSTANT 1972 $) 1981
Business Economy	100%	100%	100%	100%	$3.48	$7.59
Goods Industries:						
Farming	5.9	3.2	18.3	5.7	1.12	5.63
Manufacturing	30.4	30.7	28.9	27.6	3.64	8.45
Mining	2.8	1.9	1.8	1.7	5.32	8.43
Construction	6.6	4.4	5.1	6.6	4.48	5.15
Services Industries:						
Transportation	7.6	4.3	6.5	4.3	4.10	7.61
Communications	1.4	4.8	1.6	1.8	3.11	23.24
Public Utilities	1.5	3.0	1.1	1.2	4.50	18.99
Trade	19.5	21.4	22.9	25.2	2.96	6.44
Finance & Insurance	4.4	5.6	2.73	5.5	5.60	6.44
Real Estate[a]	6.4	5.8	1	1	1	1
Services	13.3	14.8	10.0	20.2	4.63	5.55

[a]The BLS estimate of hours paid for this sector excludes commissioned sales persons; hence human resources input is understated, and "share of hours" and "output per hour" levels would not be accurate if computed.

SOURCE: *Productivity Perspectives* (1982 Edition), American Productivity Center: Houston, Texas.

that numerous efforts have been made to model local government and private sector "technology transfer" programs after the agricultural program.

Within the manufacturing sector, there have been, of course, differential productivity performances for various industry groups. Tables 1.5a and b depict this. Total-factor productivity data reflect the relationship between the value of net outputs (total output minus intermediate goods and services purchased), in this case within a given sector and industry, and the value of inputs (labor and capital). Traditionally, it has been assumed that productivity growth is associated with increased input of capital (capital to labor substitution). However, as we can see in Table 1.5a, in periods of strong cyclical fluctuations and stress, as we experienced in general during the years 1973–81, low capital productivity growth can be expected because of relatively lower rates of capital investment. These lower rates of capital investment result from uncertainty, risk, high interest rates, and overall conservatism or a wait-and-see attitude. Capital investment, unlike labor costs, is not susceptible to easy or quick changes during dynamic periods. Thus, during dynamic and turbulent economic times we can expect most productivity growth to come from the labor side. In a nonlabor-intensive industry or in an industry with poor labor relations, this may put severe pressure on an organization.

Table 1.5a Capital Productivity Trends for Manufacturing Industries

	Average Annual Rates of Change	
	1973-79	1979-82
Manufacturing	−1.7%	−5.4%
Food	−1.9	1.7
Tobacco	−1.3	−6.8
Textiles	1.1	−3.7
Apparel	−0.3	−2.6
Lumber	−3.3	−8.2
Furniture	0.4	−6.2
Paper	−3.1	−6.2
Printing & Publishing	−2.4	−3.8
Chemicals	−1.1	−5.0
Petroleum	−4.1	−9.2
Rubber	−2.3	−3.0
Leather	−1.9	−2.7
Stone, Clay & Glass	−2.3	−7.9
Primary Metals	−4.2	−14.9
Fabricated Metals	−2.3	−6.5
Machinery, exc. Elec.	−1.3	−4.6
Electrical Machinery	0.5	−2.8
Transportation Equipment	−2.1	−12.6
Instruments	0.3	−1.1
Miscellaneous Manufacturing	0.3	−4.9

(a)

Table 1.5*b* Labor Productivity Trends for Manufacturing Industries

	Average Annual Rates of Change	
	1973-79	1979-82
Manufacturing	1.5%	1.6%
Food	0.6	5.6
Tobacco	2.6	−2.4
Textiles	4.5	4.0
Apparel	3.5	3.1
Lumber	1.4	2.2
Furniture	3.9	−0.2
Paper	1.3	0.3
Printing & Publishing	−0.8	−0.3
Chemicals	2.5	−0.8
Petroleum	−1.9	−5.5
Rubber	0.0	2.5
Leather	1.6	4.3
Stone, Clay & Glass	0.9	1.6
Primary Metals	−1.3	−2.5
Fabricated Metals	1.3	1.8
Machinery, exc. Elec.	1.3	3.7
Electrical Machinery	3.8	3.3
Transportation Equipment	1.0	−2.8
Instruments	1.2	4.4
Miscellaneous Manufacturing	2.2	3.4

(*b*)

As selected industries begin to become influenced by declining interest rates often sparked by a recession, capital investments, which have often been delayed by excessive capital costs, begin to mount. Benefits accrued from capital investments lag behind the resource investment, causing capital productivity to fall. Increases in labor productivity usually do not appear until after a recession. On the downside, managers are reluctant to lay off employees; hence, labor productivity falls. On the upside, managers are reluctant to rehire and labor productivity increases. Coupled with potential capital investments on the upswing, productivity growth potential often is good during the last phases of a recession.

So, during turbulent economic times, we see labor productivity low on the downside and high on the upside and at the peak. Capital productivity trends, in contrast, are more difficult to predict because the benefits of capital investments often lag behind the investments themselves. In addition, capital investments often take longer to implement than does a layoff or hiring freeze. A critical element associated with productivity management then becomes balancing and stabilizing these trends. Companies able to capitalize on these trends will outperform competitors during turbulent times.

The manufacturing sector of the Japanese economy has led the world in productivity growth rates for almost 30 years. In a number of key industries, Japan has far surpassed the United States in absolute levels of productivity. One factor contributing to Japan's success is its focus on improving competitive capability, efficiency, and "striking power" in strategically selected target industries.

The Japanese have an industrial policy that in essence operationalizes the popular Boston Consulting Group's theory on product mix and diversification management. In a very simplified form, this theory suggests development of a portfolio of products that "optimize" performance. The theory is based on the matrix depicted in Figure 1.4. The two axes of the matrix are productivity performance and rate of growth. The four quadrants of the matrix are high growth, high productivity; low growth, high productivity; low growth, low productivity; and high growth, high productivity. According to this matrix, then, Japan would focus on those industries in quadrants III and IV. Quadrant IV represents what is known as the sunrise industries and quadrant I represents the sunset industries.

Japan's government and industry cooperate to manage industrial policy strategically. As part of the strategy, growth, productivity, quality, and innovation

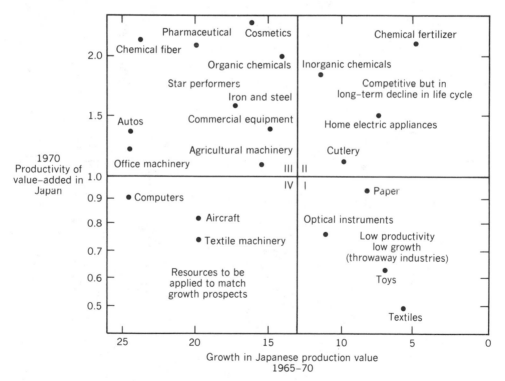

Figure 1.4 Portfolio Analysis for National Industrial Policy Planning (adapted from D. Horvath, and C. McMillan, "Industrial Planning in Japan," *California Management Review*, Fall 1980, pp. 11–21)

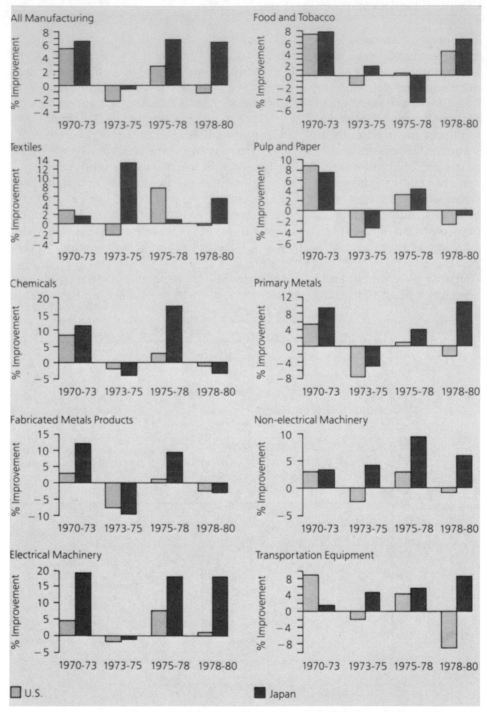

Figure 1.5 Productivity Growth for U.S./Japan by Manufacturing Industry (from *Productivity Perspectives*, 1982 Edition, American Productivity Center; Houston, Texas)

are managed systematically in target industries. In Japan, Pareto's Principle is operationalized, even at the national industrial policy level, whereas in the United States, there appears to be no national industrial policy. At a recent defense industry productivity workshop, managers from major American corporations, academicians, and defense-related government officials identified the lack of American industrial policy as a major strategic issue that would or is adversely impacting defense industry productivity (Sink, 1983).

Figure 1.5 reveals how Japan dominates numerous industries. The message that is beginning to emerge is that Japan seems to be able to dominate just about any industry it focuses its attention on. The Japanese are disciplined, dedicated, systematic, hard working, and apparently can maintain satisfactory balances between productivity, quality, innovation, and quality of work life. American industries and managers need to stop looking for excuses and begin to focus on the basics.

Productivity in Professions

Engineers, psychologists, economists, politicians, sociologists, organizational behaviorists, and managers all have different perspectives on the concept of productivity. These perspectives stem from education, training, and experiences, and surface in discussions, books, articles, interviews, and, of course, behavior. None of these perspectives is wrong. However, many confuse the terms "performance" and "productivity." Most books, articles, interviews, and discussions that claim to study productivity really focus on the broader issue of performance. The problem with this is that performance is a generic term with many attributes or components, whereas productivity should be a fairly specific concept.

If someone wishes to assess the productivity of an individual, group, department, organization, industry, system, and so forth, the assessment can be accomplished with little difficulty as long as outputs and inputs can be measured and placed in relationship to one another. However, if someone wishes to assess the performance of that same individual or group, then one immediately needs to specify what the term "performance" means. Is it effectiveness, efficiency, quality, innovation, and so forth that one is really after?

Perspectives on productivity should not really vary if definitions are accepted. However, perspectives on performance can and probably should vary significantly. The distinctions between productivity and performance are discussed in detail in the next chapter.

An Overall View of Productivity

The concept that productivity is a relationship between outputs from a given system during or over a given period in time, and inputs to that system during that same time period, should be generic and universal. Regardless of whether the system is an individual, a work group, a division, a plant, a firm, an industry, a region, or a nation, the concept is the same. Perspectives on productivity based on discipline or profession come into the picture in at least two ways. First, one's discipline will largely determine the types of systems one is interested in. For example, the psychologist is often interested in the individual or the group; the economist, in the industry or the nation; the engineer, in the product or the plant; and the manager, in either a specific function or an organizational unit. Second, one's discipline tends to narrow views not only on what productivity really means, but, more importantly, on what should be done to improve it. Myopic views on how to improve productivity accompany myopic views on what productivity is and how to measure it. Our impressions and biases on how to improve productivity shape our beliefs about whether productivity should even be measured.

This book attempts to provide a comprehensive, systematic, integrated theory of how to manage productivity by planning for, measuring, evaluating, controlling, and improving it. One of the key ideas is that the sequence of productivity management activities is critical.

Productivity management entails the following components:

1. Laying the foundation for productivity within the organization by creating awareness and a common language for performance and productivity issues

2. Developing strategic plans for productivity programs

3. Developing plans for measurement and evaluation strategies and techniques

4. Developing plans for control and improvement strategies and techniques

5. Ensuring that action planning and effective implementation of all components take place

These steps must be executed effectively, in roughly the sequence outlined, in order for productivity management to be successful.

Many managers today, however, are too undisciplined in their approach to managing in general, and to improving productivity specifically, to carry out this idea. Management is both an art and a science. There is a definite need to more effectively integrate the science into the process in the coming decade. Unfortunately, "We are more likely to act our way into a new way of thinking than to think our way into a new way of acting." Managers have to reverse this trend if they are going to compete and survive in the eighties and nineties.

Management of productivity requires pragmatic, effective, participative planning. Specifically, this planning needs to be done in a two- to 5-year strategic sense and then operationalized effectively in terms of management and employee commitments and behaviors. Measurement is an important part of management. One should never measure just for the sake of measurement, however. Measurement is a means to an end—in this case, improvement. Measurement necessarily precedes evaluation, control, and improvement. Measurement systems, existing as well as potential, need to be diagnosed, dissected, and scrutinized carefully to ensure they are achieving the desired results.

As we noted earlier, productivity is not the same as performance. In fact, productivity is presented as one of seven criteria making up performance. These seven criteria are effectiveness, efficiency, quality, quality of work life, innovation, costs and prices (profitability), and productivity. A manager has to plan for, measure, evaluate, control, and improve upon these criteria to be successful in the long run.

There are countless examples of firms that were in the right place at the right time with a product or service no one else could deliver and which ignored most of the performance criteria except for marketing innovation and profitability. (For an example of research in this area, see MacMillan, Hambrick, and Day, 1982.) However, competition forces management to focus on a broader range of criteria. Competition and a free-market environment have in theory caused improved performance in product and service from the consumer perspective.

In the United States in the fifties and sixties, we operated in essentially a closed, national free-market system; for example, General Motors competed against Ford, Chrysler, and American Motors. There existed in many industries a rather narrow concept of what performance meant because of this system. There was economic growth, and most companies succeeded in spite of themselves. Emphasis during this period was placed on marketing, innovation, and profits largely through a price recovery mechanism. Let's examine a case study to understand more clearly the situation managers found themselves in.

CASE STUDY

The owners of a florist shop in a small but prosperous town in the Midwest started a business in 1950. The business began with a small shop and two greenhouses. Initially, the management of this shop struggled to survive. The postwar economy was shaky and slow to grow. The management concentrated on basics: customer service; good quality; tight margins of prices over costs; good, tight inventory control; and a lot of hard, long work. During the early fifties, the shop's sales volume increased slowly but steadily.

During the late fifties, as the economy started to shift into high gear and productivity growth rates were averaging 3.2 percent per year, the business expanded to a second shop. Its management continued to work hard and to focus on basics. However, they also now began to sense the strength of the economy and, along with other managers from large as well as small companies, began to raise margins of prices over costs.

The inflationary spiral began. If the cost to these florists of a dozen roses went up 50 percent, they simply punched out a new price tag. In fact, if they felt they had a product people really wanted, they might even have raised the price 75 percent and increased their margin. Price recovery of this fashion happened rampantly during the fifties, sixties, and seventies, since markets for most goods and services were rather elastic. Consumers were always able to reach into their pockets and pull out the little bit of extra cash.

The management of this florist business, like managers in many lines of business, became acclimated to this easy and effective little price recovery game. A manager could just about ignore the basics during this period and still survive; in fact, many, if not most, did. They spent less time managing. They started to ignore the customer. They became lethargic except when it came to raising prices. They started "living off the fat" instead of building for the future. Costs were ignored because they usually could be recovered. Managers strayed from the basics because price recovery was an effective management technique.

During the middle to late seventies, the economic environment began to change. As early as 1967, the United States began to experience rates of productivity growth far below the "traditional" (from 1947 to 1967) 3.2 percent rate. Many markets became inelastic as consumers began to feel the pinch created by wage and price controls, concern over rates of inflation, absolute price levels, and so forth. The florist shop continued to grow during the sixties and seventies. Its management expanded to three shops and held a large share of the market for almost those entire two decades. They continued to work hard but began to become concerned that sales volume was up but profits were down.

Across the United States, managers began to realize that price recovery was ceasing to be an effective way to maintain financial performance. They began to sense that "their suppliers were putting it to them faster than they could put it to their customers." The management of the florist shop heard about the term "productivity" and became interested in the prospect that it might somehow contain a solution to their dilemma. In this case, the management went off to a three-day course entitled "The Essentials of Productivity Management: Planning, Measurement, Evaluation, Control, and Improvement." They hoped that something learned at the short course would help them return to the good old days, improve the motivation of their employees, get government off their backs, find simple ways to increase efficiency, and so forth. (Notice that none of the expectations points the finger at management. Very few managers come to this type of seminar expecting to find that they are the problem or that going back to basics is the solution.)

The management of the florist shop heard about the seven criteria of performance and began to sense that things were more complex than they had hoped. They had no idea how to manage effectiveness, quality, productivity, quality of work life, or innovation. They were comfortable with efficiency and profitability, but the others were hard to operationalize. They listened intently to discussions regarding the difference between price recovery and productivity improvement, sensing that this difference was at the heart of their problems. They impatiently sat through day two of the short course as the instructor presented alternative approaches, models, methodologies, and techniques for measuring productivity. But the message fell on deaf ears. They wanted to improve productivity, not measure it.

On the third day of the short course, the instructor presented several specific approaches to improving productivity. The instructor stressed the importance of developing a two-to-five-year strategic plan for the productivity management effort: "What we need to learn from the Japanese is not what to do, but to do it." He pointed out that productivity management requires going back to basics, working hard, concentrating on improving management effectiveness and leadership, and, most of all, persistence, consistence, and patience.

The florist shop managers in that short course were uneasy about the recommendations. They sensed the truth in the analysis and diagnosis but felt uncomfortable with the remedy. Upon returning to their company, they put the short-course notebook on a shelf with other short-course notebooks. The first week back they talked among themselves about the concepts and techniques presented in the course. They even applied some of the concepts on motivation that were presented in the course. During the second week, crises arose that required their undivided attention. For six months or so they continued to "put out fires," and they practiced price recovery as much as possible to avoid profit margin erosion.

Then one day they received a brochure in the mail for a short course on quality circles. The brochure said, "at last the technique has come along to solve all of management's productivity problems." They hurriedly signed up for the next offering. They rushed off to the short course anticipating a panacea and hoped the instructor would not let them down. He stated that quality circles are foolproof, easy to implement, cost effective, and would improve productivity and everything else. He outlined the process, step by step, gave all the participants the order form for the "cook book" complete with a "dog and pony show" for employees, and offered to assist them in setting up the program.

The managers in the seminar were, for the most part, excited. At last someone had given them a prescription that was simple, easy, and almost guaranteed to succeed. Many went back to their firms and either implemented or pushed for implementation. They designed, developed, and implemented the program in six weeks. A year down the road, disenchantment set in. Quality circles were not a panacea. They required hard work, patience, more planning, better communication between labor and management, and more time than they had thought.

Many American managers have lived through this 30-year scenario just described. For big company or small, the issues are the same. Many American managers are at the same juncture at which these managers from the florist shop found themselves—concerned about profits, concerned about survival, curious about productivity, having searched for and perhaps even experimented with one or more of the "solutions," and having experienced failures with "productivity improvement" panaceas. American managers face many challenges in the eighties and nineties, not the least of which is how to manage productivity. This book, like that short-course notebook, may end up on a shelf collecting dust as managers search for easier solutions. Our hope is that when the search for panaceas ends, management will reach back to the shelf for this and other books like it, blow the dust off, and get down to the basics.

REFERENCES

Horvath, D., and C. McMillan. "Industrial Planning in Japan." *California Management Review*, Vol. xxiii, No. 1, Fall 1980.

MacMillan, I. C., D. C. Hambrick, and D. L. Day. "The Product Portfolio and Profitability—A PIMS-Based Analysis of Industrial-Product Businesses." *Academy of Management Journal*, Vol. 24, No. 4, 1982.

Maddison, A. "Comparative Analysis of the Productivity Situation in the Advanced Capitalist Countries." Bureau of Labor Statistics, Unpublished data, OECD, "Economic Outlook," July 1982, Goldman Sachs Economics.

Multiple Input Productivity Indexes. Second Quarter Update, American Productivity Center. Houston, Texas, Vol. 3, No. 2, December 1982.

"Productivity and the Economy: A Chartbook." U.S. Department of Labor, Bureau of Labor Statistics, October 1981.

Productivity Perspectives. American Productivity Center, Houston, Texas. Annual Chartbook on Productivity, 1979 to present.

Sink, D. S. and R. Engwall (eds.), "Defense Industries Productivity Workshop," Workshop and Report Supported by the Aerospace Industries Association, August 1983. Copies available upon request from AIA, Richard Engwall, and the Virginia Productivity Center.

White House Conference on Productivity Report. *Productivity Growth: A Better Life for America,* A Report to the President of the United States. NTIS, Springfield, Va. 22161, PB 84-159136, April 1984.

QUESTIONS AND APPLICATIONS

1. Interview someone from as many of the following disciplines as possible on the topic of productivity:
  (a) Industrial engineering
  (b) Industrial psychology
  (c) Economics
  (d) Business
  (e) Accounting and/or finance
  (f) Sociology
Focus the interview on what productivity means to them, how it can be measured, how it can be improved, and its importance relative to other aspects of performance. Summarize your findings and share them in class.

2. Interview someone from as many of the following professions as possible on the topic of productivity:
  (a) Quality control
  (b) Manufacturing engineering
  (c) Upper-level management
  (d) First-line supervision
  (e) Line employees
  (f) Maintenance
  (g) Marketing
Focus your interview on the same areas listed in Question 1. Keep the interview as short and concise as possible. Summarize your findings and share them in class.

3. What functional as well as dysfunctional effects do all the varied perspectives on productivity have concerning our ability to research, manage, measure, and improve productivity?

4. Is there a perspective for productivity that all disciplines and professions can use to facilitate progress in this field? If so, develop that perspective, present it, and discuss it. What are its critical components?

5. Select a given industry and analyze, in some detail, trends in performance for that industry. Interfirm comparisons within that industry should be developed where possible. Include productivity analysis as a component of your overall performance evaluation.

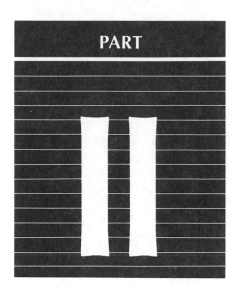

THE BASICS

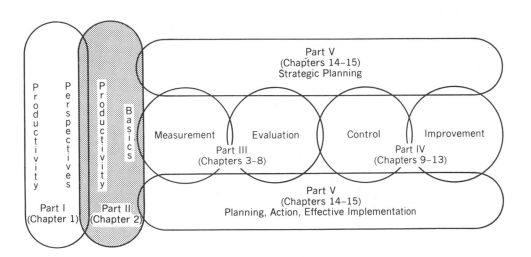

CHAPTER

2

PRODUCTIVITY BASICS

HIGHLIGHTS

- What Is Productivity Management?
    Basic Productivity Management Process
    A Synopsis of Productivity Management
- *Case Study*
- What Is the Relationship between Productivity Management and Organizational
    System Performance Management?
    Classifying Performance Control Systems
    Measures of Organizational Systems Performance
- Productivity in Theory and Practice
- Productivity Process Modeling
- References
- Questions and Applications

OBJECTIVES

- To further refine the definition of productivity presented in Chapter 1.
- To introduce the productivity management process conceptual model, and to use it to structure our thinking about productivity measurement, evaluation, control, and improvement.
- To expose the reader to the seven primary criteria that comprise performance of organizational systems.
- To challenge the reader to become more disciplined about the use and management of the seven performance criteria.
- To develop an improved understanding of the differences between the seven criteria and of the interrelationship (causal relationship) that exists among them.
- To develop an improved conceptual understanding of the dynamic and situationally variable character of the performance equation.
- To provide the performance/productivity process modeling exercise as a technique for operationalizing the concepts presented in this chapter. This exercise

is used by a number of leading organizations, is very practical, and is a necessary first step in the productivity management process.

"You know you're getting smarter when you start to call things by their right names."

Productivity is an extremely abused term and concept. This abuse occurs because there has been no disciplined attempt to develop a sound conceptual framework for productivity. The "half-truth" rhetoric being written about productivity is amazing and, at times, overwhelming to students of this subject and to managers attempting to come to grips with how they can begin to improve it. Productivity has become such a significant buzz word that almost every discipline and profession imaginable has begun to use the term in an attempt to further market and promote its own often myopic "solutions." The need for synthesis, clarification, disciplined definitions, and a conceptual framework is quite evident.

Many terms and concepts are presented to accomplish this sizable task. In addition, a model is used to efficiently communicate the meaning of terms and the relationship between them. The model that is being used here has been developed, tested, and reviewed over a four-year period. Some 400 managers and over 50 graduate students have provided critiques. This model will set the stage for later discussion and will help further your understanding of productivity.

What Is Productivity Management?

Basic Productivity Management Process

Figure 2.1 depicts a model of the "Basic Productivity Management Process." This process includes the following: (1) measuring and evaluating productivity; (2) planning for control and improvement of productivity based on information provided by the measurement and evaluation process; (3) making control and improvement interventions; and (4) measuring and evaluating the impact of these interventions.

We first define the terms used in the model and then discuss the major components of the model in some detail. The following terms appear to reflect consensus:

Input variable any controllable factor or resource that may be acquired in various quantities, types, and/or qualities (for example, energy, people, materials, and data)

Process or transformation a change in the form, outward appearance, condi-

24

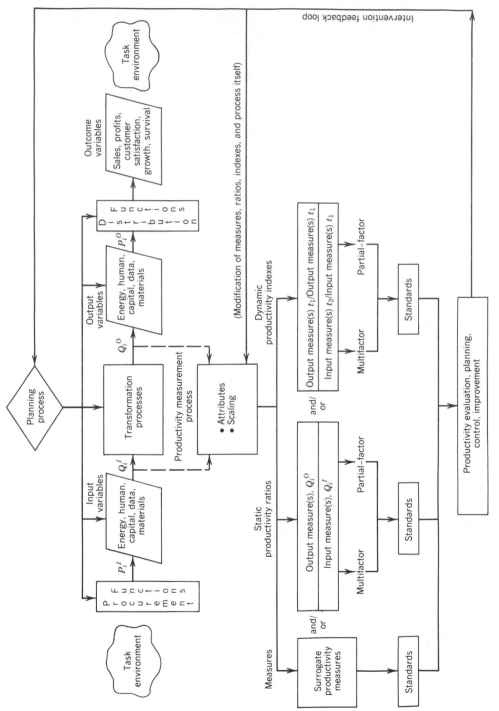

Figure 2.1 The Basic Productivity Management Process

tion, nature, function, personality, character, and so forth of an input variable (for example, manufacturing, training, and processing)

Output variable any controllable factor or resource that results from a transformation of the input variable (for example, energy, people, services, and data/information)

Outcome variable the result(s) of selling and/or delivering an output variable to persons or organizations in the environment of an organization (This element does *not* affect productivity, by definition.)

Attribute a unit of measurement or an identifiable characteristic of a variable (for example, size, color, age, personality, response time, quality, quantity, weight, and kilowatts). A given variable may be described and defined by one or more attributes.

Measure the development and/or selection of a scale with which to assign "signs" to an attribute according to some "rules." By signs, we simply mean numerals, letters, or symbols. By rules, we mean some consistent, logical, and valid matching process between the attribute and some scale.

Productivity measurement the selection of physical, temporal, and/or perceptual measures for both input variables and output variables and the development of a ratio of output measure(s) to input measure(s).

An example that will clarify these terms is the use of time cards (time is an *attribute*) to measure labor input (an *input variable*) for various activities (*process* or *transformation*). These activities, too, can be measured in terms of their quality, timeliness, appropriateness of methods, and so forth (all *attributes*). The activities, if designed correctly, create outputs (*output variable*) that in turn can also be measured in terms of quantity, quality, timeliness, and total costs or price (all *attributes*). In general, input resources, transformations or activities, as well as outputs, should have well defined and measured quantity, quality, timeliness, and cost or price attributes.

Outputs delivered to customers (internal or external) create *outcomes*. That is, customers pay for, use, and react to the outputs (goods and/or services) they receive, and we can and should monitor outcomes as measures of system performance.

In general, there are two basic categories of pure productivity measures. The first are called *static productivity ratios*. These are simply measures of output divided by measures of input for a given period of time. The second are called *dynamic productivity indexes*. These are essentially a given static productivity ratio at one point in time divided by the same ratio at some previous period in time. We end up with a dimensionless index that reflects the change in productivity from one period to the next.

There are also three *types* of productivity measures within each category; partial-factor, multifactor, and total-factor. Each of the three types represents a ratio of output to input; they differ, however, in terms of how much of the inputs is captured in the denominator of the equation. If only one class (labor, capital, energy, data, materials) of input is captured, we call this a *partial-factor measure*. If more than one class is captured, we call this a *multifactor measure*. And if all classes of inputs, are captured, we call this a *total-factor measure*.

Usually, we assume that most or all of the outputs are captured in the numerators of these measures. Certain authors differentiate between what is called total productivity and total-factor productivity (Sumanth, 1984); in this text, however, that distinction is not made. Our use of the term "total-factor productivity" will coincide with the use of the term "total productivity." The fundamental distinction between the two terms has to do with what output and input components are included in the numerator and denominator, respectively. (For further clarification on this distinction, refer to Sumanth, 1984.)

Two final terms are used in the model. These have already emerged in our discussion, and are briefly defined below.

Productivity management a process that entails strategic and action planning and a critical focus on ongoing and effective implementation. Figure 2.1 depicts the general flow of the productivity management process itself.

Productivity improvement the result of managing and intervening in key transformations or work processes. Productivity improvement will occur if any of the following conditions are made to exist:

1. Output increases; input decreases. $\dfrac{O\uparrow}{I\downarrow}$

2. Output increases; input remains constant. $\dfrac{O\uparrow}{\text{In.c.}}$

3. Output increases; input increases, but at a lower rate. $\dfrac{O\uparrow}{I\uparrow}$

4. Output remains constant; input decreases. $\dfrac{O\text{n.c.}}{I\downarrow}$

5. Output decreases; input decreases, but at a more rapid rate. $\dfrac{O\downarrow}{I\downarrow}$

In the discussion that follows, components of the model will be explained one by one until the entire model has been described. The first component is depicted in Figure 2.2. Theoretical development and support for viewing organizational systems as general systems and, more precisely, as open systems and cybernetic feedback models can be found in several sources (Bobbitt et al., 1974; von Bertalanffy, 1968; Thompson, 1967; Hall and Fagen, 1956; Bennis et al., 1976; Cyert and March, 1963; Katz and Kahn, 1966; Porter et al., 1975).

ORGANIZATIONAL SYSTEM RESOURCE FLOW COMPONENT

Figure 2.2 presents an organizational system as a process. An organizational system is defined by the elements of a system and the relationship among these

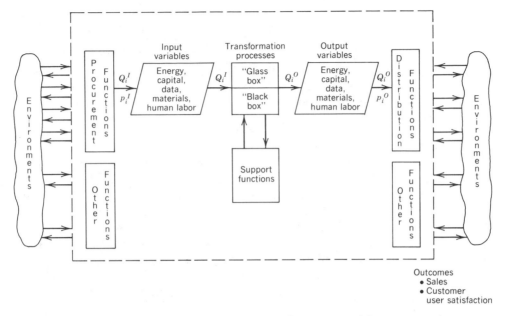

Figure 2.2   Organizational System as a General Systems Model: Resources Flow Component of Basic Productivity Management Process

elements. Work groups, functions, divisions, plants, cities, departments, organizations, firms, corporations, and stores are all examples of different organizational systems. It is essential, when discussing productivity, to clearly define the organizational system boundaries. The boundaries are defined by the places in which elements outside the system exchange energy, information, and resources with elements inside the system. When the boundaries are defined, the *unit of analysis*, or scope of the measurement system, is defined. That is, the appropriate inputs, outputs, and outcomes are clearly understood. Without this delineation, the measurement activity will be confusing, frustrating, more difficult, and less effective than it should be. Most difficulties with productivity measurement originate with a poor definition of the unit of analysis.

An organizational system procures inputs (Q_i^I—I = quantities of inputs; i = quantities of type) at price (p_i^I—I = price or cost of inputs; i = price of each type). Basic categories of input types are energy, capital, data, materials, and human labor. There are obvious subcategories for each input (for example, management labor, oil, electricity, steel, paper, and market demand forecasts). Procurement functions vary from organization to organization, but examples are purchasing, vendor relations, receiving, incoming quality control, and personnel.

Once the input variables are procured for use by the system, the costs (p_i^I) of those resources often become either obscured or temporarily irrelevent until budgeted expenditures are compared to actual costs. That is, once resources are procured, personnel responsible for transforming these resources into a good or a service tend to focus on quantities of the resources consumed and not on costs. This is reflected by the Q_i^I (quantities of inputs by type, i) entering the transformation process box. Quantities of energy, capital, data, materials, and human labor are transformed into outputs (good and/or services). These outputs are, of course, the same variables or resources that came in; only the form of the resources has changed. They have been transformed into quantities of output (goods and/or services), Q_i^O (i.e., value has been added).

This output is then distributed—sold, delivered, received, and so forth. Prior to distribution of these goods and/or services, distribution functions (marketing, sales, and so forth) assess and determine a unit price for each output, Q_i^O. This unit price, p_i^O, is arrived at using either simply a cost recovery process or a cost recovery plus profit margin process. The customer's perception of value added by the transformation processes essentially determines the p_i^O's. For more detailed discussions of this process, see any basic marketing, micro economics, or corporate finance text (for example, Weston and Brigham, 1981).

This is an overly simplified and generic view of an organizational system. There are, as mentioned, many types of organizational systems. Each system has its own unique operating characteristics, combination of inputs, combination of transformation processes, combination of outputs, environments, and so on. Yet each organizational system has all of the elements depicted in Figure 2.2. In addition, and most importantly from a productivity measurement perspective, each system uses quantities of resources (Q_i^I), pays a price for these resources (p_i^I), distributes quantities of outputs (Q_i^O), and receives a price for those outputs (p_i^O). These common elements are critical to the development of your understanding of productivity.

PRODUCTIVITY MEASUREMENT COMPONENT

Figure 2.3 depicts a second major component of the basic productivity management process. The first component, just discussed, addresses the flow of resources through an organizational system. This second component addresses the productivity measurement process. By definition, productivity is the relationship between quantities of outputs, Q_i^O, and quantities of inputs, Q_i^I. Several relationships can be developed to permit identification and evaluation of relative changes in output quantities to input quantities. The most common relationship developed is the *ratio* Q_i^O/Q_i^I. Other relationships of interest incorporate *measures* of output and input transformations and *indexes* of output changes to input changes over time. These three types of productivity relationships will be described in the next part. It suffices to say at this point that the goal of the productivity measurement process is to develop relationships between measures of output and measures of input that enable practitioners to make decisions and better manage their systems. Productivity measurement and resulting measures should tell managers something they didn't know relative to the performance of the system(s) they manage.

Figure 2.4 illustrates the concept of productivity ratios and productivity indexes. As you can see, productivity ratios are developed by placing measures of output from a given system over measures of inputs for that same system

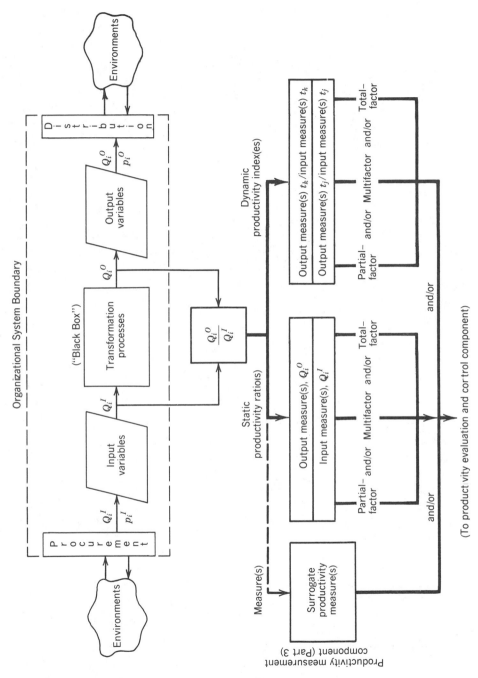

Figure 2.3 Productivity Measurement Component of the Basic Productivity Management Process

29

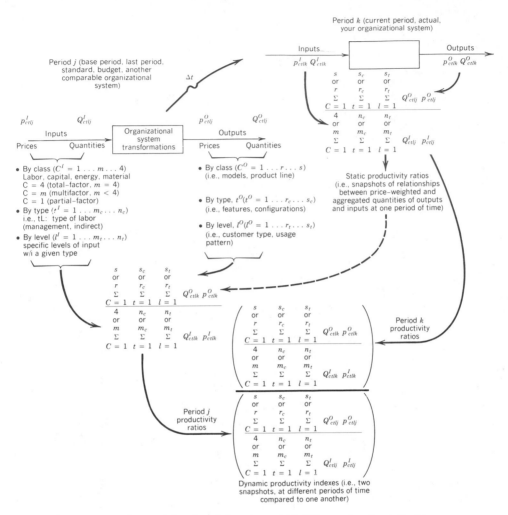

Figure 2.4   Price-Weighted and Aggregated Multifactor Productivity Measurement Model

and for a given period of time. These ratios represent "snapshots" of relationships between what came out of an organizational system and what resources were consumed to produce those outputs. Figure 2.4 depicts productivity ratio development for period k (say, the current period) and period j (say, some previous period). The period represents the length of time we leave the shutter in the camera open. The length of a period can be a quarter, a week, a year, a month, six months, and so forth. Typically, we use unit prices and costs as common denominators for aggregating unlike outputs and inputs, and this notation appears in the figure. (More will be said about this in upcoming chapters.) Note the development of productivity indexes at the bottom of the figure.

In this case, we take productivity ratios for period k and divide by the same productivity ratios for period j to obtain a productivity index.

Productivity measurement seems simple enough at the outset. One simply assigns outputs from a selected organizational system to the numerator and the inputs used to produce those outputs to the denominator. This process, however, is not as easy as it appears, for several reasons. Most organizational systems:

1. Have multiple products and/or services.

2. Are faced with continual price and cost changes.

3. Are redesigning products, services, and processes on an ongoing basis.

4. Must consider other performance measures (such as quality, effectiveness, efficiency, profitability, and quality of work life).

5. Have a variety of categories, types and levels of input resources, each of which has specific costs and other significant characteristics that must be considered.

For example, deciding what to include in the numerator and denominator of the productivity ratio can often be a major difficulty. How much detail or how much to disaggregate the data becomes another factor. For instance, do we measure consumption of all categories of labor together or break it out in more detail? In addition, the selection of the length of the period to be analyzed and what to use as a base period becomes a major consideration. Product mix changes over time can become sources of difficulty in a productivity measurement system, too. In general, any time the type or character of outputs or inputs changes over time, measuring productivity reliably and validly become harder. There are a large number of combinations of outputs to inputs which a manager might examine. The critical question becomes which ratios and indexes will give us the most insight. Another critical and often difficult issue is associated with how to link the productivity measurement system to control and improvement. Finally, in a great number of situations, certain outputs and attributes for outputs will be difficult to quantify. The outputs of service organizations and functions and the quality and timeliness of goods and services are often hard to quantify. These and other sources of difficulties cause productivity measurement to be more complicated than it might first appear.

Basic issues inherent in a measurement process—such as attribute definition, scaling, and combining or aggregating unlike scales—are also inherent in the productivity measurement process. And since costs, p_i^I, and prices, p_i^O, are necessarily excluded from the productivity definition (except for weighting, indexing, or aggregation purposes), only quantities of output and input remain. Productivity measurement is thus simply the determination of relationships between outputs and inputs of a system.

The following list sums up the major difficulties with productivity measurement. These difficulties have occurred for both academic and practical reasons.

1. Determination of the boundaries of the system or the unit of analysis to be measured.

2. Determination of what to put in the numerator (what to call outputs and how the measure them) and what to put in the denominator (what inputs to include and how to measure them).

3. Lack of consensus definitions of productivity, causing confusion as to what productivity really is, how to measure it, and how to improve it.

4. Differences in perspectives among various academic disciplines. These differences, apparent in the literature, cause confusion on the part of practitioners, who have a very focused view of productivity based on the organizational systems they are managing.

5. Technical measurement issues, such as aggregation of unlike outputs and inputs, changing output mixes, price and cost inflation contamination, work-in-process data input problems, and length of measurement period.

6. Operationalization of the productivity concept in the areas of professional, white-collar, management, service-type activities in which measurement of outputs is particularly difficult.

7. Differentiation of productivity measurement systems for control, improvement, planning, and so forth.

8. Integration of productivity measurement with other performance measurement systems.

PRODUCTIVITY EVALUATION AND IMPROVEMENT COMPONENT

The final component of the basic productivity management process is depicted in Figure 2.5. Note that this figure is essentially the same as Figure 2.1. Once a productivity measurement system is developed, the system can be operationalized, and standards can be generated. That is, the ratio(s) and/or index(es) can be implemented. Actual measurements can begin and data can begin to be produced for the system. Standards can be generated using at least the following methods:

1. Estimation

2. Engineered approach

3. Comparison, previous period, or historical (another plant or firm could also be used)

4. Normative (ask persons in the system to generate a standard for given ratios or indexes)

Actual performance is evaluated against standards and appropriate management actions are set in motion. These actions can range from doing nothing to immediate intervention based on interpretation of measurement and the perception that there is a cause-and-effect relationship to be manipulated.

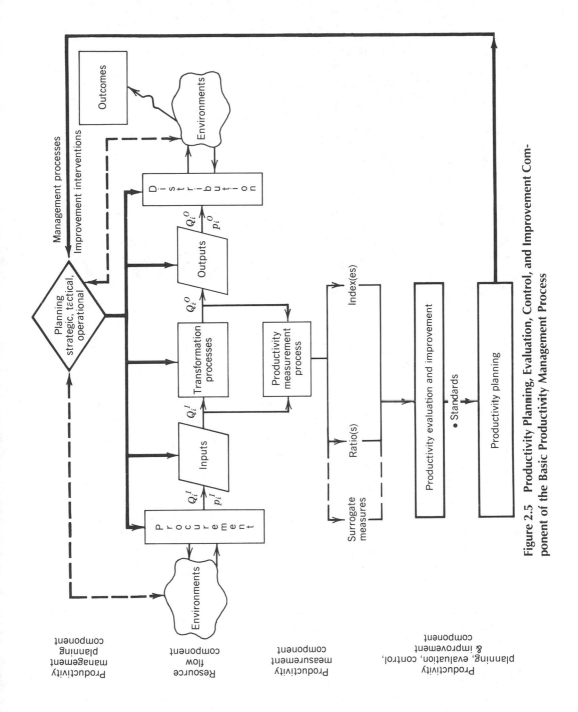

Figure 2.5 Productivity Planning, Evaluation, Control, and Improvement Component of the Basic Productivity Management Process

33

The design of the productivity measurement system itself will be discussed in much more detail in the next chapter. Let it suffice at this point simply to mention that a productivity measurement system needs to be highly individualized. There is no one best productivity measurement system, particularly for smaller, more narrowly defined units of analysis (for example, a division, a particular function, or a work group or a department). At higher units of analysis, such as a plant or a firm, it is possible to design "absolute" or one-system-fits-all productivity measurement systems. However, these systems run the risk of providing great amounts of data and computer reports while not generating much evaluation and action.

It also is important to point out that a productivity measurement system could comprise a combination of various types of productivity measurement relationships. For instance, a system might comprise several partial-factor ratios as well as several multifactor ratios, and might include corresponding indexes for those ratios. Remember that an index is simply a ratio over that same ratio at some previous period in time. The productivity ratio, therefore, has units such as cars/employee, or reports/person hours, while the index has no dimensions or units. An index might be, for example,

$$\frac{\text{Cars July 1982/employees July 1982}}{\text{Cars July 1981/employees July 1981}}$$

Note that the productivity evaluation and improvement and the productivity planning components are the feedback component in the general systems model being developed. (For theoretical development of these concepts, see Bobbitt et al., 1974, Chapter 8.) Feedback in the productivity management process is utilized in at least two fashions. First, as a result of the measurement system evaluation, the design of the measurement system itself might be modified. Perhaps the measurements are not being operationalized correctly (a possible validity or reliability problem), or the composition of measurements being considered is inappropriate. Certain measurements may be deleted or modified; others might be added.

Second, the evaluation of the measurement system may promote productivity planning. That is, results obtained from the evaluation may cause management to plan for and execute productivity improvement interventions. **The essence of the productivity management process is to direct and motivate productivity planning and action through the productivity measurement system.**

A productivity measurement system should indicate to management when there is a need to plan and act, and should provide some clues as to where in the system to intervene. It should not generate redundant information, but information that complements and supplements other organizational performance measurement systems. The productivity measurement system should provide a basis for establishing priorities, or indicate the relative degree of importance of various measurement results. In short, a productivity measurement system should indicate when to act and where to direct efforts. It will not make decisions, and alone is not sufficient to cause productivity improvement to take place.

By utilizing the basic conceptual model for the productivity management process (Figure 2.1), we can delimit and therefore precisely define the concepts of productivity, productivity measurement, productivity improvement, and productivity management. This should clear up confusion over such related terms as profitability, production, performance, efficiency, working harder, and effectiveness. The reader does not necessarily have to agree with these definitions or with this conceptual framework. However, one should feel comfortable enough with the terms and concepts to be able to apply them to a specific organization.*

The conceptual framework presented in this chapter is sufficiently generic to accommodate any type of organizational system. Therefore, a manager or a student of service systems should find these concepts equally as relevant as does a manager or a student of manufacturing systems. As this book progresses and specific techniques are presented, however, this will not necessarily be the case.

A Synopsis of Productivity Management

In this section, it may be worthwhile to restate and summarize the productivity management process. At the end of the chapter, several case studies are provided to help the reader better understand the concepts just presented.

Productivity management, as you by now see, is actually a subset of the larger management process. It involves planning, organizing, leading, controlling, and adapting, based on the relationship between quantities of output from a system and quantities of input from a system. It is neither more nor less important than the other control functions and processes. However, there is a growing concern that it has been overlooked and needs to be better developed. In other words, it is becoming apparent to some practitioners that managers have incorrectly weighted various system performance measures.

The productivity management process necessarily includes productivity measurement and productivity improvement. The productivity measurement process is evolving and is beginning to have specific, unique techniques associated with it. The productivity improvement process is also evolving although in a different way. Peter Drucker recently wrote: "What we need to learn from the Japanese is not what to do, but to do it." This quote appropriately describes the way the productivity improvement process has evolved. Certainly, we are learning how to use computer-aided technology more effectively and efficiently, are experimenting with participatory problem solving, and are reexamining the role that financial reinforcement (productivity gainsharing) can play in productivity improvement. But there really are no significant new breakthroughs. Much of the

---

*It is important to note that the remainder of this book will adhere to the specific definitions and concepts outlined and discussed to this point.

success in Japan and elsewhere appears to result from more effective implementation.

For example, Mr. Robert Lynas, vice president and general manager, Automotive Worldwide for TRW, recently addressed productivity centers from around
the world and presented personal comparisons of the relative effectiveness of
certain critical functions in American industry as compared to Japanese industry
(see Figure 2.6). This comparison highlights the need for managers in the United
States to focus on the basics, to more effectively execute what Americans already
know.

The productivity management function in an organization will necessarily
have to be interdisciplinary and broad in perspective. It will require effective
communication and an understanding of the various functions that contribute
to the creation of goods and/or services; an ability to take a large strategic view
of the system as well as a smaller, tactical, or even operational view; an ability
to justify change and to measure and evaluate the effect of change; and, perhaps
most importantly, an ability to link productivity improvement with productivity
measurement.

Productivity management is vitally interrelated with quality management (the
process that ensures quality), planning (the process that determines what it will
take to be effective), work measurement and budgeting (the process that facilitates evaluating efficiency), accounting and the comptroller (the functions responsible for evaluating profitability), and personnel (the function often responsible for ensuring quality of work life). Who is responsible for productivity
management?

> [i]n turbulent times, the first task of *management* is to make sure of the institution's
> capacity for survival, to make sure of its structural strength and soundness, of its
> capacity to survive a blow, to adapt to sudden change, and to avail itself of new
> opportunities" (Drucker, 1980).

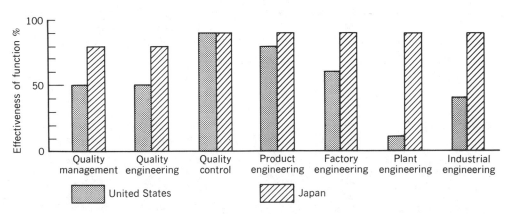

Figure 2.6 Relative Effectiveness of Critical Functions, Japan/United States Comparisons

These are challenges not only for productivity management, but for all managers. Because, in Drucker's words, "productivity is the source of all economic value," the first test of management performance is productivity.

CASE STUDY

STUDY A

1. A work group of five employees in a particular month produced 500 units of output using 880 units of labor (where the units are hours: 5 employees × 22 days/month × 8 hours/day = 880 employee hours/month). Several productivity relationships can be derived from this simple example:

(a) Productivity $= \dfrac{500 \text{ units output}}{880 \text{ units labor}} = 0.57$

(b) Productivity $= \dfrac{500 \text{ units output}}{5 \text{ employees}} = 100$

2. Assume that in the next month, this work group produced 600 units of output using 800 units of labor (5 employees × 20 days/month × 8 hours/day = 800 employee hours/month). Again, the same two productivity relationships can be developed:

(a) Productivity $= \dfrac{600 \text{ units output}}{800 \text{ units labor}} = 0.75$

(b) Productivity $= \dfrac{600 \text{ units output}}{5 \text{ employees}} = 120$

In both relationships, productivity has gone up. However, note that we are looking only at partial-factor static productivity ratios. The partial-factor dynamic productivity indexes for this example are

(a) Productivity $= \dfrac{\dfrac{600}{800}}{\dfrac{500}{880}} = \dfrac{.75}{.57} = 1.32$

(b) Productivity $= \dfrac{\dfrac{600}{5}}{\dfrac{500}{5}} = \dfrac{1.20}{1.00} = 1.20$

It is not until we examine the dynamic indexes that we get a feel for the rate at which productivity has improved from one period to the next.

3. Assume that in addition to labor, this organizational system has utilized materials to create this output. Table 2.1 depicts the data for months 1 and 2. Although we add another category of input (materials), the productivity measurement calculations are still rather straightforward. It is not until we get to multifactor ratio and index development that significant problems occur. How can we attempt to operationalize the following equations?

$$\dfrac{\sum\limits_{i=1}^{q} Q_{i,(korj)}{}^{O}}{\sum\limits_{i=1}^{m} Q_{i,(korj)}{}^{I}}$$

Table 2.1 Work Group Output and Input Data

	MONTH 1		MONTH 2	
	AMOUNT	INDEX[a]	AMOUNT	INDEX[b]
1. Output, units	500	—	600	120.0
2. Input, employees	5	—	5	100.0
3. Input, hours	880	—	800	90.9
4. Input, materials	1000	—	1250	125.00
	Productivity ratios			
5. Output per employee (line 1 ÷ line 2)	100	—	120	120.0
6. Output per hour (line 1 ÷ line 3)	0.57	—	0.75	131.5
7. Output per unit of material input (line 1 ÷ line 4)	0.50	—	0.48	96.0
	Productivity ratio			
Multifactor (line 1 ÷ line 2 or 3 + line 4)	$\dfrac{500}{880 \text{ or } 5 + 1000}$		$\dfrac{600}{800 \text{ or } 5 + 1250}$	

[a]The index for the base period, month 1, in this case, is normally not shown or understood as being 100.
[b]The index for the following period, month 2, in this case, is calculated by dividing amount month 2 by amount month 1 and multiplying by 100.

and

$$\frac{\dfrac{\displaystyle\sum_{i=1}^{q} Q_{i,k}{}^{O}}{\displaystyle\sum_{i=1}^{m} Q_{i,k}{}^{I}}}{\dfrac{\displaystyle\sum_{i=1}^{q} Q_{ij}{}^{O}}{\displaystyle\sum_{i=1}^{m} Q_{ij}{}^{I}}}$$

When q and m involve either outputs or inputs from different categories, a difficulty arises in terms of combining unlike units. To resolve this problem, we need to find a least common denominator that the various inputs and outputs share. Within certain categories, there can exist physical common denominators. For instance, in the case of energy, BTUs is the common unit of measure to which all forms of energy are converted for productivity measurement purposes. However, between certain categories (for example, energy, labor, and output types), no physical common denominator exists. So, it is conventional to utilize constant value financial measures as a common denominator (for example, dollars, pesos, marks, pounds, and bolivares).

4. We now add constant value, indexed value, base period utilized, and financial measures to each input and output. Table 2.2 depicts the data for months 1 and 2.

Table 2.2 Work Group Output and Input Data

	MONTH 1			MONTH 2		
	AMOUNT	PRICE	INDEX	AMOUNT	PRICE	INDEX
1. Output, units	500	1000	—	600	1000	120.0
2. Input, employees	5	—	—	5	—	100.0
3. Input, hours	880	15	—	800	15	90.9
4. Input, materials	1000	150		1250	150	125.0
Productivity ratios:					*Productivity indexes*	
5. Output per employee (line 1 ÷ line 2)	100	—	—	120	—	120.0
6. Output per labor (line 1 ÷ line 3)	0.57	—	—	0.75	—	131.5
7. Output per material input (line 1 ÷ line 4)	0.50	—	—	0.48	—	96.0
Productivity ratio: (Multifactor, price and cost weighted)	$\dfrac{500(1000)}{880(15) + 1000(150)}$ $= 3.06$			$\dfrac{600(1000)}{800(15) + 1250(150)}$ $= 3.01$		3.01 3.06 $= 0.98$

Note that except for the multifactor case, the ratios were calculated without indexed prices and costs as weighting factors. This is because there were no different types of labor and materials broken out. Had there been several types of labor (management, supervision, line, and so forth), then cost weighting would have been necessary to calculate aggregated labor partials. (This is discussed in more detail in the next chapter.)

In interpreting all the ratios and indexes, we find that compared to the first month, the work group in the second month produced 20 percent more output with approximately 9 percent less labor hours (paid) but 25 percent more material. Output per employee went up 20 percent; output per labor hour went up 31.5 percent; and output per unit of material input went down 4 percent. (That is, workers used 4 percent more material per unit of output in month 2 than in month 1. $(500/1000)/(600/1250) = 0.96$.) Without indexed or constant value price and cost weighting, we cannot interpret the multifactor case in item 3. However, in item 4, we see that the multifactor productivity ratio for month 1 is 3.06 and for month 2, 3.01. There is no way to know if these numbers are good, bad, or indifferent without some form of comparison standards. However, if we simply compare month 2 to month 1 in ratio form and create the productivity index, then we can see that, on a relative basis, the work group has been 2 percent less productive in month 2 than it was in month 1.

STUDY B

The unit of analysis in this study is a group of ten professional employees. They are all college graduates performing staff- and project-type work in an organization that produces business systems equipment (copiers, microfilmers, microfilm readers, and reader/printers). The department head utilizes, fairly effectively, management by objectives concepts and effective performance appraisal techniques with subordinates in the group. Group objectives as well as individual objectives are set on a periodic basis, usually once a year. Each individual is assigned and held accountable for individual job objectives and responsibilities as well as certain group-related objectives. Subgroups, committees, or task forces of the department are usually assigned group objectives with a group leader chosen to coordinate activities and progress. The manager of this group is frustrated by the literature on productivity. Generating ratios and indexes of output to input for the work group and specific employees seems to be an academic exercise.

What is the problem? Our nice, neat, disciplined model for productivity measurement seems to be in trouble. For some reason, in this application, which is not unlike any professional or service organization system example, the discussion and model for productivity management in this chapter seems constraining or even not applicable.

One of the dilemmas and major sources of confusion for many managers is how to operationalize productivity concepts for "white-collar," professional, indirect, secretarial, clerical, and managerial employees. Study A seemed relatively straightforward in terms of applying the conceptual framework presented in this chapter. However, Study B, dealing with professional employees, seems not to be so straightforward. First, it is hard to measure all the outputs of a group of engineers. Second, it is difficult to place a value on the outputs to be able to combine unlike outputs for the multifactor case. Third, it somehow seems too narrow to define productivity of these engineers as just output over input. There appears to be more to measuring the performance of professional employees than just measurable output over input. This dilemma and source of confusion is addressed in detail in the remainder of this chapter and in Chapter 3.

In an attempt to shed some light on what is probably the major source of confusion surrounding the term of the concept "productivity," we now examine the relationship between various types of organizational system performance criteria.

What is the Relationship between Productivity Management and Organizational Systems Performance Mangement?

The productivity management process has just been examined. Rigid definitions of terms were provided, and a perspective for the basic components of the productivity management process was developed. Specific components discussed were (1) organizational system resource flow, (2) productivity measurement, and (3) productivity planning, evaluation, control, and improvement.

The following points were stressed:

1. The definition of productivity as a relationship between quantities of outputs and quantities of inputs.

2. The importance of defining the boundaries (or unit of analysis) of the system to be measured.

3. The distinctions between four basic types of productivity measurement relationships.

4. The need for the measurement component to effectively integrate with the evaluation and improvement component.

It is important to place this development in proper perspective. As this book develops, more specific techniques will be presented.

Classifying Performance Control Systems

In the preceding section, we mentioned that productivity measurement is essentially a type of managerial control. Organizations have control systems for behaviors, costs, prices, information, decisions, financial performance, production, inventory, quality, and so forth.

There are many ways to classify control systems. We can classify or categorize them with respect to the resource they are supposed to manage. (Financial control systems, production control systems, and behavioral control systems would be examples of this type.) We can also classify control systems with respect to the type of "organizational system" performance they are attempting to control or manage. ("Organizational system" can again be interpreted as meaning a system comprising a variety of resources. The unit of analysis for the system could be individual, group, function or division, plant, firm, and so forth.) In general, there are at least seven distinct, although not necessarily mutually exclusive, measures of "organizational system" performance. They are

1. Effectiveness

2. Efficiency

3. Quality

4. Profitability (benefit/burden)

5. Productivity

6. Quality of work life

7. Innovation

Every organization in one way or another has systems designed to monitor, evaluate, control, and manage functions utilizing one or more of these seven measures of system performance. Note that productivity is only one measure of performance for a system, and not necessarily the most important one. We might consider these measures of system performance as a multiattribute or multicriterion measurement system. Let's now examine each of these measures in more detail.

Measures of Organizational Systems Performance

Effectiveness is the degree to which the system accomplishes what it set out to accomplish. It is the degree to which the "right" things were completed. At least three criteria need to be used to evaluate degree of effectiveness:

1. *Quality*: Did we do the "right" things according to predetermined specifications?

2. *Quantity*: Did we get all of the "right" things done?

3. *Timeliness*: Did we get the "right" things done on time?

The planning process is closely tied in with effectiveness. We decide what we will accomplish, when it is to be accomplished, and usually what kinds of quality standards to adapt. This is true at the individual level as well as at higher levels or units of analysis in an organization. We may not make these decisions objectively, systematically, or explicitly. However, in one way or another we determine goals, objectives, and activities, and work toward them.

To measure effectiveness, we simply compare what we said or intended to accomplish against what we actually accomplished. (The resources we utilized are not an issue unless stated as an objective.) We can objectively and explicitly or subjectively and implicitly determine the degree of effectiveness.

Effectiveness is therefore an output or accomplishment issue. It is a measure of an organizational system's performance which focuses on the output side of the system. One can develop effectiveness indexes that reveal the level of accomplishment in one period compared to that in another.

Efficiency is the degree to which the system utilized the "right" things. It can be represented by the following equation:

$$\frac{\text{Resources expected to be consumed}}{\text{Resources actually consumed}}$$

From this equation, we can see that efficiency is simply the comparison between resources we expected or intended to consume in accomplishing specific goals, objectives, and activities and resources actually consumed. We utilize budgets, standards, estimates, forecasts, projections, rules of thumb, intuition, and so forth to develop quantitative expressions for the numerator of this equation. We utilize accounting systems, records, estimates, and so forth to develop quantitative expressions for the denominator of the equation. If the denominator is smaller than the numerator, the ratio will be greater than 1.00, and we can, in a discrete sense, evaluate a state of efficiency. If the numerator is smaller than the denominator, the ratio will be less than 1.00, and we can, in a discrete sense, evaluate a state of inefficiency. On a continuous scale, the extent to which the final number is greater than or less than 1.00 determines the magnitude of our efficiency or inefficiency.

Efficiency is therefore a measure of an organizational system's performance that focuses on the input side. One can develop indexes that examine efficiency in one period compared with efficiency in another period.

Quality is the degree to which the system conforms to requirements, specifications, or expectations. Traditional definitions of quality incorporate the conformity to specifications and a timeliness criteron, which could be considered simply as a kind of specification. The key element of quality that distinguishes it from effectiveness is the concept of *quality attributes*. A quality attribute is a specific quality characteristic for which a product is designed, built, and tested. Quality attributes can be subjective or objective. Key quality-related questions are: Was the product built or delivered the way it was intended or required? Is the customer satisfied with the good and/or service? Will the good or service do what it was intended to do? There are several references that provide more detail on this aspect of organizational system performance (Crosby, 1979; Ishikawa, 1976; Juran and Gryna, 1980; Besterfield, 1979; Charbonneau and Webster, 1978).

Profitability is a relationship between total revenues (or in some cases, budget) and total costs (or in some cases, actual expenses):

$$\frac{\text{Total revenues}}{\text{Total costs}}$$

Profitability can be measured in a number of ways. Typical financial measures of performance are called "operating ratios" or "financial ratios." Typical ratios are (Weston and Brigham, 1981)

1. Liquidity ratios
2. Leverage ratios
3. Activity ratios
4. Profitability ratios
5. Growth ratios
6. Valuation ratios

These financial ratios can be utilized to assess the financial health of a firm. Specifically, the profitability ratios are generally

1. Profit margin on sales

$$\frac{\text{Net income (after taxes)}}{\text{Sales}}$$

2. Return on total assets

$$\frac{\text{Net income}}{\text{Total assets}}$$

3. Return on net worth

$$\frac{\text{Net income}}{\text{Net worth}}$$

Financial ratio analysis is a well-developed science with industry averages and well-delineated standards of performance. Similar development needs to occur in the productivity analysis field.

Productivity is a relationship between quantities of outputs from a system and quantities of inputs into that same system:

$$\frac{Q_i^O}{Q_i^I}$$

If we dissect these definitions, we can see that the numerator of the productivity equation contains an aspect of effectiveness.

$$Q_i^O = good \text{ output} = \text{quality quantity}$$

The denominator contains an aspect of efficiency.

$$Q_i^I = \text{resources actually consumed}$$

Quality of work life is the way participants in a system respond to sociotechnical aspects of that system.

We have known for a long time that people's psychological reactions to working in an organization are a factor affecting performance. However, management theorists and researchers generally did not recognize the importance of this factor until the publication of the reports on employee productivity at the Hawthorne works at Western Electric (Roethlisberger and Dickson, 1939). Whether the environment caused this factor to become predominant at that point in history or whether it was always a key factor but overlooked by managers is a question yet to be resolved. However, the way participants respond to sociotechnical aspects of an organizational system is an important measure or aspect of an organization's ability to perform.

Innovation can be defined as applied creativity. It is the process by which we come up with new, better, more functional products and services. Goodyear, the tire company with the blimp, has recently been running commercials that highlight quality and innovation. The National Science Foundation had, up until President Reagan's budget cuts, been researching how to spark and facilitate the innovation process in the United States through the development of Innovation Centers (Colton, 1977). However, in most of the literature on productivity, one does not read much about innovation, although it is a critical factor in the productivity equation. An organization that does not innovate in product, service, and process will likely not be able to compete favorably over the long haul.

Figure 2.7 and the following discussion address the innovation criteria and, more importantly, relationships among performance criteria. The discussion highlights the point that performance of an organizational system is multidimensional and that high performance in one or even six of the seven areas does not ensure success and survival.

At the level of the firm, output is not counted in the equation until it is distributed (sold and delivered). For example, if the product or service in Figure 2.7 is not "competitive" in features, operating characteristics, design, aesthetics, price, demand, and so forth, it will not sell. Note, however, that at the manufacturing system level, the output (square wheels) is counted for that unit of analysis when it is completed for finished goods inventory. Hence, although the person making those square wheels, Mr. Og, may be very productive (at least on a relative basis), Og and Trog Wheels, Inc., is likely not to be productive

"OF COURSE OUR PROBLEM IS PRODUCTIVITY. WHAT ELSE COULD IT BE?"

Figure 2.7   Is it possible for this organization to be effective? Efficient? Productive? Innovative? Profitable? Successful? Is it possible for this organization to produce quality goods? Is it possible there is good quality of work life in this organization? Think about it carefully!

or profitable. Og and Trog Wheels may be effective (the company could very well be accomplishing *all* it planned to in a timely fashion; efficient (it could be utilizing resources efficiently—that is, expected resource consumption/actual resource consumption ≥ 1.00); and producing a quality square wheel (conforming to requirements within an acceptable tolerance). Moreover, the company could be promoting and supporting quality of work life: Og may be very satisfied with his work, his working conditions, the company in general, and so forth. *But* Og and Trog Wheels, we must suspect, does not have a product that many people want. Therefore, on at least three of the performance measures, product innovation, productivity, and profitability, the company's performance would be low, unsatisfactory, even unacceptable.

In a way, one important job of a manager is to determine

1. What the appropriate priorities or relative weights are for each performance measure

2. How to measure, operationally, each performance measure

3. How to link the measurement system to improvement

In other words, managers must determine how to most effectively use the control system to cause appropriate changes or improvements. It is clear that the priorities or weightings for each of these performance criteria will vary according to several factors (size of the system; function of the system—marketing, manufacturing, research and development, etc.; type of system—job shop, assembly line, service, process industry, etc.; and maturity of the system in terms of employees, management, technology, organizational structure and processes, etc.). Several sources concur with these findings (Hitt, Ireland, and Palia, 1982). It is also clear that no one organizational system will equally weight the performance criteria.

Productivity in Theory and Practice

Because productivity is connected with a number of other organizational system performance measures (effectiveness, efficiency, quality, profitability, quality of work life, and innovation), it is possible for many combinations of states to exist in any given organization. For instance, it is possible for an organizational system to be productive, effective, and efficient, but not profitable; profitable but not productive; and effective but not efficient.

Specific answers to the questions of what productivity is and whether or not it is a necessary or important measure of system performance depend on the type of organizational system (public sector, private sector, manufacturing, service, oil, automotive, electronics, large, small, vertically integrated, highly differentiated and specialized, etc.) and on the specific type of organizational system or the specific unit of analysis (work group, department, division, plant, firm, cost center, industry, etc.). The weighting or importance of productivity or even the existence of productivity as a component in an organizational system's performance measurement depends at least on the above-mentioned factors. There is no research to support the propositions shown in Figure 2.8, so the reader is urged to evaluate the inferences and to modify the chart based on experience and wisdom.

Several features of this chart are noteworthy. First, it reflects, in a relative and general sense, relative weightings of various performance measures in various types of organizational systems. Second, it suggests that effectiveness is universally a critical measure of performance, hence a necessary condition for "success." Third, at the individual level, productivity and profitability are relatively less important measures of performance. Fourth, in such specific functions as R & D and service, efficiency, productivity, and profitability are relatively less important than effectiveness, quality, and innovation. Fifth, in manufacturing, the activities are so diverse and current pressures so great that performance measurement requires a highly diversified system. Sixth, in certain sectors, the critical nature of certain performance measures overrides emphasis on other areas. For instance, in the public sector, effective and quality delivery of services sometimes overshadows efficiency issues. Finally, in certain functions, sectors, and levels, the nature and character of the work renders quality of work life a relatively less critical factor. That is, the nature and characteristics of work for professors in academia are generally satisfying and, therefore, quality of work life is not normally a problem.

Managers ought to think through the relative importance of each of the seven performance criteria to their organizational units. They should analyze what is currently being done to operationalize (measure, evaluate, and manage) each of the important criteria. They should also evaluate areas for potential improvement in terms of the control system for the various performance measures. This approach requires a clear understanding of the differences among the seven performance measures as well as the overlap of certain measures. It assumes an ability to design a "differentiated" measurement and control system (that is, an independent measurement and control system for each performance measure) and to then successfully develop these systems into an integrated performance management system. (Chapter 14 discusses in more detail the productivity management process itself and the relationship between this process and the performance management process.)

In theory, this may all seem natural and logical, but in reality we have an altogether different picture. Experience shows that if you ask managers how they currently measure productivity in their organizations, you might get such responses as those shown in Table 2.3. These data represent sample results from a structured, interactive survey taken during a three-day short course on productivity measurement. The 213 respondents were primarily managers from a variety of organizations who were asked to identify ways in which they or their organizations currently measure productivity.

In a recent study, Viana developed a productivity measurement and improve-

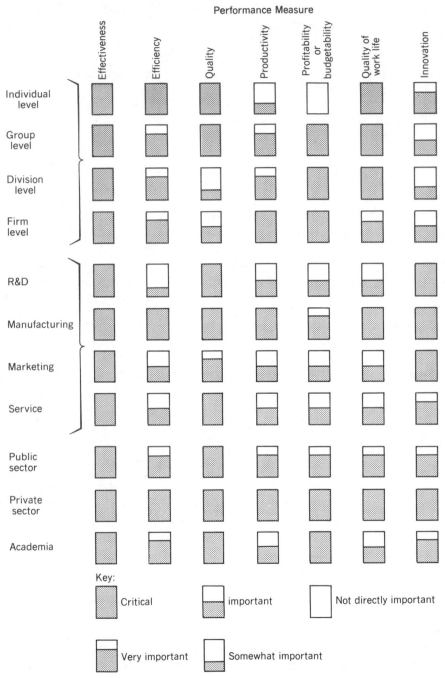

Figure 2.8   Performance Measures Relative Importance by Unit of Analysis and Function (An Example)

Table 2.3 Responses from a Sample of Managers and Organizations to the Question, "How Do You Currently Measure Productivity?"

Return on investment
Defects per 1000 units
Response time
Customer satisfaction
Costs over sales
Profits
Costs per unit
Production per employee
Sales

$$\frac{\text{Direct hours}}{\text{Standard hours}}$$

$$\frac{\text{Product cost}}{\text{Units produced}}$$

$$\frac{\text{Total operations personnel less finance}}{\text{Finance personnel}}$$

$$\frac{\text{Estimated hours on work order}}{\text{Actual hours on work order}}$$

ment program for a company in Brazil (1982). When asked to develop measures of productivity, the managers of this plant agreed on the list of ratios shown in Table 2.4.

The points to be made from these experiences are somewhat obvious. First, very few of the "measures" identified by managers are actually productivity relationships in a strict sense. In fact, based on these data, the following generalizations regarding the categories that these responses fell into can be made (see Table 2.5).

Experience suggests that when asked how they measure productivity or would develop productivity measures, managers will generate very few (typically less than 10 percent) actual productivity ratios or indexes. This suggests that they don't know what productivity is or feel that it can be managed without measuring it as precisely defined.

Second, managers apparently do not understand the basic definition of productivity, or perhaps they feel that they can measure other characteristics about the system and infer a productivity relationship. These other "measures" might be thought of as surrogate productivity measures (measures used in place of productivity but that correlate with it). Third, when managers are asked to measure productivity, they respond as if asked how to measure performance. In other words, managers evidently do not distinguish between productivity and performance.

This points out several significant problems. It has been said that one cannot manage something without being able to understand it, cannot understand

Table 2.4 "Productivity Ratios" Developed by Managers in an Oil Field Product Manufacturer

Revenue/payroll[a]
Products shipped/person-machine[b]
Revenue/capital[a]
Product shipped/energy[b]
Revenue/production cost[a]
Nonconformity cost/production cost[c]
Machine idle time/machine availability[c]
Rejected parts/inspected parts[c]
Absentee hours/person-hours available[c]
Number of accidents/number of employees[c]
20 percent good productivity ratios
30 percent potential productivity ratios that are more directly financial ratios
50 percent nonproductivity ratios, miscellaneous measures of performance

[a]These ratios, if developed correctly, can be partial-factor productivity ratios.
[b]These ratios are partial-factor productivity ratios.
[c]These are miscellaneous indexes and/or ratios. None are productivity ratios or indexes.

Table 2.5 Generic "Measures" of Productivity Typically Identified by Managers

1. MEASURES
    (a) Measure(s) of outcomes
       Customer satisfaction
       Sales
       Revenues
       Share of market
       Net dollar sales
       Cash flow
    (b) Measures of inputs
       Costs
       Quality
       New orders received

    (c) Measures of transformation
       Efficiency
       Response time
       Work standards
       Activities
    (d) Measures of output
       Units produced
       Units sold
       Students trained
       Papers written

2. INDEXES
    (a) Outcomes

$$\frac{\text{Sales 1981}}{\text{Sales 1980}}$$

    (b) Inputs

$$\frac{\text{Labor costs 1981}}{\text{Labor costs 1980}}$$

    (c) Transformations

$$\frac{\text{Conferences attended 1981}}{\text{Conferences attended 1980}}$$

Table 2.5 (*Continued*)

$$\frac{\text{Meetings held 1981}}{\text{Meetings held 1980}}$$

$$\frac{\text{Activities 1981}}{\text{Activities 1980}}$$

(d) Output

$$\frac{\text{Units produced 1981}}{\text{Units produced 1980}}$$

(e) Productivity

$$\frac{\text{Measure(s) of output } t_2/\text{Measure(s) of output } t_1}{\text{Measure(s) of input } t_2/\text{Measure(s) of input } t_1}$$

3. RATIOS
  (a) Performance ratios (outcomes/inputs)
     ROI
     Profitability
     ROE
     Net earnings per share
  (b) Cost ratios (Mali)—$\dfrac{\text{Outcomes}}{\text{Input}}$ or $\dfrac{\text{Input}}{\text{Outcomes}}$

$$\frac{\text{Labor costs}}{\text{Total revenue}}$$

  (c) Objectives ratios (Mali)

$$\frac{\text{Actual output}}{\text{Expected output}}$$

  (d) Work standard ratios (Mali)—essentially transformation system ratios. Depending on unit of analysis, they could be productivity ratios.
  (e) Productivity ratios

$$\frac{\text{Measure(s) of output}}{\text{Measure(s) of input}}$$

     Partial-factor static
     Multifactor static

something without being able to measure it. It seems the questions to be asked are

1. Can one manage productivity without measuring it directly?

2. Can one manage productivity without measuring it at all?

3. Can one measure and infer productivity directly?

4. Do managers really not understand or accept fundamental definitions of productivity, or do they simply not operationalize this concept as strictly defined?

A discussion using Figure 2.9 may help to answer these questions.

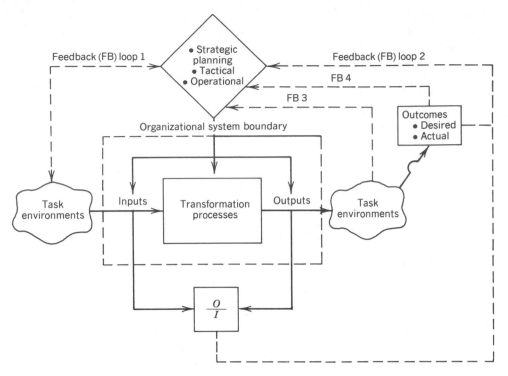

Figure 2.9   What is organizational systems performance? What are: efficiency, effectiveness, quality, profitability, quality of work life, and innovation? How are they measured and managed? What is productivity? Does it need to be measured? How can it be?

As you can see, the figure depicts an organizational system: the components of the system, inputs, transformations, outputs, outcomes, productivity measurement relationships, and several feedback loops. The questions managers need to ask themselves are identified on the figure. The basic question posed here is whether feedback loop number 2, productivity measurement to planning, is *necessary* for managing organizational performance. Note that it is reasonably clear that productivity measurement is not alone *sufficient* for managing performance. In a larger sense, the fundamental question is this: What combination of types of measures is *optimal* for managing organizational system performance?

It is not the purpose of this book to tackle the problem or question of how to measure organizational performance. There are many views and disciplines addressing the issue; and, unfortunately, in the past it has been common to treat the question in parts. Industrial engineers often concentrate only on the efficiency issue, while some of the organizational behaviorists concentrate only on the effectiveness component; certain psychologists may concentrate only on quality of work life, whereas accountants and finance people may emphasize only the profitability issue. It has been and probably always will be the responsibility of upper management to integrate these perspectives into a whole. How this is done is also not the concern of this book. Our focus is on whether productivity measurement is in fact a component in that holistic, integrated system. This book also focuses on how productivity can be strictly, accurately, and correctly operationalized.

In summary:

1. Productivity is simply one of a number of important performance measures or criteria for a given organizational system.

2. For a variety of reasons, productivity is *not* operationalized in a strict fashion in most organizations.

3. Not including productivity in certain organizational system performance measurement systems is a potential problem for management in the coming decades.

4. It appears that many organizational performance management systems are not designed and developed in a rational, systematic fashion, nor well thought-out in a design sense. Managers are often not really sure what they are measuring, or why.

Productivity Process Modeling

One of the first steps that should be taken prior to developing a productivity measurement system or program is that of productivity process modeling. The process is a very straightforward one, although it requires considerable thought and analysis. It primarily entails developing a productivity management process model for a specific organizational system of interest. Before writing this activity off as being overly academic, try it. You will find, as other managers and students have, that the exercise is (1) more challenging than it seems; (2) valuable from the standpoint of causing you to really think about the performance of the organizational system; (3) valuable as a starting point for further development of specific productivity measurement systems; and (4) valuable from the standpoint of causing the concepts presented in this chapter to make sense to you.

The steps in executing a productivity management process model are as follows:

1. Select an organizational system of interest. Clearly define the boundaries of the system.

2. Identify "task environment" organizations on the input and output side (suppliers and customers). Note that for some organizational systems, suppliers and customers are internal to the organization or firm itself.

3. Identify major goals of the organizational system (What is the mission? What are the "superordinate" goals?). This data should be available from the strategic planning process, if it exists.

4. Identify major subcategories of input resources to the organizational system

(subcategories of resources one level more detailed than energy, capital, materials, labor, and data).

5. Identify major transformations that take place within the organizational
system to convert inputs to outputs. Do not get bogged down on this step.
There are literally thousands of transformations that take place in most organizational systems. In a work measurement sense and in a hierarchy of work
sense, you want to identify major objectives of the organizational system. In
the hierarchy of work that follows (adapted from Mundel, 1983), you want to
be specific but not too specific. In this step you should identify transformations
that reflect sixth-order objectives and that occur between the third-order tasks
and fourth-order major transformation steps.

Step 3	Goals	
Step 7	8th order work-unit	Results/ Outcomes
	7th order work-unit	Total/Gross Output from Organizational System
	6th order work-unit	Programs, Objectives
Step 6	5th order work-unit	Output, Good and/or Service
	4th order work-unit	Intermediate Good Product, and/or Service/
Step 5 Major Transformations		Work in Process/Major Transformation Steps
	3rd order work-unit	Task, Job
	2nd order work-unit	Elements, Activities
	1st order work-unit	Motions

6. Identify major outputs of the organizational system. These outputs may be
goods and/or services. Do not worry at this point about the ability to measure.

7. Identify expected or desired outcomes to be realized from the organizational
system that has distributed the outputs.

8. Identify and prioritize performance criteria or measures for the organizational system. Define what performance means for this system. Prioritize the
seven basic criteria. Develop operational measures or subcriteria for each of the
six criteria (exclude productivity for this part of this step).

9. Identify and prioritize output to input ratios that, if developed, measured,

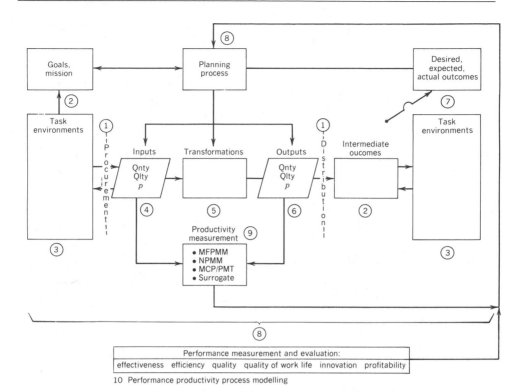

Figure 2.10   Productivity Process Model

monitored, and evaluated, would provide you or a manager with additional, useful insights as to how well the organizational system is performing or has performed.

10. Identify and discuss the evaluation, feedback control, and improvement planning process. How do, should, or will the control systems developed in steps 8 and 9 be utilized as decision support systems for productivity/perform-ance improvement?

A blank productivity process model is presented in Figure 2.10. The steps in the productivity process modeling exercise are identified on the figure.

REFERENCES

Bennis, W. G., et al. *The Planning of Change*, 3rd ed. New York: Holt, Rinehart, and Winston, 1976.

Besterfield, D. H. *Quality Control*. Englewood Cliffs, N.J.: Prentice-Hall, 1979.

Bobbitt, H. R., et al. *Organizational Behavior*. Englewood Cliffs, N.J.: Prentice-Hall, 1974.

Charbonneau, H. C., and G. L. Webster. *Industrial Quality Control*. Englewood Cliffs, N.J.: Prentice-Hall, 1978.

Colton, R. M. *An Analysis of the National Science Foundation's Innovation Centers Experiment.* NTIS, U.S. Department of Commerce, 5285 Port Royal Road, Springfield, Virginia, July 1977.

Crosby, P. B. *Quality Is Free.* New York: Mentor, 1979.

Cyert, R. M., and J. G. March. *A Behavioral Theory of the Firm.* Englewood Cliffs, N.J.: Prentice-Hall, 1963.

Drucker, P. F. *Managing in Turbulent Times.* New York: Harper and Row, 1980.

Gryna, F. M. *Quality Circles: A Team Approach to Problem Solving.* New York: AMACOM, 1981.

Hitt, M. A., D. D. Ireland, and K. A. Palia. "Industrial Firms' Grand Strategy and Functional Importance: Moderating Effects of Technology and Uncertainty." *Academy of Management Journal.* Vol. 24, No. 2, 1982.

Ishikawa, K. *Guide to Quality Control.* Asian Productivity Organization, Tokyo, Japan, 1976.

Juran, J. M., and F. M. Gryna. *Quality Planning and Analysis.* New York: McGraw-Hill, 1980.

Katz, D., and R. Kahn. *The Social Psychology of Organizations.* New York: John Wiley and Sons, 1966.

Lynas, R. "Two Views of Quality Results: Corporate View." Presentation made at 7th Annual Partners Seminar, Productivity and Quality Connection, Utah State University, Logan, Utah, 1982.

Mundel, M. E. *Improving Productivity and Effectiveness.* Englewood Cliffs, N.J.: Prentice-Hall, 1983.

Nadler, D. A., and E. E. Lawler. "Quality of Work Life: Perspectives and Directions." *Organizational Dynamics.* Winter 1983.

Porter, L. W., E. E. Lawler, and J.R. Hackman. *Behavior in Organizations.* New York: McGraw-Hill, 1975.

Roethlisberger, F. J.,and W. J. Dickson. *Management and the Worker.* Cambridge, Mass.: Harvard University Press, 1939.

Sumanth, D. J. *Productivity Engineering and Management.* New York: McGraw-Hill, 1984.

Thompson, J. D. *Organizations in Action.* New York: McGraw-Hill, 1967.

Viana, W. S. B. "Productivity Measurement and Improvement Strategy Program." Master's Thesis, Oklahoma State University, Stillwater, 1976.

Von Bertalanffy, L. *General Systems Theory.* New York: Braziller, 1968.

Weston, J. F., and E. F. Brigham. *Managerial Finance,* 7th ed. Hinsdale, Ill.: The Dryden Press, 1981.

QUESTIONS AND APPLICATIONS

1. In some detail, develop a productivity management process model for one or more of the following organizational systems:
   (a) A restaurant
   (b) An automobile assembly plant
   (c) A department of industrial engineers acting as internal consultants to a company
   (d) A secretarial/clerical pool
   (e) A maintenance department for a manufacturing plant
   (f) Custodial and maintenance function for a high school

(g) Administration for a college or a university
(h) A department, school, or faculty in a college or a university
(i) A major sports team/corporation
(j) Food service/dietary service in a hospital
(k) A job shop
(l) A hotel
(m) An accounting department
(n) A convenience store
(o) A florist shop
(p) Manufacturing engineering for a plant
(q) General Motors
(r) Delta Airlines
(s) A line work group with supervisor
(t) A fire department
(u) Research and development function
(v) A teacher and a class
(w) A McDonald's franchise
(x) A service department for a city
(y) Quality control function
(z) Any other specific organizational system of your choosing

Prepare a professional five-page report and presentation of your analysis. Focus primarily on steps 8, 9, and 10. Present and discuss in class.

2. Explain the difference between an output and an outcome.

3. Describe and define, in your own words, the following terms:
(a) Measure
(b) Variable
(c) Attribute
(d) Scale
(e) Ratio
(f) Index
(g) Parameter
(h) Statistic
(i) Surrogate

Give examples that operationalize these terms and concepts.

4. Operationally describe the following:
(a) Surrogate productivity measure
(b) Productivity measure
(c) Static productivity ratio
(d) Dynamic productivity index
(e) Partial factor
(f) Multifactor
(g) Total factor
(h) Performance
(i) Quality
(j) Effectiveness
(k) Efficiency
(l) Quality of work life
(m) Innovation
(n) Profitability

5. In the public sector or in a private sector service function, what is the correlate to profitability?

6. Analyze and discuss the meaning and implications of Figure 2.4.

7. Conceptually and operationally discuss how the productivity management process integrates with or fits into the general management process. First, what does the general management process look like? What are its components? Second, what role does productivity management play in that larger process?

8. An organizational system has produced 200 units of output A, 450 units of output B, and 1000 units of output C during a given period. It also has provided a wide variety and number of services during that period that were not measured. In addition, 50 units of output A, 10 units of output B, and 200 units of output C were partially completed during the period. The organizational system utilized the following resources:

> Labor: 10 welders, 1700 hours, $15/hour
> 5 fitters, 800 hours, $12/hour
> 15 assemblers, 2500 hours, $11/hour
> 3 supervisors, 500 hours, $17/hour
> 2 foremen, 350 hours, $19/hour
> 1 plant manager, 170 hours, $25/hour
> 2 engineers, 350 hours, $20/hour
> 3 secretaries, 500 hours, $7.50/hour
> 2 salespeople, 360 hours, $13/hour
> 1 procurement agent, 170 hours, $11/hour
> 1 accountant/comptroller, 180 hours, $15/hour
> 1 maintenance person, 190 hours, $10/hour

Materials, energy, and capital data were not collected. Plant capacity utilization was 90 percent. Develop some "measures" of productivity for this organizational system from this data. What are the weaknesses in the data? What are the weakneses in your analysis?

9. Using the same example presented in question 8, and with additional data, reanalyze the "productivity" of this organizational system. In a previous base or comparative period, the organizational system produced 150 units of A, 300 units of B, and 750 units of C. It also provided services during that period that were not measured. Ten units of A, 5 units of B, and 50 units of C were partially completed during that period. Additionally, 150 units of output D was discontinued later and was not produced at all during the period evaluated in question 8. Resources used during the base period were:

> Labor: 7 welders, 1200 hours, $11/hour
> 3 fitters, 500 hours, $7/hour
> 14 assemblers, 2000 hours, $7/hour
> 3 supervisors, 500 hours, $12/hour
> 2 foreman, 350 hours, $15/hour
> 1 plant manager, 170 hours, $20/hour
> 3 engineers, 500 hours, $18/hour
> 4 salespeople, 670 hours, $12/hour
> 1 accountant, 180 hours, $11/hour
> 2 maintenance people, 350 hours, $8/hour

Materials, energy,and capital data were available but not captured for the productivity analysis. Plant capacity utilization was 76 percent. Develop some measures of performance and productivity (ratios and indexes) for this period as well as for a comparison between the two periods. What are the weaknesses and problems with your analysis? What additional data would you like to have in order to validly assess the performance and productivity of this organizational system?

10. Select a real organizational system that you have access to. Perform a productivity management process model exercise on that system. Outline a realistic and workable performance and productivity measurement, evaluation, and control system.

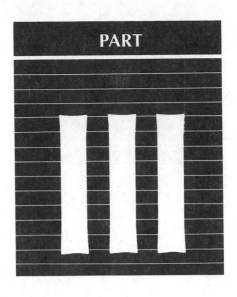

PRODUCTIVITY MEASUREMENT AND EVALUATION STRATEGIES AND TECHNIQUES

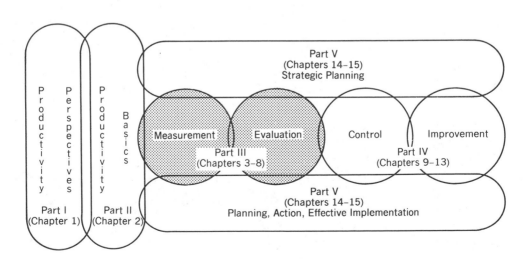

CHAPTER 3

INTRODUCTION AND STRATEGY OVERVIEW

HIGHLIGHTS

- Measurement Basics
- Measurement Components and Critical Decisions
  Planning Horizon
  Desired Outcomes
  Scope
  Intended Interface with Other Performance Measurement Techniques
  Development Plans and Procedures
  Mechanisms for Converting Productivity Measurement Planning into Productivity Measurement and Improvement Actions
- Moving from Productivity Measurement Strategies to Productivity Measurement Techniques
- Case Study
- Productivity Measurement Techniques: Introduction
  Beginning
  Three Techniques

OBJECTIVES

- To develop an understanding of
  (a) When productivity measurement is appropriate and useful.
  (b) The causal relationship between the seven criteria of organizational systems performance.
  (c) When specific productivity measurement approaches are appropriate and useful.
  (d) Underlying measurement fundamentals.
- To present a generic productivity measurement and evaluation strategy that could be used to develop productivity measurement systems.

- To acquaint the reader with the types of productivity measurement systems available and to identify their appropriate applications.
- To discuss how to evolve from a productivity measurement strategy to specific productivity measurement techniques.
- To introduce two specific productivity measurement techniques.

As discussed in Chapter 2, productivity should be thought of as one of at least seven measures of organizational system performance. It is highly related to and dependent on such performance criteria as quality, effectiveness, efficiency, quality of work life, and even innovation. If one were to construct a simple cause-and-effect diagram of the relationship between the seven performance measures, it might resemble Figure 3.1.

From a broad perspective, the purpose of productivity measurement is to assist the organizational system with assessment, evaluation, control, and improvement of its effectiveness, efficiency, quality, productivity, quality of work life, innovation, and profitability. Yet it may not always be necessary to measure productivity in order to accomplish this. In some organizational systems, measuring productivity might be critical to gaining insights into the management interventions needed to improve productivity and perhaps overall performance. In others, however, systems measuring productivity may be dysfunctional. The reason is that what you measure is often what you get. Productivity can be and often is defined operationally in a very narrow fashion. Managers may fail to communicate the broad, integrated performance picture when developing and implementing measurement and control systems. As a result, employees might misread the intent of such systems.

Before deciding whether or not to measure productivity, practitioners must first define and prioritize specifically what performance means in their organizational system. As an example, consider a work group of ten industrial engineers providing staff, internal consulting assistance to the management of a manufacturing firm. Should productivity be used directly as a measurement, evaluation, and control system? Given the intensity of interest in white-collar productivity today, one might immediately respond affirmatively. However, after carefully considering the integrated concept of organizational system performance, the response might be no. The argument could be made that a group of engineers should be more concerned about whether they are doing the right things in the right way with the right resources at the right time. It might be more important to operationalize a measurement, evaluation, and control system to manage effectiveness, efficiency, quality, and innovation. Further, it could be argued that if such a system were developed and implemented, a productivity measurement system would be redundant and perhaps even dysfunctional.

On the other hand, a case for measuring productivity directly could be made. The benefits of developing ratios of outputs to inputs could be espoused. For example, as we have seen, there exists a strong relationship between effectiveness, efficiency, quality, and productivity. When we measure and evaluate changes

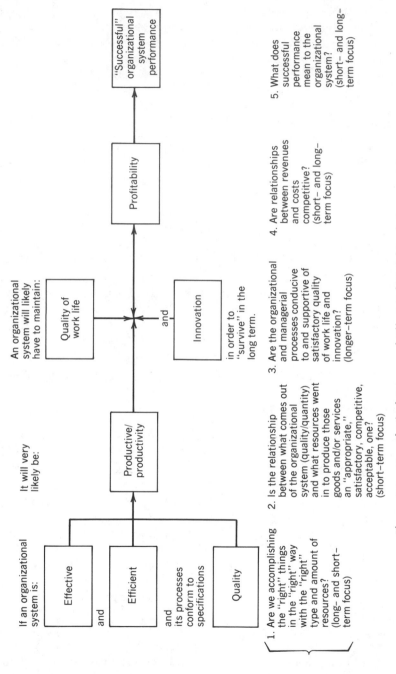

Figure 3.1 Hypothetical Cause and Effect Relationships between and among Organizational Systems Performance Measures or Criteria

64

in productivity we are actually monitoring root changes in effectiveness, resource consumption, and quality of outputs. Productivity measures represent efficient control measures in this respect. That is, we get a great deal of insight into system performance with the productivity criteria. Productivity measurement and evaluation can tell us when we are ineffective, inefficient, and when there is a potential quality problem. Certain productivity measurement and evaluation systems can even point us in the right direction in terms of control and improvement.

The point is the managers must thank critically about the appropriateness of productivity measurement. Productivity is vital to organizational system performance; it is appropriate to measure it. Yet productivity measurement alone is not sufficient to measure, evaluate, control, and improve performance, although it can play an important role in decision support systems for performance management.

Measurement Basics

If we decide to measure productivity, we must understand something about the concept of measurement. Most of us take measurement for granted. For example, when we fill up our car with a tank of gas, we rely on a measurement system without even thinking about it. We know the tank if full when gas splashes from the neck of the tank. However, we do not know how many gallons or liters went in or how much money we owe without the meter on the gas pump. Similarly, we sense whether it is hot, cold, comfortable, or chilly, but do not know exactly what the temperature is without a thermometer. We sense relative magnitudes of light but do not know exactly how much light is available without a light meter.

Measurement is a natural part of the analysis, control, evaluation, and management process. If we want to know something about a particular phenomenon, we measure certain attributes: its size, color, shape, temperature, magnitude, weight, state, quality, and so forth. If our interest is casual or not particularly critical, we do not spend much time, effort, or resources on measurement. However, if the particular phenomenon is of great interest, then typically we attempt to be precise and accurate in our efforts to measure or specify its characteristics.

Managers must measure in order to manage and improve productivity. As implied above, measurement can be casual and intuitive, or it can be specific, disciplined, and systematic. One is not necessarily better than the other, although explicit measurement is viewed as critical to improved decision making. It can be demonstrated that explicit measurement leads to more consistent decisions and that consistent decision makers will outperform inconsistent ones in the long run (Morris, 1979).

Measurement requires collection of data. In general, there are three basic ways to collect data about a given phenomenon or organizational system:

1. Inquiry
2. Observation
3. Collecting system data or documentation

All specific data collection techniques will fall into one of these three categories. Surveys, time studies, interviews, and work sampling are examples of data collection techniques.

In the case of productivity measurement, we are interested in collecting data about productivity. This necessarily means that we will need to collect data about quantities of inputs coming into a particular organizational system and quantities of outputs going out of that same system. From a strict measurement sense, that is all that is necessary. We do not need to know anything about the transformation processes. The basic measurement question for productivity is what should be and can be included.

Physical measurement simply involves the "orderly assignment of numbers to the objects, events, or properties of a system in accordance with certain rules (laws and conventions)" (Archer, 1970). The act of adopting a set of rules is known as scaling. Measurement invariably consists of either adopting an existing scale or creating a new one" (Smith, 1978). There are four basic types or levels of scales:

1. Nominal
2. Ordinal
3. Interval
4. Ratio

Table 3.1 illustrates the basic distinction between these scales. It also depicts relative "power" of the scales in terms of transmission of information about the phenomenon being examined. Specific statistical analysis techniques appropriate for analyzing data measured on the various scales are also indicated.

To the practicing manager, this discussion will not be very useful. To the student or researcher in the field of productivity, however, these issues have important implications. As will be seen later in this chapter, one of the models we will consider, the multifactor productivity measurement model, utilizes primarily ratio scales. In contrast, the normative approach, another measurement technique, potentially utilizes the entire spectrum of scales.

In particular, sociopsychological and psychophysical investigations typically are attempts to quantify certain attributes in a given system. Attitude surveys and many measures of the quality of work life are examples of this type of measurement process. Typically, the data received from these types of meas-

Table 3.1 The Four Classes of Scales of Measurement and Statistics Appropriate to Measurement for those Scales

SCALE	BASIC EMPIRICAL OPERATIONS	TYPICAL EXAMPLES	MEASURES OF LOCATION	DISPERSION	ASSOCIATION OR CORRELATION	SIGNIFICANCE TESTS
D Ratio scales (e.g., length)	Determination of the equality of ratios	Numerosity, length, density, work, time intervals, temperature (Rankin or Kelvin)	Geometric mean, Harmonic mean	Percent variation	All of the below	All of the below
C Interval scales (e.g., temperature)	Determination of the quality of intervals or of differences	Temperature (Fahrenheit or Celsius), position, time (calendar); Performance rating scales	Arithmetic mean	Standard deviation; Average deviation	Product-moment correlation; Correlation ratio	t-test, F-test
B Ordinal scales (e.g., preference)	Determination of greater or less	Hardness of minerals, street numbers, grades of leather, lumber, wool; Daywork pay classes	Median	Percentiles	Rank-order correlation	Sign test, Run test
A Nominal scales (e.g., identity numbers)	Determination of equality	"Numbering" of football players; Assignment of type or model numbers or job numbers	Mode	Information	Information transmitted, T contingency correlation	Chi-square

Scale diagrams:

D: Nonarbitrary zero — Constant unit — 0 1 2 3 4 5 6

C: Arbitrary origin — Unit assumed constant — 0 1 2 3 4 5

B: Order — No unit — 1st 2nd 3rd 4th 5th

A: Identify — No order — (No. 2, No. 4, No. 5, No.3, No. 1, No. 6 — Common membership)

SOURCE: Smith, 1978 and Stevens, 1959. Reprinted with permission.

urement systems are on an interval scale at best. They also often limit the specific inferences that can be made.

In general, the quality of a measure or a measurement system can be evaluated using the following criteria, which can be thought of as design criteria:

1. *Validity*: Does the measure or set of measures in fact measure or specify that which it purports to? In other words, are we really mesuring what we think we are?

2. *Accuracy and precision*: Does the measurement system accurately and precisely measure the "true" state of a given phenomenon? In other words, can the measurement system accurately and precisely determine the statistical characteristics of the behavior of a particular phenomenon?

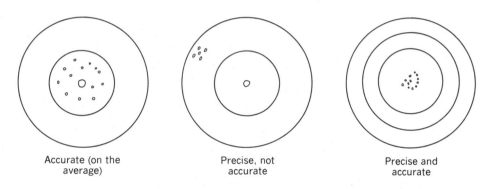

|Accurate (on the average) | Precise, not accurate | Precise and accurate |

3. *Completeness or collective exhaustiveness*: In the case of a measurement system where we are interested in *completely* specifying the behavior of a phenomenon, the total set of measures in the system should be collectively exhaustive or include all measureable variables.

4. *Uniqueness or mutual exclusiveness*: Specific measures in a measurement system should be unique. That is, unless preplanned and accounted for, there should be no redundant or overlapping measures. In general, the goal is to strive for one "good" measure for each property of a given phenomenon.

5. *Reliability*: Measures or the measurement process should consistently provide valid results. Errors in the measures or measurement process should be either consistent over time or minimized.

6. *Comprehensibility*: Measures and a measurement system should be as simple and understandable as possible and still convey the message and meaning intended. This criterion is highly affected by the intended users of the measures or measurement system, since comprehension is a developed skill. Some users will be more knowledgeable and skilled than others.

7. *Quantifiability*: In order to convey the maximum amount and quality of information, measures should be quantified. The basic reason for this is that when something is described quantitatively, we can better understand its behavior.

This is not to say that qualitative information is not important. On the contrary, qualitative information is often necessary and useful for supplementing our quantitative measures or making them more robust and meaningful.

8. *Controllability*: Measures should reflect variables, factors, relationships, or any phenomena that we have some control over. Managers are obviously much more likely to value a measurement system that measures things they can control than one that measures only things out of their control.

9. *Cost effectiveness*: Measures and a measurement system should be designed to be cost effective. The benefit-to-burden ratio should satisfy specific aspiration levels. Some measures are simply too difficult to operationalize to justify their development.

There are specific statistical methods that are used to ascertain validity and reliability. Further, there are a number of types of validity (construct, content, internal, and so forth). The interested reader is directed toward the following references: Kerlinger (1973); Dunnette (1976); and Guion (1965).

It is important to add that productivity measurement systems should or will primarily comprise ratios of output measures and input measures and/or indexes or a ratio of productivity ratios $\left(\dfrac{Q^O}{Q^I} \text{ and/or } \dfrac{Q_{ij}^O/Q_{ij}^I}{Q_{ik}^O/Q_{ik}^I}\right)$. The measures of output and input could be specific measures of quantities of any resource utilized and of quantities of any good or service produced as output. The basics for productivity measurement then boil down to (1) identifying which measure or measures to include; (2) testing the measures against the nine criteria just discussed; (3) operationalizing the specific measures of output and input (for example, which scales to use or how or where to get the data); and (4) evaluating the resulting ratio(s) and/or index(es) against the nine measurement criteria.

Measurement Components and Critical Decisions

Measurement seems like such an easy concept. Somehow, though, when we attempt to apply measurement concepts to specific applications, confusion and ambiguity tend to set in. It would appear that, in particular, this confusion stems from the following difficulties:

1. Not really knowing precisely what productivity is and, therefore, being confused about what exactly should or should not be included

2. Thinking productivity and performance are synonymous and, therefore, viewing productivity measurement as being a larger and more complex task than it really is

3. Matching, at least temporally, outputs with the corresponding inputs utilized to create those outputs

4. Identifying quantifiable outputs

5. Measuring or scaling the outputs

6. Confusing outcomes with outputs

7. Understanding how to use indexed prices and costs as a weighting and aggregating device in the productivity ratios and indexes

8. Being unwilling to accept precise definitions of productivity, thus causing ambiguity in the measurement process

9. Being unable to understand the concept of "unit of analysis" and, in particular, developing and integrating productivity measurement systems at the firm, division, plant, and work-group levels

10. Not taking the time or being disciplined enough to sit down and evaluate which *performance* criteria are important and whether or not productivity is or should be a component of the organizational system's performance measurement system.

In any case, organizations should have a strategic plan for measuring productivity. This plan will integrate or fit directly into the "Grand Strategy" for productivity management (see Chapter 14). It is a vital component of that larger picture. Further, such a plan would necessarily be a component of a larger performance measurement system for the organizational system of interest.

Productivity measurement strategies are characterized by the following parameters:

1. Planning horizon

2. Desired outcomes

3. Scope

4. Intended interface with other performance measurement systems

5. Development plans and procedures

6. Explicit mechanisms by which to convert strategy and plans into tactical and operational plans and eventually into actions

As such, the process of developing a productivity measurement strategy is not unlike that associated with developing a marketing strategy, a business plan, product planning, or any other strategic planning activity and focus. The only difference is that the focus is on how the organizational system will design, develop, implement, and maintain a productivity measurement system. Let's look at each of the parameters in more detail.

Planning Horizon

Developing a successful productivity measurement strategy requires long-range planning. What is meant by long range varies from organization to organization and is affected by a number of factors. These include type of organization, level within the organization at which the planning is taking place, whether or not this is an initial start-up or a continued development, environmental uncertainties, budget, and so forth. In general, though, strategic planning for an initial start-up effort will require a three- to five-year planning horizon. Note that the term "planning horizon" does not infer or mean the length of the planning process itself. Rather, it refers to the length of time that needs to be set aside to reach some stage of accomplishment or completion of the project itself. For a small unit of analysis, say, a department, the planning horizon may be only one or two years (perhaps less) through completion of an accepted, operating measurement system. For a larger unit of analysis, say, an organization with a number of plants, the planning horizon may be closer to five years.

In any initial start-up activity, there is typically going to be a need for considerable productivity awareness training to take place. A groundwork or foundation for grass-roots support of a productivity measurement system needs to be laid. This activity alone can require a sizeable planning effort. Without an understanding of the need for productivity measurement, most efforts in today's organizations are doomed. Figure 3.5, presented at the end of this section, depicts an exemplary plan for the design, development, and implementation of a productivity measurement system. Horizon variances are indicated on that figure.

Desired Outcomes

The statement of desired outcomes is perhaps the most critical component of the strategic plan. Management's ability to clearly explicate visions of where an organization should be going and the levels of excellence it should be achieving is a critical leadership requirement. Too often management sets a plan in motion without having clearly thought through what the plan should accomplish. In Albert Einstein's words, "Perfection of means and confusion of ends seems to characterize our age."

The following is a possible set of desired outcomes for a plan to design, develop, and implement a productivity measurement system. (Many of these statements look like goals. There is such a close relationship between desired outcomes and goals that to a great extent differences are semantic.)

1. To have successfully implemented a productivity measurement system that satisfies certain explicitly stated design criteria by the specified target date

2. To end up with a productivity measurement system that accurately identifies areas for productivity improvement

3. To end up with a productivity measurement system that is well understood and accepted by all personnel involved with its application

4. To have included a measurement activity that enhances a commitment and motivation to act on the indicated areas for improvement

5. To end up with a measurement system that is flexible and dynamic so that over time, it reflects the true state of productivity in the particular organizational system

6. To end up with a measurement system that does not focus solely on a standard set of measures created by experts and imposed on specific organizational systems, but that also incorporates a method by which organizational systems can create each specific productivity measurement system suited to their own inevitably special circumstance (Morris et al., 1977)

7. To end up with a productivity measurement system that, where feasible and appropriate (perhaps depending on unit of analysis), utilizes participative techniques. "The greater the participation in creation of the productivity measurement system, the greater the acceptance and commitment, the greater the ease of implementation, the greater the resulting productivity change, and the greater the ease of implementation of future changes based on productivity measurement" (Morris)

8. To end up with a productivity measurement system that, especially at lower units of analysis (work-group level), results in a vector of productivity measures, ratios, or indexes and does not necessarily attempt to achieve a single measure, ratio, or index (Morris)

9. To end up with a productivity measurement system development process that, especially at lower units of analysis, utilizes "analysts" or outside assistants to facilitate or stimulate the creation of measurement systems by those involved in the organizational systems themselves (Morris)

10. To end up with a productivity measurement system that is as simple as possible (not simpler) and cost effective (Morris)

11. To end up with a productivity measurement system that does not appear to those concerned and involved as simply another passing fad that will soon disappear (Morris)

12. To end up with a productivity measurement system that clearly indicates how it fits into existing management processes and how it can function as a decision aid (Morris)

13. To end up with behavioral consequences of a productivity measurement system that are anticipated and reflected in system design (Morris)

14. To end up with a productivity measurement system that is viewed by those whose behaviors and outputs are being scaled or measured as being essentially nonmanipulative (Morris)

15. To have made sure that any possible dysfunctional consequences of a productivity measurement system were explicitly identified and that contingency plans were drawn up (Morris)

16. To have made sure that during the early development of the productivity measurement system, expectations were managed. It is important not to claim

too much, to identify clearly the possible shortcomings, and to emphasize deferral of early judgments (Morris)

17. To have used self-observation, self-reporting, self-analysis, and self-design of jobs and programs so that the productivity measurement system will be cost effective (Morris)

18. To have employed a combination of "decoupled" and "coupled" productivity measurement systems, especially for a large unit of analysis (a corporation or at the firm level). *Decoupled* productivity measurement systems are those developed by specific work groups for themselves. *Coupled* productivity measurement systems are those developed for the aggregate firm itself. The productivity measurement system will have to address how to integrate and coordinate the "decoupled" productivity measurement systems.

This list presents a set of desired outcomes that might be utilized in a strategic plan for the design, development, and implementation of a productivity measurement system. As you will see in Chapter 4, many of these desired outcomes were taken into consideration in the design of a productivity measurement system for ACI services.

Scope

The scope of the plan refers essentially to the units of analysis to be covered by the development of the productivity measurement system. The scope can range from the very macroscopic (national, industry, regional, and so forth) to the very microscopic (work group). For instance, the scope could be defined as the firm. A desired outcome or goal statement might infer this by specifying what areas, levels, functions, and so forth of the firm are to be included. The scope would further state if only a "coupled" productivity measurement system, only a "decoupled" productivity measurement system, or perhaps both kinds of systems need to be developed.

Figure 3.2 depicts a much simplified organization structure and a hierarchical system of performance measurement and evaluation. Note that for each level of the organization, for each major function, for each work group, there can exist a vector of performance indexes, ratios, and measures. If each subgroup in the organization develops its own vector(s) with which to measure performance and productivity and no attempt is made to systematically integrate these systems, then we would call this a "decoupled" performance measurement system. The implicit assumption in this strategy is that each subgroup can best measure and manage performance independent of integration with other systems. This does not imply that the planning system does not coordinate goals and objectives. It implies only that the performance measurement system itself will be decentralized. There might be some centralized procedure for how the performance measurement system is developed. However, the results of those

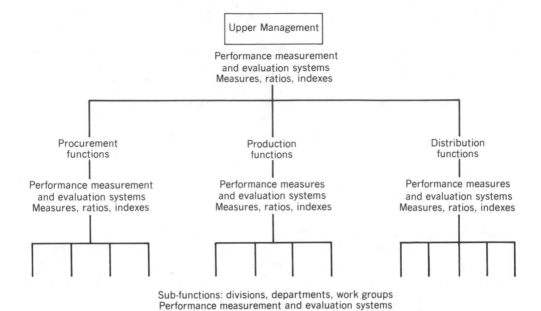

Figure 3.2   Organizational Systems Performance Measurement and Evaluation Hierarchy

efforts would not be aggregated in any fashion. Figure 3.3 depicts the relationship among various performance measurement and evaluation subsystems.

On the other hand, a productivity measurement strategy might call for the development of a "coupled" or integrated system. In this case, productivity indexes, ratios, and measures might be developed for a variety or even all subgroups in the organization with the intent of aggregating ("coupling") the results, somehow, into a more macroscopic measurement system. This strategy would still focus on subgroup-level productivity measurement systems. However, the thrust of these developments would be directed toward eventual aggregation and integration.

A third strategy with respect to scope would be a hybrid of the first two just mentioned. An organization could develop fairly microscopic (say, down to the work-group-level) productivity measurement systems that are "decoupled" and, on top of those systems, add a larger (say, firm-level) aggregated productivity measurement system. Productivity measurement systems would then be executed in a decentralized, "decoupled" fashion throughout the organization at a number or all units. A centralized, aggregated (not necessarily a "coupled") productivity measurement system would also be developed at the firm level. Figure 3.4 depicts this.

Many combinations of productivity measurement system coverage could ob-

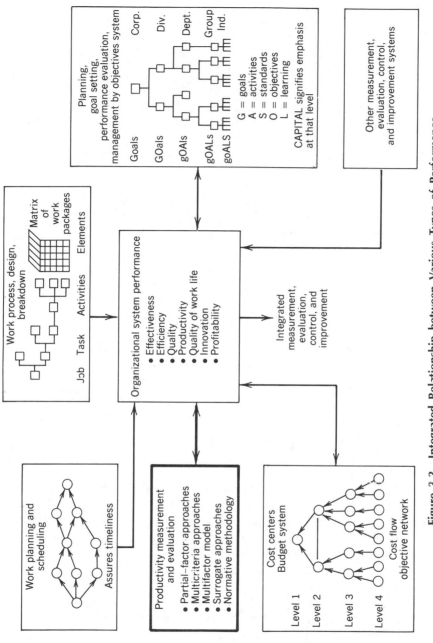

Figure 3.3 Integrated Relationship between Various Types of Performance, Measurement, Evaluation and Control Systems

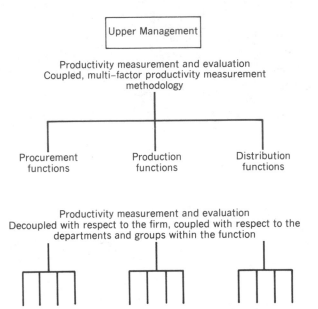

Productivity measurement and evaluation
Coupled, multi–factor productivity measurement
methodology

Procurement   Production   Distribution
functions   functions   functions

Productivity measurement and evaluation
Decoupled with respect to the firm, coupled with respect to the
departments and groups within the function

Productivity measurement and evaluation
Decoupled at the work–group level: NPMM, PFPMM, MCP/PMT

Figure 3.4   Decoupled and Coupled Productivity Measurement and Evaluation Systems

viously take place. These three basic strategies just mentioned have been the most common.

Intended Interface with Other Performance Measurement Systems

As discussed previously, there are a variety of other performance measurement systems that managers utilize to monitor, evaluate, and manage organizational system performance. The strategic plan for a productivity measurement system will have to identify and define how the specific productivity measurement system will interface with and supplement and enhance these other systems. Desired outcome 12, mentioned above, specifically addresses this design parameter for the strategic plan.

There are no specific guidelines that can assist one with the process of satisfying this parameter. As the specific productivity measurement system is being designed and developed, specific indexes, ratios, and measures will need to be examined in light of what management is currently measuring. The interface will not only need to check for redundancy but also to integrate productivity measures into such management control systems as the reward mechanisms and the performance appraisal processes. This is a critical design parameter and a step quite frequently omitted from the planning process.

Development Plans and Procedures

The strategic plan will obviously need to contain system development plans and procedures. This planning will need to include how the specific measures are determined; how the data will be collected; how the measures will be built into a management information system or a decision support system; whether "decoupled" measurement systems will be developed and, if so, how they will be integrated or brought back together; and whether or not macro, aggregated, multifactor models will be utilized and, if so, how that information will be utilized to direct specific productivity improvement activities. The plans will also need to focus on the type and amount of training and development that will be required. Let's examine each of these aspects of the planning process.

HOW SPECIFIC PRODUCTIVITY MEASURES WILL BE DETERMINED

There are a number of ways for identifying the specific productivity measures:

1. An expert (consultant, manager, and so forth) can designate specifically what should be measured to evaluate productivity in a particular organizational system

2. One could simply use what others have used in similar systems

3. Select persons in the organizational system under study could be asked to identify the things that should be measured

4. The measurement model itself might specify precisely what will be measured and analyzed.

Each of these basic approaches reflects specific techniques for measuring productivity. In particular, the third approach reflects the use of structured group processes (such as the nominal group technique and the Delphi technique) to develop normative indexes, ratios, and measures of productivity. This technique is useful at the microscopic and intermediate levels, such as the work group, department, and perhaps function. The fourth approach reflects the use of a productivity measurement model that specifies the ratios and indexes that are examined. The multifactor model, which is used at a more macroscopic level, is an example of this approach. Both of these techniques will be detailed in the upcoming chapters.

HOW THE DATA WILL BE COLLECTED

In the strategic plan, and in particular with respect to development plans and procedures, general data collection issues should be addressed. How the data will be collected, where the data can be obtained, who will collect the data, and what is then done with this information are all key questions to be addressed. The data requirements and the specific answers to these questions depend on the specific approach taken and the particular technique selected.

HOW THE MEASURES WILL BE BUILT INTO INFORMATION OR DECISION SUPPORT SYSTEMS

Once approaches and techniques are selected and data collection issues are addressed, planning must be directed to specifying how the measurement system can be developed into a decision aid. The designer needs to focus on such user requirements as documentation preferences, timing of feedback, formatting, and support documentation. Emphasis should be placed on significant communications with end users. If the normative approach (the third method of identifying productivity measures) has been taken, then this step should be less difficult. If, however, a consultative approach (method number 1) or a model approach (method number 4) has been chosen, then this step might require considerably more attention, communication, training, and sensitivity.

USE OF DECOUPLED PRODUCTIVITY MEASUREMENT SYSTEMS

If it is decided that productivity measurement systems should be developed in a decoupled, or decentralized, fashion, then a decision as to whether or not to also couple, or centralize, these measurement systems has to be made. This remains one of the more difficult aspects of productivity measurement in organizations of any size. Attempting to integrate a decoupled productivity measurement system appears to be a major stumbling block for many managers. Many of the problems arise from either a weak statement of desired outcomes or a complete absence of any planning itself.

Decoupled systems have advantages. If productivity awareness and basic education and training are achieved prior to the actual development of the productivity measurement system in a given organizational system or unit, the quality of the resulting measurement system should be high. Decoupled systems place measurement activities, accountability, and design logic with the people who actually manage the system. A decoupled system will typically be closer to actual cause-and-effect mechanisms so as to strengthen the linkage between measurement and improvement. If decoupled systems are legitimized and supported from top to bottom and from the grass-roots levels themselves, they will more likely be an integral part of the way an organization and its various subgroups of organizational systems do business.

Decoupled systems also have disadvantages. Control of actual measurement techniques and applications is diminished. Specific units may bog down and become discouraged or even turned off to productivity without stronger guidance and support. Productivity measurement may be applied inconsistently from unit to unit. If the overall business planning process is ineffective, the productivity measurement system may end up being confused as the only goals, objectives, and targets of the unit. This could cause dysfunctional outcomes. For instance, a particular work group could focus on improving certain productivity measures at the expense of other critical measures. Or, one unit's productivity mesurement system could in fact interfere with or cause another unit's productivity to decrease. In decoupled systems, there is in general less control than if there is a plan to couple. This can be either an advantage or a disadvantage depending on a number of factors.

In general, if the organization type, structure, processes, and management style are compatible, a decoupled system is recommended. The advantages seem to outweigh the disadvantages if certain key factors are managed correctly (for example, overall planning system, a flexibility and adaptive attitude, good communications between critical interfaces in the organization, and effective productivity basics education and training). The preference would be for an effective and efficient, well-planned, and well-executed decoupled productivity measurement system with an aggregated multifactor measurement system at the firm level. Figure 3.4 depicts this preferred strategy.

USE OF MACRO, AGGREGATED MULTIFACTOR PRODUCTIVITY MEASUREMENT MODELS

The development of multifactor productivity measurement models for firm- or organizational-level analysis has been taking place for over 20 years, with most of the effort occurring in the past 7 years or so. The multifactor productivity measurement model is a computerized and primarily accounting-based approach used primarily to measure and evaluate changes in productivity, price or cost recovery, and profitability at the company or plant level over time. The model automatically constructs partial-factor productivity ratios as well as partial- and multifactor indexes for productivity, price recovery, and profitability. The focus of the model is on tracking changes in productivity and price recovery over time and providing information as to what factors are causing these changes.

Multifactor productivity measurement models are being continually refined into measurement systems that can support decisions directed at productivity improvement. They are reasonably simple and easy to understand, and they create information that managers have typically not had at their disposal. In addition, they have been programmed to allow for compilation of large amounts of data in an efficient fashion. Moreover, there are a variety of instructional or training programs available that teach how the models work and how to use them. In general, they observe and integrate the basic definitions and concepts of productivity presented in Chapter 2.

These models, although widely applicable, are not for all organizations and certainly not for all types of organizational systems within a given organization. They are primarily, as indicated before, useful at more macroscopic units of analysis (the firm, the corporation, the plant, the division, and so forth). The firm's requirements for data input as to outputs, quantities, and prices, as well as inputs, quantities, and prices make the models difficult to apply in certain areas. In general, though, development plans and procedures should consider the multifactor, aggregated models for productivity measurement.

So, a key component of a productivity measurement system strategy is development plans and procedures. How productivity measures are to be determined and by whom is a critical step in this component. How the productivity data will be collected is a second major step. A third step is how the measures can be built and integrated into existing management systems. A fourth step considers the use of decoupled, decentralized productivity measurement systems. Finally, appropriateness and utility of a multifactor, aggregated productivity measurement model will need to be examined. The desired relationship between measurement and improvement brings us to the final major component of a productivity measurement strategy.

Mechanisms for Converting Productivity Measurement Planning into Productivity Measurement and Improvement Actions

The five major components of a productivity measurement system strategic plan discussed to this point (planning horizon, desired outcomes, scope, intended interface with other performance measurement systems, and development plans and procedures) have fairly well defined the productivity measurement plan itself. We know how much time the plan covers. We know what we expect or desire the resulting system(s) to accomplish. We know how much of the organization is involved and whether we have a bunch of small productivity measurement systems, one big productivity measurement system or a bunch of small ones all added together. We also know how the productivity measurement system(s) is to be viewed and utilized in relation to other management systems. Finally, we know specifically the type of productivity measurement methodologies or techniques that will be utilized and how they will be executed.

There is at least one last parameter, however, that the strategic plan should specify. This parameter is perhaps one of the most critical since it deals directly with converting productivity measurement planning into productivity measurement and improvement actions. It is one thing to tell people what productivity is and to convince them it is important. But it is quite another thing to get managers and professionals to actually measure productivity. A strategic plan for developing productivity measurement systems in an organization *must* incorporate explicit mechanisms for facilitating the conversion of productivity measurement plans into productivity measurement and improvement actions.

The primary and perhaps most effective way to do this is to involve the persons responsible for managing productivity and those who "cause" it in the productivity measurement process itself. (Note desired outcome 7 on page 72.) Particularly in organizations today, managers, professionals, and other employees want (and sometimes demand) to be involved in the development of systems in which they work and for which they are held accountable. The basic question facing managers in the eighties and nineties is not whether to involve employees, but how to do it effectively and efficiently. Specific methodologies for accomplishing effective and efficient involvement in productivity measurement systems will be described later in this part.

The primary goal to be achieved by paying attention to this parameter is an awareness and perception of the direct relationship between individual and group behaviors and productivity. This goal can be achieved through involvement in the design, development, and implementation of productivity measurement systems. This involvement, if effective, can at the individual- and group-behavior levels create positive expectations for the relationship between efforts, actions, and programs taken in response to results from the measurement system(s). It also can cause individuals and groups to refocus their attention on what in fact they should be doing and how productivity performance can be measured. Measurement systems developed participatively can focus efforts. Participatively developed measurement and evaluation systems often target problem and/or opportunity areas for the organizational system. In this respect, they are perhaps more proactive than other approaches. This approach also tends to promote the linking of improvement to measurement.

The essence of an effective productivity management system is for our efforts, actions, beliefs about cause-and-effect relationships, and perceived problems and opportunities to be effectively integrated with the productivity measurement system(s). When this takes place, productivity measurement facilitates productivity improvement efforts and vice versa. This synergy between measurement and improvement at the individual and group levels (work-unit level of analysis) is the key to successful productivity management. The only way to achieve this is to have the productivity measurement system for the organization incorporate a decoupled system.

Moving from Productivity Measurement Strategies to Productivity Measurement Techniques

A productivity measurement strategy is a long-range plan. It is developed by taking the time to sit down and think through these questions:

1. What do you want your productivity measurement system to accomplish?

2. Who you think should be involved in its development?

3. What are your specific measurement technique alternatives?

4. What specific steps or actions will be required to implement the plan?

5. What is the logical sequence and timing for each step, activity, or stage in the plan?

6. How much of the organization do you want involved in or incorporated by the productivity measurement system (scope)?

7. What level of aggregation do you desire for the data?

8. What are the (intended) consequences?

9. What are the critical and perhaps necessary linkages between this performance measurement system and others in your organization?

10. What specific steps can be taken to ensure that plans are converted into actions?

Planning is thinking, and thinking is hard work and dealing with ambiguity and uncertainty.

Planning is distinctly different from doing. Doing involves putting things into motion, action, involvement. Most managers like doing better than planning. In fact, many managers like doing so much better that they "never plan to fail, they just fail to plan." (We discuss in more detail planning techniques and how to improve planning effectiveness in Chapter 14.)

Planning is a critical function in the productivity measurement process. If we take the time to do it right, we won't have to take the time to do it over or to have to accept lower quality than was possible. How do you become a better planner? Take a short course? or a seminar? hire a consultant? Not likely! Planning is a skill. Skills take discipline, hard work, a commitment to excellence, and perfect practice. "Practice doesn't make perfect; perfect practice makes perfect," said Vince Lombardi. Somehow we seem to have lost sight of the basics. There is no magic, no secret, no panacea. If you really *want* to improve productivity in your organization, there is a price. Developing a productivity measurement strategy is part of that price.

Every strategy or long-range plan, if approved and accepted, must somehow lead to the development of tactical and/or operational plans. The strategy maps out where we want or should go; the tactics specify how we are going to get there. There are many components of the tactical and operational plans: objectives, activities, responsibilities, accountability, resources expected to be consumed, timetables, milestones, and techniques to be utilized. Many of the concepts involved in project management will pertain to this stage in the development of a productivity measurement system. There is not sufficient space to go into project management techniques. However, essentially, the goal of these techniques, approaches, and concepts is to monitor and control the progress of a particular project, in this case, the development of productivity measurement systems. Gantt charts, critical path method, and/or program evaluation and review technique are examples of such techniques.

Although these are all formalized techniques, there is no reason that management of the productivity measurement system development project needs to utilize them. Again, the goal is to develop effective and efficient mechanisms for dealing with these questions:

1. Are we doing what we set out to do?

2. Are the people we wanted to be involved actually participating?

3. Which specific measurement techniques are being developed?

4. Are the specific steps being executed on time and in the proper sequence?

5. Is the scope of the measurement system as specified?

6. Are the specific measurement techniques being developed and implemented appropriate based on the specified level of aggregation for the data?

7. Are the resources being consumed within budgeted or estimated figures?

8. Is the productivity measurement system being integrated into other performance measurement systems that managers currently utilize?

The project management mechanism(s) does not have to be sophisticated; it just
has to work.

CASE STUDY

An organization has decided to develop a productivity measurement system. It is a
service department providing maintenance services for a small city of population
40,000. Services include street repair, traffic control, hardware maintenance, city park
and building maintenance, city vehicle maintenance, sewage and drainage system
maintenance, and other miscellaneous maintenance and repair activities. The service
director reports directly to the city manager. The service director has one assistant,
three foremen, and 18 employees of various labor skill classifications.

The city manager wants to develop a productivity measurement system while also
improving productivity in the process. He decides to approach a local university for
assistance. A graduate student in Industrial Engineering who is studying productivity
measurement for a thesis topic discusses the concepts presented in Chapter 2 and
this chapter with the city manager. Based on the student's suggestions, the city manager agrees to ask the service department to conduct a "pilot study" to develop a
productivity measurement system.

The city manager, service director, assistant service director, foremen, and the
student/consultant discuss and outline the elements of the strategic plan mentioned
earlier in this chapter. Critical elements of the plan are identifying the scope of the
measurement system, deciding which productivity measurement technique to utilize,
anticipating consequences, and linking this system to other existing performance measurement systems presently in use. Several other issues are raised that the student/
consultant had not anticipated.

The service director is concerned about the use of the term "productivity." He is
not sure his staff will understand the term. In fact, he is not sure he understands the
term. He is also concerned about the time that it will take to develop the system. He
is already too busy and is not sure this is a "priority" activity he and his people should
take time out for.

Later in the discussion, the service director asks the city manager why his department
has been singled out for the pilot study. He feels his department is as productive as
the rest of the organization. Why pick on him? The city manager explains that he
selected the service department because he felt it would be the most likely to cooperate
and to succeed with the project. He also explains that the decision to proceed with
the pilot study is strictly up to the department itself.

The city manager suggests that the service department take a week to think about
the opportunity. He also asks the department to develop a tentative rough outline of
a plan to move forward based on the discussion that took place between the city
manager, the service director, the assistant service director, the foremen, and the
student/consultant. He asks the department to focus on the elements outlined by the
student/consultant but not to limit its attention to only those elements. If there are
other items to consider in developing the plan, he wants to know about them. He

then asks the department members to develop a "realistic" timetable for the plan and to pay particular attention to any necessary training and development.

The city manager stresses that the department must view this effort as its program and regard it as beneficial. He points out that the major goal of the program is to redesign the control system that the managers of the department are using to guide and direct the organization and the employees. He uses an aircraft example. The managers are the pilots and the weather is getting rough outside. What instruments would they like to have to keep the plane on course and performing at peak? He adds that he has recently heard a well-known management speaker say that an organiza-

Table 3.2 Modes of Professional Functioning

Acceptant, Listener—The consultant encourages client expression, listens effectively, makes no evaluative responses. The consultant exhibits the clinical skill of concerned listening without being judgmental, thus encouraging the client to clarify problems and to relieve anxieties and frustrations.

Expert, Solution Provider—The consultant functions in the traditional mode of the expert with special knowledge and experience, solves the client's problem, and produces a recommended solution or design for a new system.

Structured Group Process Coordinator—The consultant coordinates and facilitates groups of client people working on well-defined tasks using structured group processes such as brainstorming, brainwriting, the nominal group technique, or the Delphi method. The consultant is careful to stay out of the substantive considerations and play the low-profile role of structuring an efficient and effective group process.

Teacher, Skill Developer—The consultant seeks to enhance the competence of client people by teaching tools, techniques, methods, and skills that have direct application to immediate client concerns. The consultant may also teach principles, theories, or bodies of knowledge that are expected to increase client competence in a much longer run sense.

Data Gatherer, Information System Designer—The consultant undertakes to collect or plan the collection of information believed by either client people or the consultant to be useful in some phase of the client change process. Note that in this mode the consultant is acting as a data gatherer only, leaving the analysis and interpretation of the data to the client or to some joint effort involving both client and consultant.

Challenger—The consultant challenges or confronts the client with data, information, standards, comparisons, and "tough" questions. The consultant seeks to make client people "face up" to opportunities, shortcomings, inconsistencies, and evidence that may suggest unfavorable interpretations.

Collaborator, Team Member—The consultant becomes a member of a team of client people, working on some fairly well-defined task. The consultant may bring special skills and experience to the work of the team but is careful to recognize the special knowledge of the client members as well. The consultant avoids leading or dominating the work of the team, seeking consensus decisions.

SOURCE: Morris, 1979.

tion's best source of consultants is its people. This is the reason for going to the people to find out what they think should be measured to assess and evaluate the system's performance.

Note several features about this example of the initial planning process. It went through several stages. First, the topic was introduced in a nonthreatening way to the lower-level managers who would ultimately be responsible for maintaining the program. Second, the productivity basics and productivity strategic planning process were presented by the student/consultant in a very much simplified and succinct fashion. No fancy theories or terms were used. All of the concepts presented in Chapter 2 were restructured and relabeled in a way that the managers could relate to. Third, the city manager did not push the program on his managers. He did not present it as a new fad or as the newest technique to come along. He stressed that this process was simply part of an ongoing attempt to improve the control mechanism for the city. Finally, he did not push for a commitment in that meeting. He gave his managers time to think about the pilot study. At the same time, though, he asked them to develop a tentative plan. This prevents them from going away and forgetting about it for a week. He forces them, subtly, to become involved in the planning process before they even commit to the pilot study itself.

Morris has identified what he calls professional modes of functioning. Table 3.2 presents the modes. Note that it is relatively easy to map what modes the city manager and the student/consultant were in at each stage of this meeting.

We have discussed the concept of strategy formulation for the development of a productivity measurement system. Planning for any specific management activity is certainly more an art than a science. So many factors come into play in determining what kind of planning process will work that it is difficult to prescribe the one best way to develop a strategic plan for a productivity measurement system. Instead we have attempted to outline the elements such a plan should include. Now we can turn our attention to specific productivity measurement techniques that might be used to operationalize the plans and procedures outlined in the strategic plan.

Productivity Measurement Techniques: Introduction

Productivity, strictly defined, is a relationship between quantities of outputs that a particular organizational system produces (goods, services, cars, reports, boats, graduated students, inventions, new products, and so forth) and quantities of inputs that the same organizational system utilized to produce those outputs (energy—gas, electricity, and so forth; labor—direct, indirect, management, line, staff, and so forth; material—steel, plastic, paper, fasteners, wood, and so forth; capital—land, buildings, equipment, cash, investments, and so forth). To measure productivity, one simply needs to quantify outputs in relationship to corresonding inputs. In other words, one just has to put what came out of a system over what went in.

This seems simple enough, and for some organizational systems, it is. However, in most organizations, a multitude of outputs exist. Some of the outputs are goods and easy to quantify; some are services and very difficult to quantify. In many organizations there exists tremendous variety across outputs, even within a model or a type of output.

For example, consider the auto industry. A given model of a typical American car can come with a tremendous variety of options. A 1981 Pontiac, for instance, can be "outfitted" with over 50 accessories. What do you count as output? A basic model, or each uniquely "outfitted" car? In organizations with a multitude of product and/or service mixes, the question of how to quantify the output data can be a difficult one. The decision is not unlike one a statistician makes when strata of data or class intervals are determined. The amount of detail desired and that can be afforded must come into consideration. Also, variability in product mix is a factor to consider with respect to the overall strategy of the productivity measurement program. If a type of output exists one month and is absent the next, this will obviously cause variation in productivity. How that variation is handled and interpreted can be a major issue for certain organizations.

As mentioned, some outputs may be services rather than goods or products. Most productivity measurement analysts have considerable difficulty quantifying services. For example, what is the output of a hotel/motel? Many would sat it is satisfied customers or sales or profits. Let's go back to basics! Think back to the definitions provided in the model discussed in Chapter 2. Recall the productivity process modeling exercise. For a hotel/motel, what should satisfied customers, sales, and profits be called? Assuming a "firm-level" unit of analysis, then satisfied customers, sales, and profits are clearly outcomes. They are a result of a system output.

The question again is, what are the system outputs? A hotel/motel is in the business of providing rooms. Therefore, in a precise analysis of productivity, one could contend that occupied rooms is the major output of a hotel/motel. (For the purpose of simplicity, the restaurant and bar portions of the business are excluded.) This precision and simplicity in definition of actual output assists not only with the productivity measurement process, but also with the management process in general. In service-type organizations, disciplined productivity process modeling can be of great benefit. Managers should identify desired outcomes, identify the intended output from the system, and then work back through the cause-and-effect relationships that will create desired levels of output, outcomes, and output/input ratios. This is a management basic. It is the type of basic that Peters and Waterman (1982) suggest is present in the best-run American companies.

The point to be made is that if you go back to the kinds of basics presented in Chapter 2, even organizational systems that have traditionally been labeled as being hard or impossible to measure become measurable. It seems that we want to make it more complex than it needs to be. In the process we confuse ourselves, become frustrated, and give up. Or perhaps it seems too simple. For example, in the hotel/motel case (firm level), if occupied rooms are the major output, then there are just a few productivity ratios to examine. An initial reaction would be to contend that ratios like (1) occupied rooms/labor input; (2) occupied rooms/energy input; or (3) occupied rooms/indexed costs are too simple. One might contend that they don't tell us enough about the "true" productivity of the hotel/motel. A manager might suggest looking at occupancy rates or at occupancy rates in relation to industry cost standards. There is obviously nothing wrong with these measures of organizational systems performance. But let's call things by their right names. Occupancy rates are not productivity measures. They may tell us quite a bit about the effectiveness of the organization, but they are not a measure of productivity.

One cannot expect productivity measurement to satisfy all the requirements and needs of an organization's control system, just as an airplane's airspeed indicator doesn't convey all the information one might need to know about the performance of the airplane. Again, productivity measurement may be a much needed instrument on your organization's control panel. You will never know until you try it out.

Successful American firms have a bias for action. "The most important and visible outcropping of the action bias in the excellent companies is their willingness to try things out, to experiment. . . . Our experience suggests that most big institutions have forgotten how to test and learn" (Peters and Waterman, 1982, p. 134). Effective and efficient control systems for complex organizations need to evolve in a systematic way. We are beginning to learn that most control systems have been sloppily designed and that few, if any, incorporate productivity.

Is productivity measurement as defined in this book important? To be quite honest, the answer to this question is not as clear as it might seem. For some organizational systems, the answer is definitely yes; for others, no. That is, as a measure of control in some organizations, other measures of performance are clearly more critical. This book is not intended to give you the answer to your experiment with control system design, only to help you design an experiment to determine if productivity should be on the panel. As stated earlier, there is growing evidence that productivity is a missing control system element for many organizations and organizational systems.

Is productivity measurement difficult? Sometimes yes and no. However, in most cases managers make it difficult if not impossible simply because they fail to first master the basics presented in Chapter 2. You cannot manage something if you cannot measure it. And you cannot measure it if you do not understand it. In most cases, productivity measurement is very simple if management clearly defines and understands productivity basics. It is, of course, particularly important to have a firm grasp of the basic definitions, both conceptually and operationally, for the seven basic measures of performance (effectiveness, efficiency, quality, productivity, quality of work life, profitability, and innnovation.)

Productivity measurement can, however, be operationally difficult. Phenomena such as product mix changes, long product cycle times, accessibility of data, matching outputs to inputs (in other words, specifically linking resources consumed to outputs produced), aggregation of data, and determining which measures to evaluate can cause the productivity measurement process to be less easy than it might initially appear. So, although a precise and delimited definition of productivity makes measurement easier, pragmatic, real world aspects of your organization will undoubtedly make it more difficult.

Beginning

When beginning to design and develop productivity measurement and evaluation systems certain factors should be considered at the outset. For instance, data are required to measure productivity. Both output data and input data are needed and should be matched. That is, the input data should refect all actual resources consumed to produce the measured output. The scope of the measurement system or the period length to be analyzed will determine how much data we bring together. The relationship between "unit of analysis" and "scope" is depicted in Figure 3.5. Period length could be annual, quarterly, monthly, and so forth. The data should be given in terms of physical quantities or units. For reasons to be explained later, input and output data on prices and costs for each output and input might be needed depending on the measurement model utilized. If unit prices and costs are not feasible, then total revenues and total costs for a category of outputs or inputs will be necessary.

As mentioned, data can be collected from at least three different sources or in at least three general ways: (1) inquiry or soliciting input from persons in the organizational system under study; (2) observation and documenting characteristics of the organizational system; or (3) utilizing and collecting existing system documentation, records, accounts, and so forth. In developing productivity measurement systems, these three sources of data or approaches to collecting data are available. As shall be seen, different techniques rely on different sources of data. In general, when studying or researching something as complex as productivity, it is advisable to utilize as many sources of information as possible. Therefore, it will be important to gain as much perspective as possible on the productivity measurement task.

Another factor to consider when beginning a productivity measurement and evaluation effort is that for any given organizational system, there are typically at least three unique perspectives on productivity: (1) that of the users or recipients of the goods/products and/or services produced; (2) that of providers or persons in the system itself, managers as well as employees; and (3) that of management of the organizational system. Each of these three groups of persons has unique and important views on how to measure the productivity of the organizational system. It is interesting to note that Peters and Waterman (1982) have found that "the better companies are better listeners." Lew Young, editor-

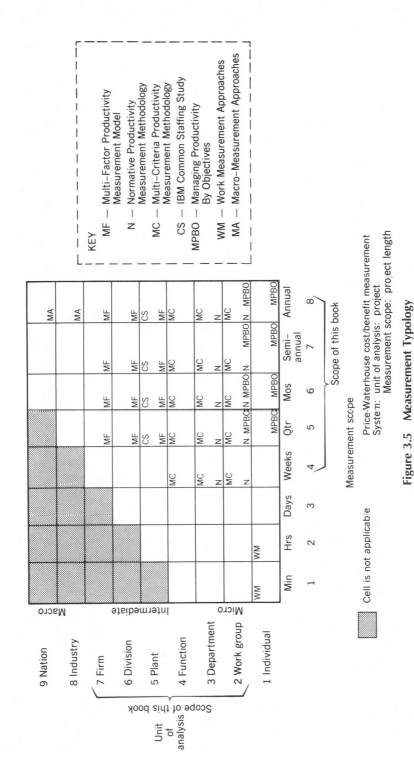

Figure 3.5 Measurement Typology

		Min	Hrs	Days	Weeks	Qtr	Mos	Semi-annual	Annual
		1	2	3	4	5	6	7	8
Macro	9 Nation								MA
	8 Industry								MA
Intermediate	7 Firm					MF	MF	MF	MF
	6 Division					MF CS	MF CS	MF CS	MF CS
	5 Plant					MF MC	MF MC	MF MC	MF MC
	4 Function			MC	MC	MC	MC	MC	MC
Micro	3 Department			MC	MC	MC	MC	MC	MC
	2 Work group			N MC	N MC	N	N MC	N MC	N MC
	1 Individual	WM	WM	N	N MPBO	N MPBO	N MPBO	N MPBO	N MPBO
						MPBO	MPBO	MPBO	MPBO

Measurement scope

Scope of this book

Price-Waterhouse cost/benefit measurement
System: unit of analysis: project
Measurement scope: project length

Unit of analysis — Scope of this book (7 Firm, 6 Division, 5 Plant, 4 Function, 3 Department, 2 Work group)

☐ Cell is not applicable

▨ Scope of this book

KEY

MF — Multi-Factor Productivity Measurement Model
N — Normative Productivity Measurement Methodology
MC — Multi-Criteria Productivity Measurement Methodology
CS — IBM Common Staffing Study
MPBO — Managing Productivity By Objectives
WM — Work Measurement Approaches
MA — Macro-Measurement Approaches

in-chief of *Business Week*, writes of the importance of considering the users' perspective:

> Probably the most important management fundamental that is being ignored today is staying close to the customer to satisfy his needs and anticipate his wants. In too many companies, the customer has become a bloody nuisance whose unpredictable behavior damages carefully made strategic plans, whose activities mess up computer operations, and who stubbornly insists that purchased products should work.

Note that in regard to productivity measurement, we are not necessarily talking about "external" customers. Every organizational system imaginable has clients or vendors (suppliers if external, and support groups if internal) on the input side and customers or users on the output side. These people and organizational systems can have important viewpoints on how productivity might be measured, what productivity ratios are important, and even what productivity standards should be achieved. American manufacturing firms are learning this lesson the hard way from the Japanese in the area of vendor relations.

Participation in problem solving, problem analysis, decision analysis, decision making, and even implementation has matured and evolved to such an extent that in one form or another, it is quite widespread in the United States. In the fifties and sixties, management theorists provided us with the underlying philosophy supporting participative management styles. McGregor, Likert, Argyris, and many others presented strong, rational arguments for "tapping the human resource" in more effective ways. Throughout the sixties and most of the seventies, managers, academicians, consultants, and others struggled to operationalize the theories and concepts of participative management. As Peters and Waterman point out, excellent companies have exhibited for some time significantly better communication and involvement on the part of and between customers, management, and employees at all levels.

When quality circles hit the United States in 1974, participation and involvement took shape and form for a greater number of managers and companies. The technique (cookbook style) for making participation "work" was neatly spelled out—step 1, step 2, and step 3. Participation came off the bookshelf, out of a theoretical context and onto the shop floor. There are, to be sure, still skeptics—those who think employees should do what they are told, when they are told to do it, and exactly how they are told to do it. And, of course, many managers have learned that a leader's style must vary according to a number of factors, such as maturity of the followers, the task, and the situation. But the point to be made is that the legacy quality circles will leave to American management will be that participative management is indeed a possibility. Quality circles have convinced and shown a larger group of American managers that involving employees, at all levels, in participative problem solving has substantial benefits. There are many lessons being learned and many more to be learned about how to incorporate involvement as a natural management process. However, the positive fact remains that American managers are now experimenting more aggressively with new patterns of employee involvement.

It is important that the reader not "lose sight of the forest for the trees" at this point. An overall and very general strategy for developing a productivity measurement system has been presented. Certain basic system design specifications have been outlined and discussed (planning horizon, desired outcomes, scope, intended interface with other performance measurement systems, development plans and procedures, and explicit mechanisms by which to convert strategy and plans into tactical and operational plans and eventually into actions.) At this juncture, we are now prepared to specifically examine and learn alternative approaches and techniques for measuring productivity of an organizational system.

Three Techniques

In the next three chapters, we will examine the three most developed techniques for measuring productivity. In fact, it would not be stretching too far to state that these are the only three explicit, systematic, and pure productivity measurement techniques. The reason is that the techniques presented here are the only three approaches that adhere strictly to the definition of productivity as developed in Chapter 2.

We will first take a detailed look at the normative productivity measurement methodology (NPMM), a methodology incorporating involvement as a major component in the approach. This methodology was first studied in 1976 by Drs. William Morris and George Smith at The Ohio State University. These two researchers headed a project sponsored by the National Science Foundation in which they tried to develop innovative productivity measurement systems for administrative computing and information services. Two important processes used in NPMM are the nominal group technique (NGT) and the Delphi technique (DT). Both are utilized to develop consensus measures of productivity for a given organizational system. The NGT will be described and detailed, and then the methodology using these techniques will be outlined. The Delphi technique is very similar to the NGT. However, it is primarily executed in a survey fashion. It will not be discussed in detail, although specific references are provided for those interested. Case studies are provided to clarify steps.

In Chapter 5, we will examine in some detail a multifactor productivity measurement model (MFPMM). This model does not incorporate involvement in any major form in the collection of data. It is a more macroscopic measurement approach. It also structures the data input in such a way as to adhere, automatically, to strict definitions of productivity outlined in Chapter 2.

The Multicriteria performance/productivity measurement technique (also called the objectives matrix) is presented in Chapter 6. This technique allows for measurement and evaluation of performance (the broader issue) or productivity (output/input) and, most importantly, it provides a mechanism for developing an aggregate performance or productivity index. The technique has roots in multiattribute decision analysis. It is very compatible with the NPMM, and we will discuss how these two techniques can be integrated.

The techniques presented differ primarily in terms of what data are collected and how the data are collected. They also vary fundamentally with respect to professional mode of functioning. That is, the techniques vary in terms of what role the participants in the system to be measured play in the development of which measures to look at. The NPMM allows certain participants to determine which measures will be part of the measurement system. In contrast, the MFPMM prescribes and dictates what measures will be provided. There is some input on the part of the analyst as to which measures will be examined. However, this input takes place only through the decision of what data can be or will be collected. The MCP/PMT has sufficient flexibility to allow for data collection in a variety of ways. It is primarily an evaluation technique that allows for the aggregation and analysis of performance against a variety of criteria.

The techniques also differ in terms of units of analysis that are appropriately handled. The NPMM has been found to be particularly useful at the work-group level and at the department or division level. It becomes cumbersome at higher units of analysis. In contrast, the MFPMM has been found to be most appropriate at the firm level. It can, however, be utilized at lower units of analysis, such as for cost centers, profit centers, plant, division, and function.

Because of the nature of the data and the different methods of data collection, the three techniques also differ significantly in terms of their uses and their ability to behaviorally *link* measurement activities to improvement actions. Since the NPMM is participatory, it has proven to be more effective at specifically motivating action from the measurement activities themselves. In contrast, the MFPMM has proven to be more of a management diagnostic tool, pointing the direction for improvement but not necessarily motivating specific actions. Note that this is not a criticism of the MFPMM. The MCP/PMT has the flexibility to be developed either participatively or autocratically. The three techniques have simply been designed to do different things. One is primarily a firm-level diagnostic tool; the other two are group-level awareness and planning tools. They all have a place and a purpose depending on specific desired outcomes determined in the strategic plan.

It also is important to point out that one is an "absolute" measurement technique while the other two are primarily "relative" measurement techniques. That is, the absolute MFPMM creates essentially the same set of measures regardless of organization, whereas the relative NPMM and MCP/PMT generate measures that are very much specific to the organization and group developing the productivity measurement system. An absolute system's basic structure will remain fairly constant over time, whereas the relative system's basic structure will be much more dynamic since people's perceptions are involved. Again, one technique is not better than the other's; they have just been designed to do different things.

REFERENCES

Archer, B. L. *Technological Innovation: A Methodology*, Royal College of Art, London, (unpublished working paper), 1970.

Dunnette, M. D., ed. *Handbook of Industrial and Organizational Psychology*. Chicago: Rand McNally, 1976.

Guion, R. M. *Personnel Testing*. New York: McGraw-Hill, 1965.

Kerlinger, F. N. *Foundations of Behavioral Research*, 2nd ed. New York: Holt, Rinehart and Winston, 1973.

Morris, W. T. *Implementation Strategies for Industrial Engineers*. Columbus, Ohio: Grid, 1979.

Morris, W. T., et al. *Measuring and Improving the Productivity of Administrative Computing and Information Services: A Manual of Structured, Participative Methods*. The Productivity Research Group, The Ohio State University, NSF Grant #APR75-20561, Columbus, Ohio, 1977.

Peters, T. J., and R. H. Waterman. *In Search of Excellence: Lessons from America's Best Run Companies*, New York: Harper and Row, 1982.

Smith, G. L. *Work Measurement: A Systems Approach*. Columbus, Ohio: Grid, 1978.

QUESTIONS AND APPLICATIONS

1. Develop a two- to five-year plan for the design, development, and implementation of a productivity measurement and evaluation effort.

2. Discuss the difference between data and information.

3. Discuss the difference between a management information system (MIS) and a decision support system (DSS).

4. How does one ensure that a productivity measurement and evaluation system will in fact be both a DSS and an MIS?

5. Differentiate between a productivity measurement system that is intended primarily for evaluation and control and one that is intended primarily for development and improvement. Are these two purposes compatible? How would one design a system that could accomplish both purposes?

CHAPTER

4

THE NORMATIVE PRODUCTIVITY
MEASUREMENT METHODOLOGY (NPMM)

HIGHLIGHTS

- The Background
- The Methodology
    Design Principles
    Methodology
- A Brief Description of the Nominal Group Technique
    The Process
    Composition
- Using the Nominal Group Technique Effectively
- References
- Questions and Applications

OBJECTIVES

- To instruct the reader how to develop and implement productivity and performance measurement and evaluation systems using structured participative approaches.
- To acquaint the reader with the development work associated with the use of the nominal group technique.
- To present sufficient instructional material on the NGT to support minimum learning requirements for this technique.
- To expand the reader's appreciation for the range of applications for the NGT.

The Background

In 1975, the National Science Foundation, Division of Research for Applied National Needs, funded a number of research projects focused on investigating innovative ways for measuring productivity in certain "hard-to-measure" or not traditionally measured functions and industries. Administrative computing and information services, as well as banking, were two of the more notable areas examined. The Ohio State University was among a select group of institutions to receive a grant to study productivity measurement. Drs. William T. Morris and George L. Smith, researchers in the Department of Industrial and Systems Engineering, directed a two-year study that focused on the development of a normative productivity measurement methodology, for Administrative Computing and Information (ACI) services. The investigation operated under several specific sets of assumptions or premises regarding the nature of ACI services and the special characteristics of the productivity measurement process relative to ACI services. The results of that research project and spinoff research and development that ensued form the basis for this section of the book. Here are the assumptions and premises of that project, taken directly from a research document entitled *Measuring and Improving the Productivity of Administrative Computing and Information Services: A Manual of Structured, Participative Methods*, (1977).

ACI services have evolved with considerable rapidity, have undergone continuing evolution driven by high innovation rates in the underlying technology, and have often been imperfectly assimilated into the operating and management structure of the organizations they serve. Three basic problems confront those who seek to design and manage ACI services:

1. Beyond some obvious and limited measures, there is little in the way of methodology for assessing productivity, little cost-effectiveness data, and virtually no explicit evidence on which management and planning decison for ACI services might be based.

2. As ACI services evolve toward a greater concern for the management information system function, basic conceptual problems are encountered. The "users," the clientele, the managers whose decision making these systems seek to enhance, are often not clear about their real information needs. Nor is there a well-developed method of determining these needs at present. Understanding the effectiveness of a management information system requires an appreciation in detail of management decision processes and a clarification of the goals of the organization itself. These considerations lie behind the results of recent surveys which indicate that computers have so far had only a modest impact on management above the lower levels.

3. Most ACI service systems have been "designed" or brought into being using some mix of three basic design philosophies:
    (a) Mechanize the existing manual system.
    (b) Utilize the best data processing equipment one can afford.
    (c) Ask managers what information they want and provide as much of it as the budget will permit.
These design philosophies have usually produced less than satisfactory results, thus there are few outstanding models to which one might look for guidance in management, design, or productivity comparison.

Productivity Measurement for ACI Systems

Productivity for ACI services must be thought of, not in terms of a single measure, but rather in terms of a series, list, or vector of measures. The vector will include individual measures which attempt to assess efficiency and effectiveness from the viewpoints of

- ACI service staff and operating management
- the various clients of the service, including those who receive its information and documentary output, as well as the managers whose decision making it seeks to enhance
- the overall organizational objectives as enbodied in the planning and dicision making of top managers who determine ACI service budgets and policies.

Efficiency measures seek to assess how well available resources are used, or how well the inputs are utilized in performing the functions undertaken by the ACI service. They tend to be relatively inexpensive, "objective," unobtrusive, and suited to portray only a very narrow view of the service's "real productivity." Efficiency measures often involve techniques well established in industry, such as work measurement or the direct analysis of operating data. Typical efficiency measures for ACI services include throughput time, turn-around time, machine availability times, CPU utilization times, quantities of documents, reports or data produced in relation to time, cost, and staffing levels, operating error rates, and so on.

Effectiveness measures attempt, on the other hand, to assess the quality of results of ACI service activities, what is accomplished with respect to overall organizational goals, the impacts of ACI services on the intended purposes of the organizational goals, the impacts of ACI services on the intended purposes of the organization and its various clienteles. Effectiveness measures tend to be relatively expensive, poorly understood, controversial, mistrusted, qualitative rather than quantitative, and "subjective" rather than "objective." They are a strong function of one's perception of the purposes of the ACI service and they represent attempts to portray the "real productivity" of the service.

Behavioral Consequences of Productivity Measurement

It has been clear since the Hawthorne studies that if productivity is measured, the process of measurement is almost certain to be accompanied by a productivity increase. It seems clear as well that the more people involved in the productivity measurement process, the greater the associated productivity change and the greater the acceptance of the resulting measures as being "fair" or "representative."

Participation in the process of designing a productivity measurement system captures viewpoints, permits the expression of concerns, and creates an involvement that enhances the process of implementing changes resulting from management actions or system redesign based on productivity measures. Participation in the design of productivity measurement systems seems likely to facilitate the trend in public sector management away from resource-oriented, budget-conforming, management decision making and toward results, impact, or program-oriented management planning.

The Role of Analysts in Productivity Measurement

The strategy of implementing a set of productivity measures created by analysts, technicians, or "experts" has a long record of the most limited success. The results have often been controversy, dissatisfaction, and the gradual decay of the productivity measurement process. The role of analysts, it would now appear, can best be that of facilitators of communication and change, of structure-giving, of potentiality-indicating, and of designers of systems by which organizations can participate in the creation of their own productivity measurement systems.

Major Considerations

The major considerations in creating productivity measurement systems for ACI services include the following:

1. What is needed is not a standard set of measurements created by experts and imposed on organizations, but rather a method by which ACI groups and their various clienteles can create productivity measurement systems suited to their own inevitably special circumstances.

2. The greater the participation in the process of creating a productivity measurement system, the greater the resulting productivity change, and the greater the ease of implementation of future changes based on productivity measurement.

3. Any system should result in a vector of productivity measures, not attempting to achieve a single measure. Much of the controversy and lack of acceptance stem from attempts to make a very complex problem appear simple.

4. Most productivity measurement systems have attempted unsuccessfully to rely on analysts for their design and implementation. The role of analysts can more effectively be that of facilitators and stimulators of the creation of measurement systems by those involved in the organization.

5. A successful system (one that is actually used in the management process and leads to behavioral change) must be simple and cost effective.

6. A productivity measurement system must not appear to those concerned as simply a passing fad that will soon disappear.

7. A system must indicate clearly how it fits into the management process and the ways in which it can function as a decision aid.

8. The behavioral consequences of productivity measurement must be anticipated and reflected in system design.

9. A useful productivity measurement system should be aimed strongly at identifying the "horror stories," the really bad aspects of any operation, the places where the greatest potential for productivity improvement exists.

10. A useful system must be seen by those whose behavior is being scaled as essentially nonmanipulative.

11. The possible dysfunctional consequences of a productivity measurement program must be explicitly identified. One must be open about the possibilities for "gaming" the system.

12. Every system will be the target of criticism. It is important not to claim too much, to identify clearly the possible shortcomings, and to emphasize the role of overriding judgments in order to blunt this criticism.

13. A cost-effective productivity measurement system involves a large amount of self-observation, self-reporting, self-analysis, and self-design of jobs and programs.

These comments, assumptions, premises, or statements laid much of the foundation for the research that followed. However, an additional set of "basic principles" also played a critical role in the research plan. These basic principles focused on the role that participation and group processes might play in the development of productivity measurement systems. If we look back on these basic principles, it is quite easy to underestimate the insight and foresight it took to envision the importance of group activity in productivity programs in the eighties. In 1974 and 1975, quality circles were not a household word, and the prominence of small-group activity in Japan was not widely recognized. Further, it is quite certain that the Ohio State study is the first to have considered the role of participation and structured group processes in the design and development of productivity measurement systems.

The following pages are excerpts from the manual just mentioned and reflect the "basic principles" associated with the role that participation or small-group activities can play in the design and development of productivity measurement systems.

We are in the midst of a slow diffusion process involving an increasing tendency for managers to use participative strategies for decision making, problem solving, planning, and design tasks. Groups are being used for these undertakings, rather than individuals, for several reasons:

1. Groups bring together knowledge and skills not possessed by an individual.

2. Groups are more effective than individuals in purging their work of errors and avoiding mistakes.

3. The product of a group is more likely to be accepted by those who must act on it, than is the product of an individual.

4. If the members of the group must themselves act on the basis of their group efforts, not only are they more likely to accept the group's findings, but they are also likely to be more effective or productive in the ensuing actions.

5. Group members learn through group processes and hence participative strategies are chosen for their educational effects. These effects presumably enhance the qualifications and future effectiveness of the group members.

6. There is a "group effect" through which group members learn from each other, stimulate each other, and supplement each other's knowledge and skills. The product of the

group is thus in some sense greater than the sum of the individual contributions. Groups operate in ways which are synergistic or superadditive.

In the experience of many managers, the validity of these reasons is such as to make participative strategies appealing in many important instances of design, decision making, planning, or problem solving. Group processes, however, are not without their shortcomings. These shortcomings or "group process losses" make participative strategies expensive almost always and seriously ineffective sometimes. Both experience and research testify to the following sorts of difficulties that arise in the functioning of groups seeking to accomplish a fairly specific task:

1. There appear to be some tasks that are simply not well suited to group methods. The pooled output of noninteracting individuals turns out to be more effective.

2. Groups fall into a pattern of overlooking or not dealing seriously with information or ideas that are inconsistent with a socially acceptable solution or product. Groups develop preferred ways of looking at problems that inhibit their abilities to look at innovative approaches.

3. The synergistic or superadditive effects are simply not present. "Brainstorming," for example, has a long record of popularity as a creative method, but research has shown that group creativity using brainstorming does not in fact exceed individually produced and combined results.

4. Politics, power, and organizational position are associated with the dominance of the group's methods and results by some members and the corresponding suppression of contributions by others. Such domination influences adversely the information that is available to the group, the weighting that the group puts on various member contributions, and the strategies used by the group to accomplish its task.

5. One of the functions of groups is to fulfill the social needs of the members. Some socializing is clearly functional but the group seldom has ways of regulating the amount that occurs.

6. It is a fairly reliable aspect of group behavior that the discussion falls into "ruts" that persist well beyond the point of being useful in task performance. Groups get "off the track" and get stuck there.

7. Groups tend to have relatively low aspiration levels with respect to the quality of solutions accepted. Once a solution or product is achieved that appears to have some implicitly understood level of member support, the result is accepted and there is little further search for better solutions.

8. The way in which the group utilizes and communicates the relevant information possessed by its members is at least as important as the nature of this information itself. Groups typically lack both the concern and the methodology for dealing with this problem.

9. Real groups (as opposed to those created in the laboratory) exhibit behavior that is strongly influenced by the cultural norms of the members. It would appear that in many ways these norms detract from the effectiveness of the group. In natural groups the members avoid uncomfortable, deviant behavior. They tend to be conservative, to avoid arousing anxiety, to take few personal risks, and generally maintain a level of circumspection.

10. If members find their contributions valued by the group, they will be increasingly reinforced in their efforts and the level of energy contributed to the group task will rise.

If, as is often the case, their efforts do not find reinforcement, they may reduce the level of energy devoted to the group task.

11. It appears to be generally true that as group size increases, the level of effort contributed by an individual member decreases.

12. Groups reliably exhibit norms against devoting time to planning their methods or strategies. Instead, they move immediately to attacking the main problem at hand, relying on implicitly shared methods of procedure. Thus, there is considerable likelihood that the group's method will be inflexible, poorly adapted to the task, and only modestly effective. Further, groups seldom have the ability to change their methods when things are not going well.

Efforts by group members, typically leaders, and outsiders to deal with these sorts of shortcomings constitute a particular class of interventions known as "process consultations." The objective of a process consultation effort is not to deal with the substantive knowledge possessed by group members, nor to contribute substantively to the solutions considered by the group. The objective is rather to intervene in the composition, size, group norms, and processes used by the group in ways aimed at increasing the effectiveness of the group's performance. Three forms of process consultation may be broadly distinguished, although each of them seeks to help groups find more effective methods of task accomplishment.

1. *Ad hoc interventions:* As practiced by Jacques (1952) and Schein (1969), these interventions involve a consultant working closely with a group, feeding back to the group his analysis of their methods and behaviors, and subtly leading the group to develop more effective working methods. These interventions are called "ad hoc" because they do not involve any method that is readily transferable, they depend very heavily on the experience and clinical skill of the consultant, and there is no systematic evidence on how to carry them out. They are described typically in the form of case studies that while rich in detail, seem to hold relatively little generalizable evidence on the methods or effectiveness of the interventions.

2. *Team building:* This type of intervention aims at the long-range modification of the social climate in a group and the development of more effective interpersonal relationships. This is necessarily a long-range strategy that requires the group members to focus their attention on a variety of interpersonal experiences that constitute the team building process.

3. *Structured group-process consultations:* The consultant or group leader involves the participants in a carefully designed strategy for the effective accomplishment of the task. The process consultant seeks also to influence the composition and size of the group and to carefully structure the task presented to the group in order to enhance the effectiveness of the process.

Two of the most widely known examples of structured group processes are the Nominal Group Technique and the Delphi Method, which are described below. As in other process consultations, the structured group-process approach gives the consultant a low profile role as a provider of methodology or strategy, rather than as a substantive resource for the group. In this approach, however, there is little that is casual or unplanned about the group process. It typically consists of a set of clear, well-defined activities that the participants are asked to undertake at specific times. So far as process is concerned, the group has little discretion, although it may sometimes be asked to make deliberate process choices by the consultant.

The considerations involved in the design of a structured group process and the factors that may influence its effectiveness may be usefully organized according to the model indicated in Figure 4.1. The "group process" includes those activities in which the participants engage at the direction of the consultant. In the Nominal Group Technique, for example, they include silent generation of ideas, round-robin listing, clarification, ranking, and voting. The process outcomes are products and effects that are available immediately on completion of the group process. A ranked list of new project proposals would be a typical example of a product that would be available immediately after the group process. Better distribution of task relevant information among the participants would exemplify an effect that might be achieved by the time the group process is completed.

Post-group-process outcomes occur days, months, or perhaps even years after the process and include behaviors that are believed to be in some measure related to the group's participation. They might include such behaviors as implementing the findings of the group, continuation of the group's efforts without the consultant, changes in intrapersonal relationships, as well as structural, technological, or process changes in the organization itself.

The relationships between the group process and these two classes of outcomes are influenced in complex and imperfectly understood ways by a large number of factors that have been roughly classified in Figure 4.1 as individual, group, and organizational factors. Individual factors that may be significant include the knowledge, skills, attitudes, needs, and interpersonal styles of the individuals in the group. Group factors likely to be of some importance in determining the relationship between process and outcomes include the size of the group, its homogenity with respect to organizational level and type of job, the group norms, and its degree of legitimization within the organization. Organizational factors are likely to include the reward structure of the organization, the importance of the group's task with respect to organizational goals, the stresses under which the group must operate, and the general tendency of the organization's management to be either authoritarian or participative in decision making.

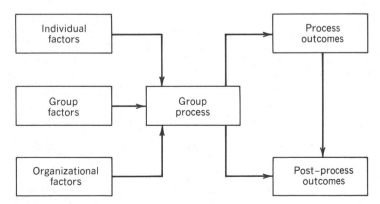

Figure 4.1   Simple Model Depicting Factors Influencing Process Outcomes
(Adapted from J. E. McGrath, *Social Psychology: A Brief Introduction*. New York: Holt, 1964)

We will shortly come to speak of the effectiveness of group processes, and it is important to clarify immediately some of the issues involved in assessing process effectiveness. Nearly all of the research in Social Psychology has considered effectiveness exclusively in terms of process outcomes. The typical approach here is to consider the quality of the group product, the quantity of product, and participant satisfaction with the group process and its immediate outcomes. If, for example, a group undertook to produce methods for measuring the productivity of a research team, effectiveness would perhaps be thought of in terms of the originality, sophistication, and organizational acceptability of the measures produced; the number of measures produced; and the expressed satisfaction of the participants with the process of working together and their accomplishments.

If the product is to be turned over to someone else for use, their quantity and quality may be sensible measures of effectiveness. If the group must accept or implement its product, then there is a presumption that if the participants liked the process and are satisfied with the product, they will be likely to accept it and act on it. Participant satisfaction is thus often supposed to illuminate the relationship between process and post-process outcomes. There is virtually no research in which effectiveness is considered in terms of post-process outcomes. Often, however, the objective of the group process is not simply to produce a product for someone else's use, but rather to begin a series of steps designed to lead to post-process outcomes. Effectiveness in these latter cases will need to be defined in ways that are specific to each situation and will necessarily include behaviors outside of, and subsequent to, the group process. Does the group, for example, having produced some measures of research team productivity, go on to refine these measures, implement them, and ultimately play a role in the improvement of productivity?

We next outline briefly the possible advantages of structured group-process consultation, its underlying general strategy, and some of the complexities that confront the group process designer.

Considered in relation to such interventions as ad hoc process consultation and team building, the structured approach is likely to offer several significant advantages.

1. It reliably leads immediately to a useful product. Process outcomes are obtained with a modest investment of participant effort.

2. The group leader is more easily trained than is the case with the other types of interventions.

3. The group leader need not depend heavily on extensive clinical skill. While clinical experience is certainly useful and some skill in facilitating group efforts is probably essential, the consultant's task is primarily that of maintaining the group's adherence to the well-defined process structure.

4. The method is robust (in the sense of reliably leading to useful process outcomes) with respect to the type of task that the group undertakes. Decision making, problem solving, problem finding, planning, and evaluation tasks may all be readily accomplished.

5. The process is generally robust with respect to group size. For example, the Nominal Group Technique has been used successfully with groups of from 13 to 14 persons.

6. The process is generally robust with respect to group composition. Providers and clients, managers and working people, line and staff, general and technical people may be included together in the same group. There are, however, instances in which group effectiveness may be improved by using more homogeneous groups.

7. The process is robust with respect to the initial motivation of the participants. Process

outcomes have been effectively achieved by participants having little initial interest in the group task and attending the session only because they were directed to do so.

8. The process reliably leads to participants being satisfied with the group process and with the process outcomes.

9. Participants learn something of the importance of their interaction process and its effect on group productivity. They experience a readily reproducible method for improving their interaction processes. The group could, if it wished, immediately adopt the structure with little further need for the consultant.

It must be emphasized that these advantages will be present to a greater or lesser degree, depending on the design of the process structure and on the constellation of factors indicated on the lefthand side of Figure 4.1. Clearly they would be expected to differ between . . . the Nominal Group Technique and the Delphi Method. These advantages, which are supported by both research and field experience, are primarily concerned with process outcomes.

Expectations with respect to post-process outcomes are subject to considerably greater uncertainty. In those cases where post-process outcomes are explicitly sought, the designer must be concerned with several questions about which we have only limited knowledge at present.

1. To what extent will the structured approach lead to the modification of group norms about task strategies?

2. In the absence of a consultant, to what extent will a group continue to work on a task, develop implementation strategies, and acutally seek to translate its process outcomes into action?

3. To what extent will the structured group-process approach succeed in moving the organization away from an authoritative-exploitive style of management and toward a participative-supportive style?

It seems probable that post-process outcomes are more likely to occur if the participants are involved in an extended learning process shaped by feedback from experience. A single structured group session can hardly be expected to make more than a dent in the norms, attitudes, role perceptions, and organizational climate. Yet there are an interesting number of instances in which a small number of group sessions have led to surprisingly effective post-process outcomes. It may be that the consultant can take advantage of interesting tradeoffs among effort invested in the intervention, the probability of desired post-process outcomes, and the persistence of these outcomes over time.

With respect to implementation in the sense of post-group-process behavior changes, one may take a fairly subtle position:

1. The group processes may be conducted so as to maximize the probability of subsequent behavior change, subject to the constraint that the consultant maintain the role of very low-profile facilitators.

2. Post-group-process activities are to be the mildest possible stimuli to implementation.

3. Under circumstances where the process would be conducted in response to a perceived need by top management and with a commitment from top management to behavior

changes, various management-guided strategies would pick up the implementation problem after the group processes were completed.

4. This subtle implementation strategy might be called a "self-actuating intervention."

A self-actuating intervention may be characterized by

1. A strictly low-profile, facilitating role for the consultant

2. A tendency to attach more importance to the behavioral consequences of the group
process experiences than to a literal, analytical, systematic evaluation of their products.
While one would very much like to see the group products meet preconceived standards
of reasonableness and usefulness, this is not the primary concern in evaluating the group
processes.

3. A recognition that whether or not any *lasting* behavioral changes result from *any*
intervention depends strongly on organizational circumstances that are well beyond the
scope of the intervention and the influence of the intervenors

4. A tendency to see behavioral outcomes as depending only in part on the intervention
as opposed to pressing hard to achieve some predetermined results

5. A tendency to design intervening activities at least as much with their predicted behavioral consequences in mind as with a concern for the literal, analytical nature of the
group products

6. The nature of the group-process outcomes is seen primarily in terms of their influence
on participant behavior. What is important is how the participants see their products and
their subsequent actions; if they decide to act they will certainly learn through immediate
experience of the need to modify, improve, elaborate, and develop the products of their
group processes.

7. An underlying assumption that a low-pressure, facilitating process that seeks to stimulate and release forces is already present in an organization. (Call this a type 1 intervention.) This type of intervention is a significantly different alternative to an intervention that
is high pressure, explicitly goal oriented, and conventional. (Call this a type 2 intervention). The rationale of the conventional, type 2, intervention is to create forces and mo-
tivations that are not already operating in an organization, or not already present in an
organization.

8. A basic hypothesis that states

Although the probability of observable behavior change is lower for type 1 than
for type 2 interventions, behavior changes resulting from a type 1 intervention are
more likely to be *lasting* than those generated by a type 2 intervention.

This type of intervention assumes that things will happen if forces that exist in an
organization are released, and it is these forces that lead to behavior change. One will
not see implementation where these forces do not exist and probably should not for the
ultimate good of the organization. Interventions should not be forced, should not strive
for 100 percent occurrence of behavior change, and should take a stimulative rather than
a manipulative point of view. This is a more realistic more honest approach.

The structured group-process consultant seeks to design in such a way as to take full
advantage of individual superiority in some activities and group superiority in others. The

underlying design strategy aims at making individuals and groups more effective by understanding and overcoming the shortcomings or process losses we have outlined above. The basic design method is to enhance group effectiveness by

1. Focusing participant effort and attention

2. Coordinating participant efforts

3. Defining and clarifying (or having participants clarify) the group task

4. Supplying and facilitating the execution of a carefully designed strategy for task accomplishment

5. Filling to a limited extent the social needs of the participants

6. Largely avoiding the dysfunctional effects of power, position, and politics

7. Assuring both the recognition and the evaluation of participant contributions

8. Permitting, encouraging, and requiring deviation from group norms

9. Arranging group size and group composition to enhance the resources available to the group and the process of making use of these resources

10. Insuring the efficient production, sharing, and evaluation of participant knowledge and skills

11. Clarifying areas of consensus and of disagreement

12. Providing task variety, identity, significance, autonomy, and feedback so as to reinforce individual commitment to the group effort

13. Permitting the group to deal reasonably with anxiety arousing, with deviant behavior, and with differences in competence among the participants

14. Helping group members to learn from each other

15. Enhancing participant satisfaction, commitment to the group's product, and consensus

The factors that influence the relationship between the group process and its outcomes (see Figure 4.1) are apparently numerous and complex. Social psychologists are not optimistic about achieving a general theory of group effectiveness in the near future. Yet, as is often the case in practical affairs, the designer must proceed as best he can, using the available research results and field experience. The result is that all group process designs are subject to greater or lesser uncertainty with respect to their effectiveness, and their use must be seen as experimental and developmental. We have set out below some of the major hypotheses that have some support in both research and field experience. They may be useful, if uncertain, guides for the structured group-process designer.

Two caveats must be made explicit in connection with these hypotheses.

1. As noted above, nearly all laboratory research involves artifically created, as opposed to natural, groups. Significantly missing are group norms, the tensions of real potential consequences, and the possibility of the occurrence of post-group-process outcomes.

2. So far as the available research goes, effectiveness is considered almost exclusively in terms of process outcomes. For applications where post-process outcomes are explicitly sought, these process-outcome measures of effectiveness are of limited relevance.

In stating these hypotheses, we will necessarily remain somewhat vague about the operational definition of effectiveness. It may be thought of generally as the quality, quantity, and efficiency of production of both process and post-process outcomes.

The ultimate aim of the process designer would be the ability to measure the independent variables in these hypotheses a priori, and thus be able to make some predictions as to the effectiveness of a particular intervention in a particular situation.

Individual Factors

Increases in the following factors are likely to be associated with increases in group-process effectiveness:

Saliency of the task

Perceived potency of the participants with respect to the task

Experience with participative methods

Consistency of conventional roles with the group task

Higher order need strengths (need for achievement, growth, variety, participation, challenge, feedback)

Group Factors

Increases in the following factors are likely to be associated with increases in group process effectiveness:

Homogeneity with respect to background, responsibility, position

Amount of task-relevant knowledge available to participants

Top management sanction of the group

Norms with respect to explicit strategy consideration

Organizational Factors

Increases in the following factors are likely to be associated with increases in group process effectiveness:

Saliency of the task

Availability of time, lack of urgency

Pressure for change

Clarity of organizational goals

Conventional use of participative methods

Tendency toward participative, supportive management style

Clearly there will be other variables not included in the above list, and the effects of those mentioned are subject to considerable uncertainty. The inevitable result is that the designer must see any specific group and group process as a highly idiosyncratic phenomenon. This is not to suggest that the available research and field exprience is useless, but rather to indicate that the design process must be a developmental one.

From a somewhat broader viewpoint, there are reasons for optimism about the success of structured group-process consultation. A study of a large number of organization development interventions suggests that there are several characteristics associated with success (Friedlander and Brown, 1974):

1. Strong internal and external pressures for change

2. Gradual involvement of many levels in diagnosis and change activities

3. Shared decision making rather than unilateral or delegated decisions

4. Shared conceptions of problems and appropriate action steps

5. Introduction of a model to guide changes

6. Establishment of linkages with others inside and outside of the target system

The first of these is situational, but the remaining factors are explicit, deliberate aspects of the structured group-process approach.

One criterion for ultimate success is the achievement of a climate in which groups become self-managing and self-developing. This is unlikely to occur spontaneously, and a consultant must start the process. The ultimate hope of the designer is that the process will be rewarding, that it will be assimilated by participants as a part of their behavior, that groups will come to regard modified interaction processes as their own, and not the work of the consultant.

These sets of assumptions, premises, basic principles and concepts, although developed between 1975 and 1977, are still highly valid and useful today. In particular, it is clear that this underlying thinking about successful design, development, and implementation of productivity measurement systems is generally applicable for all types of organizations and functions.

During the period 1975–77, the methodology that was developed was tested in four major public sector organizations: two state governments, one city government, and one large, specialized federal agency. Some 350 persons, from cabinet and top management levels down to operating personnel, have participated in nominal groups and Delphi panels. The methodology yields certain kinds of outcomes with high reliability and other kinds of outcomes that depend on the individuals, groups, and organizations involved.

Reliable outcomes are listed below.

1. The basics for a comprehensive productivity measurement system are established. The measures can reflect the viewpoints of a variety of relevant persons in the organization. Further, the measures are prioritized so as to achieve consensus on which are more and less important.

2. For the priority measures, participants in the process are involved in efforts

to operationalize the measures themselves. Most measures obtained are immediately operational. That is, most measures identified by participants in the process are either already being measured or data exist by which to specify them.

3. If outcomes 1 and 2 above are considered by management to be the basis for a productivity measurement system, the resulting system will likely have at least three significant attributes:

   (a) *Acceptance:* Those who must manage with or be evaluated by the system are more likely to accept the system because it is in reality their system.

   (b) *Acceptable quality:* The system typically includes and incorporates most of the measures typically recommended by experts, as well as measures that reflect the specific nature of the organization involved. Therefore, the research has shown, quite conclusively, that quality is not sacrificed with this approach.

   (c) *Reasonable cost:* The resulting system will be achieved with a modest amount of participant time and effort. In general, the longer term benefit-to-cost ratio is significantly greater than 1.0.

4. The organization, through this measurement approach, actually educates and develops an awareness for productivity. This outcome creates a reservoir of experience, energy, and abilities for maintaining productivity growth in the future. In a period of time when many organizations are spending considerable amounts of time and energy to create productivity awareness on the part of managers and employees, this spinoff outcome should be very valued.

5. The methodology provides a learning experience that is quite compatible with a trend toward the development of participative problem-solving techniques and programs, such as quality circles and productivity action teams. The people involved in this process become exposed to what may be evolving as a new method of problem solving, decision making, planning, and design. The methodology may represent a significant operational technique or approach for American organizations to begin to evolve toward Theory Z styles of structure and management.

Morris et al. found that although the outcomes discussed above are reasonably predictable, there are other "post-process outcomes" that are less predictable because they rely or depend on certain other preconditions, moderating variables, and so forth. In general, post-process outcomes deal with the extent and nature of follow-through and implementation of the prioritized productivity measures. In some situations, the work begun by the group-process participants was taken over and implemented by others specifically designated by management to continue the process of productivity measurement system design and development. In other cases, the participants themselves continued the process of developing and implementing their own productivity measurement system. And, in some cases, no further work took place after the initial NGT session. Keep in mind that the desired post-process outcome is to develop and/or improve productivity measurement systems in the organization. The basic reason for using a "normative" or participative approach is to ensure adequate motivation, commitment, and accountability on the part of key participants for implementation and acceptance of the resultant productivity measurement system.

Morris et al. found, not surprisingly, that the following factors strongly influenced the degree to which desired post-process outcomes occurred:

1. The degree to which top management continued to understand, maintain a commitment toward, legitimate, support, and pay attention to the "program." This sense of commitment to some master plan and the acknowledgment that there is some continuity to planning and ensuing actions is critical.

2. The salience or importance of the function for which the productivity measurement system is being developed.

3. The maturity of the organization, management, specific function, analysts, coordinators, and consultants. More important was the match between the maturity of the organization and the maturity or sophistication of the plan for designing and developing the productivity measurement system.

4. The degree to which the plan for designing, developing, and implementing the productivity measurement system is thought through by management—the quality of the planning process. Also, the extent to which the "master plan" is communicated and shared with key participants.

5. The amount and quality of basic productivity education, training, and development provided prior to start up of the program.

These factors, and probably others, are apparently critical preconditions to the successful design, development, and implementation of a productivity measurement system using this normative approach.

The Methodology

The important background development material for this methodology has been presented. The methodology itself will now be outlined and discussed. The methodology will be detailed as it was designed and executed during the Ohio State studies. Refinements to the basic approach will be introduced and discussed later. The nominal group technique will also be described later, although extensive reference to the NGT will be made in the description of this methodology.

The Ohio State University Productivity Research Group (1975–78) developed a set of guiding "design principles" that should be considered when developing a productivity measurement system. These principles follow directly from the background assumptions, premises, and hypotheses just presented. They also have previously been presented as "desired outcomes" in Chapter 3. There appear to be several broad-based fundamental design principles that, if incorporated, will yield a high probability of successful implementation of productivity management systems. These principles, though, are not particularly easy to execute. Although they are practical and intuitive, operationalizing these principles seems to be a major difficulty for management.

Design Principles

1. What is needed is not a standard set of *measurements* created by experts and imposed by organizations, but rather a method by which groups can create productivity management systems suited to their own inevitably special circumstances (OSU-PRG, 1977).

2. The greater the participation in the process of creating a productivity *management* system, the greater the resulting productivity change and the ease of implementing future changes that result from productivity management. (This principle evolves from the professions involved with helping individuals and groups change. We now know that change occurs when persons accept the need for it, make their own decisions in favor of it, and determine for themselves the directions it should take.) (Morris, 1979)

3. Any productivity *measurement* system should result in a vector of productivity measures and/or ratios, not an attempt to achieve a single measure and/or ratio (OSU-PRG, 1977).

4. A successful productivity *management* system (one that is actually used in the management process and leads to behavioral change) must be simple and cost effective (OSU-PRG, 1977).

5. A productivity *management* system must clearly fit into the ongoing management process/systems.

6. A successful productivity *management* system should operate under Pareto's Principle. It should provide the organization members with the ability to identify critical problems or opportunities. Further, it should incorporate a system by which only the critical or priority problems or opportunities are tackled (that is, resources committed).

7. Successful productivity *improvement* systems depend largely on the ability to establish a track record of successes early in a program in order to create appropriate expectations and incentives.

8. Successful productivity *measurement* systems are as concerned with how they measure, whom they measure, and who decides what to measure as with what is measured.

9. Successful productivity *management* systems incorporate involvement, as mentioned in the second design principle. Beyond that, successful systems incorporate *effective* and *efficient* involvement as perceived by the participants.

10. Successful productivity *management* systems are not considered a substitute for sound, disciplined management. In fact, in the absence of a disciplined

management environment, such systems or programs have little chance of success.

11. Successful productivity *management* systems recognize that productivity measurement and improvement can occur relative to a number of key input variables (human, technology, materials, work processes, manufacturing processes, and so forth). Recognition of where the "leverage" exists for a given organization or group is critical. Further, there is a complex relationship between the input varibles that must be considered (technology and humans; work process and humans; technology and work processes).

12. Finally, successful productivity *management* systems are perceived by the organization as being systematic, explicit (while still respecting the necessity of intuition), consistent with management style, and action oriented.

The Ohio State University Research Group devoted a considerable effort to the task of designing a productivity management system that would successfully incorporate these design principles. The result is a methodology that is depicted in Figure 4.2 and outlined on pages 112–121.

Methodology

The fundamental applied research methodology that Ohio State developed is shown in Figure 4.2. It is an action research model (formulates objectives, research questions, etc.; gathers data; analyzes data; diagnoses; acts; evaluates; formulates objectives, etc.) which is not overly sophisticated. Operationalizing concepts presented in Chapter 2 provides a useful plan of action that managers could use to design their productivity management systems. The methodology is a "how-to-do-it" model that builds on basic concepts and incorporates the principles discussed above. The methodology has been executed in more than 25 organizations, and it has succeeded in developing productivity measurement and improvement systems.

The following discussion will analyze and interpret Figure 4.2 from left to right on the flow process chart. On the far left side of the chart and at the outset of a productivity measurement system development, it is assumed that a necessary precondition would be some reasonably strong desires, interests, and motives for beginning such a program. Perhaps a grass-roots movement has brought to management's attention the need for such an effort. Or, as may often be the case, someone in upper management has recognized the need to examine this area. These antecedent conditions *must* lead to and be strong enough to cause sincere, comprehensive, persistent, and well-informed legitimization and support from top management. Who the top managers are depends totally on the scope and unit of analysis of the study. Relative to the scope and intended unit of analysis is the support and legitimization required from top or upper management.

Another early necessary precondition is the existence, selection, and involve-

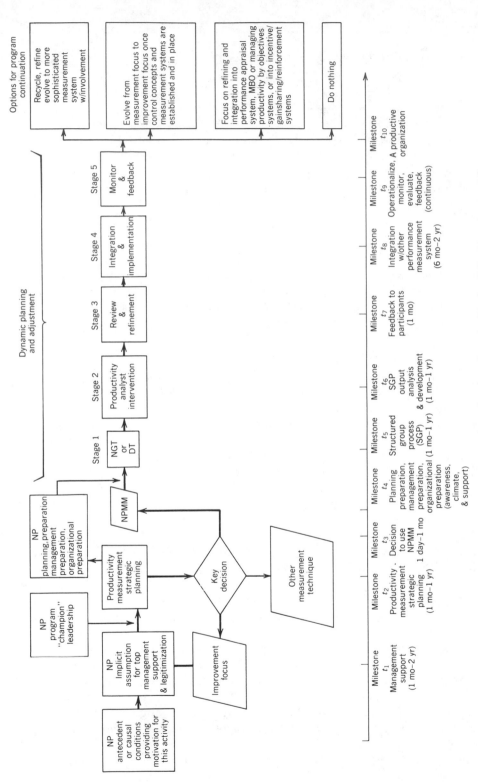

Figure 4.2 Design, Development, and Implementation of a Productivity Measurement System Utilizing the Normative Productivity Measurement Methodology

ment of a program director, leader, coordinator, and so forth. This person should obviously be well informed as to productivity perspectives and basics. In recent articles, I have presented a case for the role of the industrial engineer in these types of programs (Sink, 1982). There is no one best discipline for this position or role, although there likely is a best person within your organization. Perhaps most importantly, this person should be capable, willing, and have the opportunity to become a product or program "champion." Peters and Waterman (1982) discuss the importance of and use of "champions" in successful American firms. Basically, a champion is someone who believes; has a vision of what can and should be; is persistent; accepts failures as learning experiences and challenges; understands the concept of experimentation as a mechanism for growth and progress in oranizations; has low fear of failure; has high need for achievement; understands and can shape and facilitate change in organizations; and thoroughly understands, in this case, productivity measurement strategic planning for his or her particular organizational system.

On the bottom of the figure, a time line has been drawn in. This time line attempts to provide you with general guidelines as to milestones or critical events in this program development and a range of duration for a particular organizational system to reach or achieve this milestone. It is important to understand that the values indicated for accomplishing specific milestones depend on a wide range of factors, including (1) the scope of the program; (2) the stage of development of the specific organizational system with respect to performance measurement systems and more specifically with respect to productivity concepts; (3) staffing and other resources allocated to the program development; (4) the commitment by interface personnel, users, management, and so forth, particularly in terms of time they will actually devote to working on the project; and (5) the overall stage of development of the organization with respect to productivity management. (See Chapter 14 on the "grand strategy.")

The first milestone in this program, t_1, has to do with level of management support achieved. This is a particularly critical milestone. More than one program development has failed "down the road" because management support was not comprehensive or thorough enough. Unrecognized "pockets of resistance" severely impair and impede many programs. Attention to creating widespread awareness through the communication of appropriate expectations, realistic goals, real commitment; a willingness to legitimize the program upward, downward, and horizontally in the organization over the long run; and enthusiasm for the potential outcomes is critical. It is extremely easy to underestimate and misappreciate the time and effort it takes to accomplish this milestone. And it is also important to note that it is not enough to simply establish this milestone. This support and legitimization must be maintained.

This milestone could, perhaps in the best organizations, be established in a month. In an organization just encountering productivity as a performance measure and at an early stage in the "grand strategy" or in the productivity management program evolution, this milestone might take two years or more before it is accomplished successfully. If this were the case, you can see the benefit of realistically developing a three- to five-year strategic plan and in getting management to accept that plan. There are far too many cases where upper management has tried to install "programs" in six months without establishing certain necessary preconditions. Unfortunately, when these "programs" fail, it is the program that is criticized rather than management's ineptitude with successful implementation strategies.

Milestone t_2 represents development *and* approval of the productivity measurement system strategic or master plan. The components of this plan have been presented and discussed in Chapter 3. As a brief review, it would necessarily include and address focus and scope; unit of analysis; integration of decoupled measurement systems; purpose and objectives; desired process outcomes; desired post-process outcomes; key persons who will participate; types of involvement techniques to utilize; preliminary and ongoing training and development; specific action plans; and overall project management plans. Involvement, understanding, commitment, and approval to and for this plan is important from all appropriate key decision makers.

At this stage in the design, development, and implementation of a productivity measurement system, the decision as to what specific type of productivity measurement technique(s) to employ will be made. It is possible that an organization might decide on at least three, not necessarily mutually exclusive, options. At this juncture in the development process, management might opt to forego measurement and move toward a more improvement-oriented focus. In fact, the nominal group technique is an integral part of the productivity or participative action team process to be discussed in Parts IV and V. Management may also determine at this stage that the existing performance measurement system is sufficient, but that a more proactive improvement effort is in line.

The second option might be to decide that other productivity measurement techniques are more appropriate. For instance, as a result of the strategic planning process, management may choose the multifactor productivity measurement model for its performance measurement system needs. Development of this technique or model will be detailed in the next chapter.

And, of course, as indicated in the flow process diagram, management could decide that the NPMM is appropriate. The remainder of Figure 4.2, from the NPMM decision, milestone t_3, outlines the sequence of activities that actually constitute the five-stage NPMM.

Milestone t_4, specific NPMM planning preparation, appropriate management preparation, and organizational system preparation is a critical "necessary precondition" that will strongly moderate the success of the NPMM application. This activity essentially entails successfully "setting the stage" for the program. Awareness, plan communication, expectations, training and development, goals, objectives, desired outcomes, and specific roles are all components of this milestone. Training of facilitators for nominal group technique sessions will be required. The classic book on structured program planning techniques by Delbecq et al. is highly recommended for this preparatory stage (Delbecq, Van de Ven, and Gustafson, 1975). This book is essentially a handbook and a "cookbook" on how to execute the nominal group and Delphi techniques.

Stage 1 of the NPMM, milestone t_5, involves execution of the NGT or DT as a mechanism for generating a prioritized list of measures for each specified unit of analysis. In the case of the NPMM, it would be assumed that the unit of analysis would be the work-group level or representatives of work groups, divisions, functions, and so forth, that have been formed into small groups of 6 to 12. See Figure 4.3 for graphic clarification of alternative formulations of participative groupings. Note that it is not likely, particularly in the case of NGT application, that the unit of analysis will be very large since the NGT is a structured, *small*-group ($6 \le n \le 12$) process. As such, the development of a productivity measurement system using the NGT is, at best and almost by definition, a decoupled productivity measurement system with potential for coupling. Delphi technique characteristics, in contrast, are such that it might be and has been utilized for larger units of analysis and so forth if the task statement is appropriately phrased.

Specifically, in the case of the NGT, small groups of approximately 6 to 12

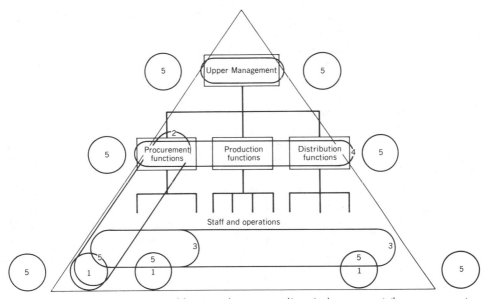

1. Work group composition (like a circle on a quality circle program) (homogeneous)
2. A functional slice from top or middle management down to line (heterogeneous)
3. Intra- or interfunctional first-line supervisor composition (heterogeneous)
4. Intra- or interfunctional horizontal slice of managers (heterogeneous)
5. Customers/suppliers can be involved in the process

Figure 4.3   Alternative Strategies for Group Composition Depending on Goals, Objectives, and Unit of Analysis

persons are formed either naturally (through work groups, departments, and so forth) or are brought together from various groups as representatives for the process. These groups are provided with a carefully designed "task statement" that is utilized to delimit and define the group's task for the NGT session. In the case of the NPMM, they are asked to "identify and list either measures, ratios, and/or indexes of productivity" for the focal organizational system.

The NGT is utilized as a mechanism for which the results are obtained efficiently and effectively; participation might foster commitment to follow up, follow through, and/or accept results as post-process outcomes; there is equality of participation in the process; there is a mechanism for forming closure on the process (priorities, preferences, and/or relative importance between measures, ratios and/or indexes are set); and the potential for group process synergy is in fact achieved. In a sense, both the NGT and DT are simply participative data collection and aggregation devices. However, for the NGT in particular, the true strength lies in the ability to act as a catalyst for action and change. The techniques facilitate group consensus, a task that may come naturally to the Japanese but that is very difficult and unnatural for most Americans.

The output of Stage 1 is a prioritized list of productivity measures (surrogate), ratios, and/or indexes. Recall from Chapter 2 that what many might term as productivity measures (for example, user satisfaction, quality levels, communication, leadership, quality of supervision, production, turnover, and absenteeism) are by definition, at best, surrogate "measures" of productivity, since productivity is strictly and correctly defined as a ratio of quantities of output to quantities of input. In fact, what most people call productivity measures are actually other measures of performance (efficiency, quality, and so forth), characteristics of the management process that influence productivity, or characteristics of the organization itself that are related to productivity. If we ask participants in an organizational system to tell us how to measure productivity without defining the term for them, they are likely to respond with factors that will relate to or correlate highly with productivity. As mentioned earlier, most people do not think of productivity as a ratio. In reality, then, when we use the NGT or DT to develop measures of productivity and do not just educate the participants, we really get measures of organizational system performance, or at best "surrogate" measures of productivity. Examples of task statements that will give different results are depicted in Table 4.1. Several actual outputs from an NGT and a DT session will be presented in the case.

Stage 2, milestone t_6, requires intervention from "productivity analysts." Persons in charge of the productivity measurement system development program need to expect to have either a productivity analyst or a team of analysts work at operationalizing measures obtained from the structured group process(es). Experience has shown that in general, the step associated with converting the prioritized measures (surrogate), ratios, and/or indexes into a workable, functioning productivity measurement system requires "consultative" intervention. In the great majority of cases, we have found that the skills, ability, and motivation for progressing from Stage 1 to Stage 4, integration and implementation,

Table 4.1 Various Task Statements for NGT Applications

APPLICATION	TASK STATEMENT
Productivity measurement (NPMM)	Please develop a productivity process model for our organizational system. Identify (1) inputs to our system, (2) outputs from our system, and (3) significant output/input ratios that you feel would give us insights into our productivity.
Goal setting	Please identify 2–5-year (planning horizon is a variable) goals and objectives we should be moving toward.
Collaborative management by objectives	Please identify specific objectives and action programs we should devote resources to in the next 6 months to 1 year in order to move toward our long- range goals and objectives.
Productivity action teams	Please identify ideas for productivity improvement for our organizational system (work group, company, plant).
Performance action teams	Please identify ideas for performance improvement for our organizational system.
Multicriteria productivity measurement methodology	Please identify criteria or measures of performance/productivity for our organizational system.

will simply be missing and must be "driven." Certainly, we have found that without strong leadership intervention, the program will very likely stall out after Stage 1. It is interesting to note that this is particularly true for measurement system development, although less predictable for productivity improvement system development. That is, when the task statement focuses on identifying and prioritizing ideas for productivity improvement or objectives for improvement, there appears to be more built-in motivation to proceed into further stages. Perhaps this occurs because improvement is action oriented and something more of us understand and can do.

Stage 2 requires actually determining how and where to collect the necessary data; how to interpret results; how to integrate consensus measures (surrogate), ratios, and/or indexes with other existing control systems; and how to couple or bring together results from more than one group's session results. Whenever considering whether or not to utilize participation, one needs to consider at least four factors: (1) Who has the knowledge to make a quality decision? (2) What level of quality is desired? (3) Is participation critical to acceptance of the decision? and (4) What are the timing constraints, and who has the time? In this stage of the methodology, it is assumed that we have significantly enhanced our chances for acceptance of the final measurement system as a result of Stage 1. Further, it is assumed that some expertise not resident with all participants is required to accomplish Stage 2. Hence, the decision to shift from a structured group-process coordinator to a consultative professional mode of functioning is appropriate (Morris, 1979).

The output of Stage 2 is a draft of a workable productivity measurement system that is congruent with both the strategic plans for this program and the results of Stage 1 of the NPMM. Again, the NPMM assumes that involvement of participants in the focal organizational system(s) is the desired and appropriate mode of data collection. It presumes that consensus about what a productivity measurement system should look like and look at is a desired outcome. This approach does not preclude the use of consultative or managerial intervention in the final product. However, the NPMM begins with a strong orientation toward involvement (hence the use of the term "normative"). So, although this methodology requires some "expert" attention during Stage 2, during the next stage it also requires review and "approval" of the draft operating system by the participants in Stage 1. In other words, the methodology is participative in Stage 1, consultative in Stage 2, and participative again in Stage 3. The methodology attempts to use the "appropriate" mode of professional functioning or leadership style at various stages in the program.

Stage 3, milestone t_7, requires a briefing, review, discussion, potential revision, and eventual approval of the draft operating system for the productivity measurement program. This is a necessary feedback loop to participants in the NPMM prior to implementation of the final productivity measurement system. As indicated, it is expected that this stage would take about a month. Again, this estimated duration is based on certain assumptions about scope and also on communication that might have been occurring during Stage 2. In particular, if Stage 2 ended up taking several months to a year, status reporting to participants would certainly facilitate Stage 3 approval. The basic desired outcome of Stage 3 is to maintain commitment to and acceptance of the final productivity measurement system. We hope that participants continue to see it as their productivity measurement system, as something that provides useful feedback to them in terms of how they, as a group or organizational system, are performing in terms of productivity. This is a critical desired outcome that cannot be overlooked if one hopes to successfully link or cause productivity improvement as a direct result of productivity measurement.

Stage 4, milestone t_8, again requires consultative or managerial intervention. During the first part of this stage, the productivity measurement system has to be integrated with other performance measurement and/or control systems. It is impossible to prescribe or even describe what will take place at this point in your organization; however, generally speaking, the focus is on integrating the resultant productivity measurement system with other existing performance measurement systems. The intent is to (1) look for overlap or redundancy; (2) link the productivity measurement system to management by objectives, performance appraisal, merit evaluation, and incentive-type systems; and (3) ensure that at least informal steps are being taken to begin to make this system an integral part of the way the organizational system does business. As indicated, this stage could easily require six months to two years.

The other element to be accomplished during Stage 4 has to do with implementing the resulting productivity measurement system, which consists simply of a set or vector of productivity measures (surrogate), ratios, and/or indexes. Again, it is difficult to prescribe or even describe what implementation will look like in your organization. For instance, it is difficult to generalize regarding who will be responsible, involved or affected, for or by the implementation. It is clear, however, that the resulting measures, ratios, and/or indexes will need to be operationalized; that is, the measurement system will need to be activated. Data will be collected, analyzed, interpreted, and then "fed back" to the participants in the focal organizational system.

Keep in mind that the measurement system is supposed to be a control system for everyone to use and benefit from, including but not exclusive of management. As such, the feedback system or the reporting system should be kept simple, succinct, and "obtrusive" or visible. There may be four or five ratios that were viewed as being critical (priority) by the group that are now being tracked quantitatively and graphically. Periodically (note the reporting period is a design variable in the earlier strategy session), the results should be posted. It is advisable that the focal participants (management and employees) in the focal organizational system meet periodically to discuss and interpret the results, as well as ways to either improve or maintain productivity performance. This feedback, evaluation/diagnosis, action process is critical to effectively linking measurement to improvement. More will be said about this process in Parts IV and V.

During this stage of development, it is reasonable to expect that many adjustments to the system will be made. In fact, in many cases, the final productivity measurement system may look nothing like the skeleton system present at the completion of Stage 1. Stages 1 through 5, as indicated in Figure 4.2, will require considerable dynamic planning and adjustment. The important things to remember at this stage of the program are the elements in the strategic plan. That plan is your road map. It reflects where you wanted to head and end up. Again, it can be adjusted because it is dynamic. However, it is all too easy to forget the plan entirely.

In particular, it is a good idea to review your desired outcomes periodically. An underlying and overriding desired outcome for the NPMM is to develop a productivity measurement system that establishes, supports, and maintains a commitment to change based on results from that system. At this stage, you have a productivity measurement system that was logically developed by participants in the focal organization. They have "bought into" that system. The assumption is that they have identified measures, ratios, and/or indexes of productivity that very likely provide excellent insights as to the productivity performance of their organizational system. Further, we assume that for many of these measures, ratios, and/or indexes, they have some or all control over factors, actions, and behaviors that cause the numbers to go up and down. Moreover, we assume that since the methodology tends to create commitment and a feeling of ownership or attribution, once productivity performance becomes visible, the participants will be motivated to cause productivity performance to increase. This trail of assumptions just outlined is reasonable and, in fact, has been shown to be highly justified. Of course, a variety of other factors and conditions has to be orchestrated and maintained in order for the full potential of this approach to be realized. Stage 4 is necessary in order to ensure that the management system successfully becomes a way of doing business in the focal organizational system.

Stage 5, milestone t_9, represents the "continuous" monitoring and feedback process. Many organizations create "control rooms" in which measurement systems, such as the one created in this methodology, are posted and are highly visible. A number of organizational systems (work groups, departments, or even functions) might use a control room for their evaluation, diagnosis, and prognosis meetings. The rooms often also serve as meeting places for productivity improvement programs, such as productivity action teams, quality circles, and participative action teams.

Milestone t_{10} represents a decision mode of sorts. Once the productivity measurement system has been "up and running" for a period of time, there will be a tendency for "entrophy" to take place. That is, the program will likely stagnate unless action is taken to impart additional momentum. There are several strategies that can be taken to do this.

As indicated in Figure 4.2, one could actually recycle the NPMM. Assuming preconditions still exist, one could begin with Stage 1 and regenerate a new list of consensus measures (surrogate), ratios, and/or indexes. Stages 2 through 5 would then be executed again for those new consensus productivity measures. There are several advantages to this recycling process. The second time through the process, participants of the organizational system are bound to be more informed, experienced, and sophisticated concerning productivity measurement and improvement. This maturity certainly should enhance the quality of not only the process, but also the final results. It is also clear that most organizational systems are dynamic; hence, priorities, problems, opportunities, performance, goals, objectives, personnel, and so on may change significantly over time. It is only reasonable to suspect, therefore, that a productivity measurement system must have a mechanism by which to evolve accordingly.

A second option for program "reenergizing" or continuance is to evolve from a measurement and control focus to one that is clearly more improvement oriented. For example, Stage 1 could be repeated, only this time with an improvement-oriented task statement such as

Please list below ideas for productivity improvement in our organizational system (i.e., work group, department, division, organization, and so forth).

In this case, Stages 2 through 5 would be replaced by stages focusing on clarification, specification, development, justification, and so forth, of the ideas for productivity improvement. This option will be elaborated on in Parts IV and V.

A third option would focus on refining the productivity measurement system by closely integrating the results with management by objectives, performance appraisal, merit evaluation, incentive, and other performance measurement and control systems. This option assumes that the productivity measurement system is valid, accurate, and comprehensive. Of course, this option does not preclude the possibility of refining the productivity measurement system should this appear necessary.

The fourth option is to do nothing. This could be interpreted to mean terminating the program or just maintaining Stage 5 in a status quo mode.

Achieving a productive organizational system is obviously not an easy task. Developing a productivity measurement system regardless of the approach taken is only one component in the process of ensuring acceptable and desirable levels of productivity. And, of course, a productive organization is not necessarily a successful one. In the long run, an organization must achieve all seven criteria of organizational performance. The NPMM has great potential, if implemented properly, for improving productivity. Reasonably simple and minor modifications to this methodology have potential for improving efficiency, effectiveness, quality, quality of work life, and innovation. All in all, the NPMM or approaches like it will prove to be extremely valuable management techniques for the eighties and nineties.

A Brief Description of the Nominal Group Technique

The nominal group technique (NGT) is one of many structured group processes that have been designed and developed. It is a special-purpose technique useful for situations where individual judgments must be tapped and combined to arrive at decisions that cannot be reached by one person. The NGT is a problem-solving or idea-generating strategy, not typically used for routine meetings.

The NGT was developed by Andre L. Delbecq and Andrew H. Van de Ven in 1968. It was derived from social-psychological studies of decision conferences, management-science studies of aggregating group judgments, and social-work studies of problems surrounding citizen participation in program planning. Since that time, the NGT has gained extensive recognition and had been widely applied.

The NGT takes its name from the fact that it is a carefully designed, structured, group process that involves carefully selected participants in some activities as independent individuals, rather than in the usual interactive mode of conventional groups. It is a well-developed and tested method that is fully presented in the work of Delbecq, Van de Ven, and Gustafson, (1975).

The Process

The NGT has four-phases in addition to an introduction and a conclusion. The participants are physically present in groups of 8 to 12, and the session is controlled by a process consultant or facilitator and an assistant.

During the introduction, the facilitator attempts to familiarize the participants with the process and make them feel at ease with what will transpire during the next two hours. The facilitator usually discusses very briefly at least the following items:

1. The purpose of the session and the importance of the process
2. The steps of the NGT
3. How the results will be used and the next steps

The facilitator then reads a carefully worded task statement. This is the task the participants should respond to during the structured group session. It is usually simple and direct. If the facilitator is asked what is meant by the task statement, he or she usually avoids introducing bias that occurs by giving examples. Instead, the facilitator often asks several participants to give their interpretation of the task statement. Additionally, the facilitator often asks several participants to directly respond to the task statement. The process of forcing the participants to clarify the task statement themselves is called *self-priming* and has been found to be very effective. When the responses appear to coincide with the objective and the remainder of the participants appear to have grasped the task, the facilitator proceeds to the first basic step of the NGT.

This first phase is called *silent generation* and typically takes about 10 to 15 minutes. During this phase, the group members are instructed to write their responses to the task statement. Both the facilitator and the assistant also write during this period. Even if a majority of participants appear to stop writing before ten minutes has elapsed, the period is not shortened. If some talking occurs, the facilitator tactfully asks for cooperation in permitting others to think through their ideas.

Like each of the steps in the NGT process, silence is purposefully designed. Research has shown that for creation, generation, and production of ideas, individuals are more effective than groups. Thus, for this portion of the session, individual behavior is sought. Silent generation focuses attention on a specific task, frees the participants from distractions, and provides them with an opportunity to think through their ideas rather than simply to react to the comments of others. In this sense, it is a search process that yields contributions of greater quality and variety. Participants are motivated by the tension of seeing those around them working hard at the group task. They are forced to attend for a longer time to the task, rather than rushing immediately to consider the first idea that is suggested to the group. They are freed from all of the inhibiting effects of the usual face-to-face interaction of unstructured groups. Judgment of ideas cannot take place during this early and crucial portion of the group process.

Next comes the *round-robin phase*. The facilitator interrupts the process, yet emphasizes that there is no need to stop generating. (Any additional ideas should be added to the silent generation lists.) The facilitator calls on participants one-by-one to state one of the responses they have written. Participants may pass at any time and may also join in on any subsequent round. A participant may propose only one item at a time, and either the facilitator or an assistant records each item as it is offered. The only discussion allowed is between the facilitator and the participant who proposed the item. The discussion is limited to seeking a concise rephrasing for ease of recording. As each participant responds, the facilitator repeats verbatim what has been said and the assistant records the concise phrase on a sheet. This phase goes on until all the ideas generated by the group are listed and displayed.

The round-robin phase permits the leader to establish an atmosphere of acceptance and trust. He or she does not unduly rephrase or evaluate the contributions, and they are equally and prominently displayed before the group. Leader openness and nonevaluative behavior are essential here. Each idea and each participant receive equal attention and acceptance. There is little opportunity for the process to be dominated by strong personalities, inhibited by possible sanctions or conflicts, or suppressed by status differences. The process separates ideas from their authors and permits conflicting and incompatible ideas to be explicitly tolerated. It provides a written record of the group's efforts as a basis for any next step.

The third phase is called *clarification*. Once all the items have been recorded, the facilitator goes over each one in order to ascertain that all participants understand the item as it has been recorded. Any participant may offer clarification or may suggest combination, modification, deletion, and so forth of items; however, evaluation is avoided. The facilitator moves rapidly from one measure to the next, keeping up the pace of the process. During this step, the underlying logic behind items may be brought out, there may be some expressions of differences of opinion, and the group may conclude that some items can be eliminated or combined because of duplication.

Pace is important to this step, and the facilitator's job is to keep the group moving rapidly through the list of items. Although in this phase the group is more like an interacting one, the facilitator seeks to control lengthy discussions, arguments, and "speech making." Again, the effort is to separate ideas from their authors, to clarify rather than to evaluate, and to ensure full opportunities for participation.

It is important to point out that the clarification aspect of the NGT is perhaps the primary determinant for the resulting quality of the list of items. If there is a great deal of overlap from item to item and if there is ambiguity on the part of the group members as to exactly what each item means, the next step, which involves voting and ranking, will be invalid. Experience has indicated that a certain amount of combination is necessary.

The fourth phase, *voting and ranking*, provides the participants with an opportunity to select the most important items and to rank those items. The participants are provided with between five and nine blank 3" × 5" cards. Usually, participants are provided with eight cards, but the number can vary depending on how many responses are generated during the round-robin phase. Each participant is asked to select the eight most important items from the list displayed before him or her. Typically, the list will contain 20 to 30 items. To avoid any confusion in handling their judgments, participants are asked to write the items out, one per card, in abbreviated fashion, in the center of the blank cards. They are also asked to write the sequential list number of the item in the upper lefthand corner of the card. When all have completed this step, they are asked to spread the eight cards out in front of them and to rank and weight the items. Typically, they are given the following instruction:

> From the eight cards, choose the *most* important item, write the number 8 with a circle around it in the lower righthand corner of the card, and set the card aside.

Another way of phrasing this to assist some in deciding which is most important is as follows:

> Which of the eight items would you use to guide future actions relative to this topic if you could only use one?

The ranking process continues:

> From the remaining seven cards, choose the *least* important item, write the number 1 with a circle around it in the lower right-hand corner of the card, and set the card aside.

Another way of phrasing this to assist some in choosing the least important item is as follows:

> If you could use only six of the seven items in front of you, which one would you drop off?

The ranking process continues:

> From the remaining six cards, choose the *most* important item, write the number 7 with a circle around it in the lower right-hand corner of the card, and set the card aside.

The process continues in this fashion until all the cards have been ranked.

At this point of the process, tabulation of the votes takes place. The facilitator has three alternatives:

1. Invite the participants to take a ten-minute break (possibly for refreshment) while he or she and the assistant tabulate and display the results.

2. Invite the participants to watch the tabulation process take place.

3. Invite the participants to fill out a brief questionnaire that has been prepared by the coordinator for the specific purpose of evaluating the reaction of the participants to the process, obtaining suggestions from the participants as to next steps, determining the likelihood of implementation, and so forth.

The tabulation process involves sorting the cards by sequential item number from the original list and recording the weights given to each. An example of results appears as Table 4.2.

This fourth step permits the participants to express their individual evaluations of the items in a way that is free of social pressure. It provides a constructive method for dealing with conflicts, and leads to a clear expression of whatever

Table 4.2 Productivity Measures Generated: Site D Maintenance Function—July 25, 1978

Management (n = 6)
Task Statement: "Please identify measures of productivity for the Service Department of our city."

MEASURE	VOTES	TOTAL
Response time and task completion within acceptable limit for a particular maintenance function	8-6-4-4 4	5/26
Effective job completion: quality of the job (lack of waste and repetition)	8-7-6-4 1	5/26
Set or establish maintenance standards	8-8-5-2	4/23
Effective supervision	8-5-5-5	4/23
Appearance of public facilities	7-7-2-5	4/21
Scheduling of preventative maintenance	7-3-3-3 2	5/18
Task communication: control of personnel	8-3-2	3/13
Task scheduling	7-4	2/11
Management of emergency or fire-fighting situations: response and completion	6-4	2/10
Effective task reporting: status	6-1	2/7
Maintaining tools and equipment in good working condition	7	1/7
Prioritization of jobs by *need*	6	1/6
Worker pride in task completion	6	1/6
Task safety: worker and public	5	1/5
Street clean (snow and leaves)	3	1/3

degree of consensus there may be with respect to the importance of items generated. It provides a strong sense of closure, a feeling of group accomplishment, and a high level of interest for future steps in the activity being examined. Although participants may not individually agree with the final product, they will typically support it as the achievement of their group.

The session closes with a brief discussion of the results of the voting process in which the facilitator emphasizes those items for which there is strong consensus. He or she may ask the group if they wish to eliminate from further consideration any items that received no votes. Again, this is done only if there is complete consensus. No participant is overridden here. At this point the facilitator may wish to comment on the future steps or to discuss the group's feelings about future action.

Composition

Selection of the appropriate participants for structured group activities is crucial. The quality of the eventual results depends directly on selecting the right personnel to participate.

Group effectiveness is strongly related to the facilitator's ability to operate the method smoothly and confidently. The following minimum logistic preparations are essential:

1. The facilitator should have a detailed agenda of group activities, resources needed, and time durations for the group activities.

2. A trained assistant should be available whose duty is simply to record participants' ideas on large sheets of flip-chart type paper, to display these sheets, to tabulate and record votes, and to provide participants with necessary materials.

3. A packet should be prepared for each participant containing the materials needed for the session. For example, the packet should contain
    (a) A card displaying the participant's name on both sides, folded so as to stand on the table
    (b) A sheet of $8\frac{1}{2}''$ × 11" paper with the task statement typed at the top
    (c) A sufficient number of 3" × 5" cards for ranking and voting (a convenience would be to have them in packets according to number, one for each participant)
    (d) Marking pens for the assistant
    (e) Masking tape to be used to tape up sheets of measures

4. A conveniently located conference room with a table that will comfortably accommodate the group while writing. Excessively large or small rooms are distracting. The room must permit the taping of the large sheets on the wall.

5. One or two large display easels on which the pads (approximately 27" × 34") can be mounted.

6. The group task should be written on one of the large sheets of paper.

7. The following simple visual aids, while not essential, have proven very useful in communicating quickly and effectively with participants:
   (a) A display of the steps in the NGT
   (b) A series of displays to supplement the facilitator's introduction to the purpose and method for the group session.

Part of the logistics is the actual execution of the NGT.

Using the Nominal Group Technique Effectively*

Since the Nominal Group Technique was first tested, interest in and applications of this structured group process have grown exponentially. What began as a technique to enhance the effectiveness and efficiency of program planning in health services has rapidly expanded into the areas of productivity measurement systems development, strategic planning and strategy implementation, participative problem solving, and many others.

The NGT has come along at a particularly appropriate time in the evolution of management thought, technique, and practices. Most American managers are reexamining basic philosophies and practices, and for many this has involved giving increased attention to group processes and techniques. Improved commitment, understanding, communication, coordination, and cooperation are viewed as valuable outcomes. Quality circles, productivity action teams, quality-of-work-life programs, team building, and productivity gainsharing plans such as Improshare, Scanlon, and Rucker all place important emphasis on group processes and behavior.

Experience is proving that the quality, effectiveness, and efficiency of specific group processes play a significant role in the overall success of these programs and techniques. And since managers often spend as much as 80 percent of their time in meetings, the quality of the group processes utilized in these meetings has a great impact on managerial productivity.

Unfortunately, group processes too often leave participants exhausted and discouraged because of the seemingly endless meanderings into unfruitful byways in what has been called "reactive search"—a search focusing on initial responses rather than a continuing creative flow; the focus effect that occurs when a group is unable to extricate itself from one channel of thought; or the mixing of solutions with problems and problems with solutions. What group has not rushed to "solutions" before the problems were clear? Felt frustrated by overbearing extroverts who dominate the sessions? Suppressed disparate or conflicting ideas because of differences in authority, prestige, age, race, sex, or levels of professionalization? What group has not experienced the general lack of creativity and absence of a sense of closure or accomplishment that leaves participants feeling impotent, bored, and frustrated?

In a management era where participative decision making and problem solving are increasingly common, techniques like the Nominal Group Technique have been quite

*Reprinted with permission of publisher. © Executive Enterprises Publications Co., Inc., New York, NY 10023.

welcome. For while the NGT is not itself a program but a participative data collection and consensus-forming device, it can be an important component of participative, group-oriented programs.

In short, the Nominal Group Technique is a real and very timely social-science break-through that managers can relatively easily and successfully apply.

NGT Results and Participation Programs

One way to conceptualize the linkage of results from an NGT session to the goals of a larger program, such as a quality-circle effort, a productivity measurement process, or a planning activity, can be depicted in Figure 4.4. The figure suggests that in those cases where programs involving participation are desirable and appropriate, the NGT can play a central role.

Programs listed in box A have been growing in popularity and, probably, in appropri-ateness in the United States. All of these programs entail a relatively high degree of participation and therefore are prime candidates for application of the NGT. In fact, the NGT has been utilized successfully in all of the types of programs listed. It has been found to be a very effective and efficient mechanism for structuring early stages of participation. It has also been found to be highly reliable for producing desirable process output (box B). That is, the probability is quite high that a correctly managed NGT session will create a high-quality list of prioritized ideas.

The generation of certain post-process behavioral outcomes (box C) is also highly likely. Experience has shown that satisfaction with the session itself, willingness to follow up, actual follow-through, commitment to future involvement and future steps, and general employee development are all reliable post-process outcomes. In a sense, the NGT process sets up certain critical psychological states deemed necessary for the achievement of many of the desired outcomes (box E) from the larger program effort. Critical psychological states reflect job-related attitudes, such as a felt responsibility for outcomes associated with the job, experienced meaningfulness of the particular job, and perceived significance of the task. Job design and job enrichment theory and research suggest that any process that can positively affect these and other critical psychological states has a high probability of causing one or more of the desired outcomes to occur.

Regardless of the success of the NGT in terms of creating process output and post-

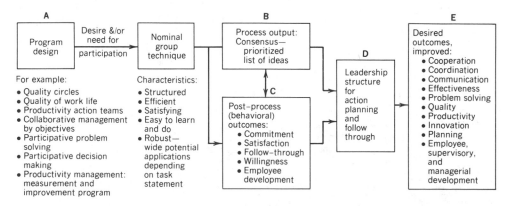

Figure 4.4   Role of NGT in Employee Participation Programs

process outcomes, it cannot be utilized in larger-scoped programs without management leadership and support. Note that box D, leadership structure for action planning and integration of NGT results into the larger program effort, reflects a critical component in the process of utilizing NGT results. It influences, and in some cases even directly determines, the extent to which desired outcomes are actually achieved. Experience suggests that without leadership intervention, the desired outcomes from such programs as those listed under box A are less likely to take place.

In the past six years, the author has been involved in numerous applications of the NGT in a variety of organizations. In each case, management utilized the NGT as an early, sometimes even initial component of a program designed to achieve one or more of the desired outcomes listed under box E. Three specific case applications of the NGT are briefly described in the following section. Their purpose is to assist the reader in thinking through applications that might be appropriate in his or her organization. These case studies do not represent empirical research on the effectiveness of NGT applications. They do represent specific systematic yet pragmatic attempts to improve the effectiveness and efficiency of employee involvement activities in organizations.

Case Studies

In the roughly 15 years that the NGT has been available to managers, educators, and researchers, the range of its applications has increased significantly. But perhaps the most important areas for application of the NGT in American organizations have been the following:

1. *Planning processes:* NGT is used to increase the amount and quality of participation in and thus, it is hoped, commitment to the planning process, to link strategic planning to strategic as well as tactical and operational actions.

2. *Productivity measurement system design and development:* NGT is used to generate "normative" productivity measurement systems composed of measures, ratios, and/or indexes.

3. *Participative problem solving:* NGT is used in programs like quality circles and productivity action teams as a mechanism for increasing the effectiveness and efficiency of identification and selection of priority opportunities, problems, causes, alternatives, and so forth.

Brief descriptions of an actual case study for each of these three applications will now be presented.

NGT APPLICATIONS IN THE PLANNING PROCESS

An academic department at a major university initiated a planning process several years ago. The basis for the planning process came from work done by Peter Drucker in 1954.

Phase I of the planning process implementation focused on getting the faculty acquainted with planning concepts and comfortable with thinking in longer-range terms. The stress was on developing department goals that were congruent with overall university goals and were acceptable to the faculty. Several years were spent on this early phase.

When the department head concluded that the faculty was ready to move to Phase II of the planning process, attention began to focus on linking strategic planning to strategic, tactical, and operational actions.

In the first year of Phase II, the stress was almost entirely on developing objectives, targets, and responsibilities for each major goal. At an annual planning retreat during the second year of Phase II, the department head decided to incorporate the NGT as a component in the planning process. Specifically, the NGT was used to generate one-year objectives for the department. The faculty received a brief explanation of the process and the technique and were then asked to respond to the following task statement: "Please list below specific objectives that our department should be working on during the next year." Each faculty member had received information on the university's as well as the department's long-range goals. Therefore, short-range planning incorporated thinking about strategic actions. Results of this NGT session are shown in Table 4.3.

Evaluation

This particular academic department has now been utilizing the NGT as an integral part of the planning process for four years. Initially there was some skepticism concerning the technique, particularly regarding its highly structured nature. But subsequently, acceptance of the technique has been so high that over a third of the department's faculty has used it in other applications outside the department itself.

Prior to inclusion of the NGT to set one-year objectives for the department, the planning process was significantly more top-down. Goals and objectives for the department were generated by its head and then cursorily discussed by faculty. And the general perception was that substantive follow-through on objectives was minimal.

Since the inclusion of the NGT in the planning process, setting one-year objectives for the department is significantly more bottom-up and participative in character. General perceptions are that faculty are significantly more involved in the short-range planning process and as such have tended to "buy into" the objectives themselves. In the four years this process has been used, the department of 14 faculty members accomplishes an average of between 10 and 20 significant department-related objectives per year. Note that these objectives are group objectives and are in addition to individual faculty position responsibilities.

The department for a number of years has earned a reputation as one of the better in the country from an administrative and academic standpoint. There is considerable agreement among the faculty that a primary reason for this success is the quality of the planning process. The NGT has been a significant component in this process.

DESIGN AND DEVELOPMENT OF PRODUCTIVITY MEASUREMENT SYSTEMS

A supervisor of a department of 12 engineers in an equipment service division of a major American corporation returned from a three-day course on productivity measurement and improvement. She was particularly impressed by her exposure to a multifactor, firm-level productivity measurement model and what had been called a "normative" productivity measurement approach. The latter approach, she had been told, was appropriate at the group level and, in particular, was very useful for hard-to-measure applications.

Members of the organization were assigned the task of developing the components of a productivity measurement system. The approach appeared to have been quite successful in other organizations at linking productivity measurement to productivity improvement actions. She was very excited about implementing it, as her greatest difficulties in the past had been precisely the linking of planning for improvement to specific individual engineer

Table 4.3 Academic Department Application of NGT for Setting One-Year Objectives

Group: Industrial Engineering Department
Facilitator Name: Scott Sink
Date: January 27, 1978

Number of participants = 14
n (number of objectives/action programs asked to vote for) = 8

Task Statement: "Please identify specific action programs or objectives our department
should be working on and accomplishing during the next year."

Key: O—Programs which are aggregates as result of clarification
S—Programs for which subprograms were identified in subsequent sessions
D—Programs which became topics of discussion during open session
I—Programs for which implementation was discussed

ACTION PROGRAM/OBJECTIVES		VOTES RECEIVED 8 = Most important, 1 = Least important	TOTAL (number of votes/ total vote score)
ISO	1. Faculty Development	8-8-8-8-8-8-8-7-6-5-4-4-1	13/83
O	2. Graduate Recruitment	8-7-7-7-5-5-5-4-3	9/51
DSO	3. Industrial Liaison	8-8-7-6-6-4-4-3-2-2	10/50
DO	4. Faculty Support	7-7-7-6-6-3-3	7/39
SO	5. Intra, Communication	7-6-5-2-2-2-1	7/25
	6. Working Papers	8-6-4-3-1	5/22
	7. Minors Ph.D.	6-4-4-2-1-1	6/18
O	8. Course Content Coordination	4-3-3-3-2-1	6/16
	9. National Conference	5-5-5-1	4/16
D	10. Coop Program	8-3-2-2	4/15
	11. Interdisciplinary Research	7-6-1	3/14
	12. Graduate Support	7-4-3	3/14
	13. Timeliness of Info on Funding Opportunities/RF	6-5-3	3/14
O	14. Begin-End Courses Evaluation	6-5-3	3/14
	15. Manufacturing Institute	5-4-2	3/11
O	16. Indices	7-3	2/10
	17. Material Handling/Plant Layout	6-4	2/10

(Note: There were 41 objectives; 33 received votes;
the 17 top vote getters are listed in the table.)

behaviors. The risks and costs, she had been told, were that the quality of the measurement system initially might not be as good as she would like and that her engineers would have to spend some time participating in the development of the measurement system. She was reluctant to involve her engineers in activities that were not directly productive but decided it was an investment that could have significant direct as well as indirect benefits.

Based on what she had learned at the course and on further suggested readings, she developed the following plan of action:

1. Hold productivity basics seminar with engineers (one day, outside consultant).

2. Make informal assessment of response to the subject with individual engineers.

3. Present proposal to engineers for development of productivity measurement system.

4. If accepted, run NGT session to develop consensus list of productivity measures, ratios, and/or indexes.

5. Review results and discuss next steps with group.

6. Integrate and operationalize results with current control system.

The plan was well received by the engineers, and an NGT session was run to develop productivity measures, ratios, and indexes. Results of the NGT session are presented in Table 4.4.

Evaluation

The engineers were highly pleased with the NGT session and its results but were unclear as to what came next. They voiced concern over how the measurement system would

---

Table 4.4 Results from NGT Session to Develop Productivity Measures, Ratios, and/or Indexes

Group: Engineers
Facilitator: Scott Sink
Date: 6/25/82

Number of participants = 12
n (number of measures asked to vote for) = 8

Task Statement: "Please list below measures, ratios, and/or indexes of productivity that you feel this engineering group should use to monitor, evaluate, and control productivity performance for the group."

MEASURE, RATIO, INDEXES	VOTES RECEIVED 8 = most important, 1 = least important	TOTAL (number of votes/ total vote score)
Number of times our clients ask for our help	8-8-6-4-8-3-1	7/38
Measurable output	8-8-8-8-8-8	12/81
Resources utilized	8-7-4-8-5-1	
Projects completed on time and within budget	7-7-6-7-5-3-8-7-5	9/55
Customer satisfaction	1-3-7-6-6-6	6/29
Percent of group objectives accomplished on time	7-7-7-6-6-7-4-3-5-4-5	11/61
Percent of successfully implemented projects	7-6-6-6-5-4-3-1	8/38

integrate with the existing individual performance appraisal and merit evaluation system.

In response to their concern, the supervisor asked them if they would like to go a step further and become involved in using the NGT-generated results to design the department's productivity system. Their unanimous response was that they wished to be involved so long as she felt they could contribute and so long as their participation did not interfere with their projects and duties. The result has been that as she has developed the productivity measurement system—guided by the NGT results—she has periodically submitted it to the engineers for review.

The group has since used the NGT to generate consensus ideas for productivity improvement. Action teams are set up and subgroups of engineers work on specific priority projects identified by the department itself.

The supervisor and the engineers are pleased with the role the NGT has played in structuring group activity in both the area of productivity measurement and improvement. They sense substantial progress within the department regarding group goals and objectives. The engineers seem pleased that they have been able to contribute to the design of a productivity measurement system. They are particularly pleased with the opportunity to work together as a group on common problems and opportunities.

PARTICIPATIVE PROBLEM-SOLVING PROGRAMS

A plant manager for an oil-industry-related manufacturing and fabricating company with 250 employees implemented an employee involvement program with the help of a local state university. The program began with a pilot-study group of 11 welders plus their supervisor. The NGT was an integral component of the program.

During the first meeting of the group, the NGT was used to identify and set priorities for ideas on productivity improvement. The results of this session are shown in Table 4.5.

After the first session, a management review committee screened the list to verify the quality of the ideas. The group of welders was then broken down into smaller action teams of three to five workers and assigned a specific idea to develop. Once taught how to develop alternatives for carrying out the ideas, each action team presented a proposal to a management review committee for approval. The committee then decided how to proceed.

Evaluation

The plant manager has noted progress in all of the four top priority areas targeted by the welding group for productivity improvement. Perhaps even more important is the improvement in worker attitudes that he has observed. In fact, the plant manager has commented that the major gain achieved through the participative problem-solving program has been the consensus and commitment generated by the NGT at the outset of the program itself. This consensus and commitment has created a willingness to identify and resolve long-standing productivity problems, which no one had ever taken the time to tackle.

Conclusion

The Nominal Group Technique can assist management in operationalizing participative management philosophies, concepts, and approaches. A structured group process, each of its five stages was developed to fulfill specific design specifications (i.e., equality of participation, functional search behaviors, closure to the decision process, and so forth).

Table 4.5 Results from NGT Session to Develop Ideas for Productivity Improvement

Group: Welders on shop floor
Facilitator: Scott Sink
Date: 6/17/82

Number of participants = 11
n (number of ideas asked to vote for) = 8

Task Statement: "Please list below ideas for improving productivity, effectiveness, and/or efficiency of your work group."

IDEA	VOTES 8 = most important, 1 = least important	TOTAL (number of votes/ total vote score)
Interdepartment and intershift cooperation and coordination	8-8-8-8-8-8-7-7-6-4-3	11/75
Labor-management relations	8-8-8-8-8-7-7-7-7	9/68
Unnecessary noncomformances, improve quality control procedures	7-6-5-4-4-3-2-2-2-2-2	11/39
Improve scheduling, reduce set-up time	7-6-4-3-3-2-2	7/27

(Note: The welders identified 22 ideas for productivity improvement. The NGT helped them reach the level of consensus shown above in an hour and a half.)

When viewed as a component that can be designed into participative activities, it will assist in improving the effectiveness and efficiency of group decision-making processes.

The eighties and nineties will present tremendous challenges to management. If management develops the skills necessary to take full advantage of the NGT, its ability to regulate group behavior and hence organizational behavior will play a key role in helping it to meet these challenges.

REFERENCES

Delbecq, A. L., A. H. Van de Ven, and D. A. Gustafson. *Group Techniques for Program Planning: A Guide to Nominal Group and Delphi Processes.* Glenview, Ill.: Scott, Foresman, 1975.

Friedlander, F., and D. L. Brown. "Organization Development. *Annual Review of Psychology*, 1974.

Jacques, E. *The Changing Culture of a Factory.* New York: Dayden, 1952.

McGrath, J. E. *Social Psychology: A Brief Introduction.* New York: Holt, 1964.

Measuring and Improving the Productivity of Administrative Computing and Information Services: A Manual of Structured, Participative Methods. The Productivity Research Group, The Ohio State University, Grant No. APR 75-20561, Columbus, Ohio, 1977.

Morris, W. T. *Implementation Strategies for Industrial Engineers.* Columbus, Ohio: Grid, 1979.

Peters, T. J. and R. H. Waterman. *In Search of Excellence: Lessons from America's Best-Run Companies.* New York: Harper & Row, 1982.

Productivity Measurement Systems for Administrative Computing and Information Services, Sec-

tion I: Introduction and General Overview, Section II: Products of Structured, Participative Methods, Section III: Summaries of Activities at Research Sites, The Ohio State
University, Grant No. APR 75-20561, Columbus, Ohio, 1977.

*Productivity Measurement Systems for Administrative Computing and Information Services: An
Executive Summary.* The Productivity Research Group, The Ohio State University, NSF
Grant No. APR 75-20561, Columbus, Ohio, 1977.

Schein, E. H. *Process Consultation: Its Role in Organization Development.* Reading, Mass.:
Addison-Wesley, 1969.

Sink, D. S. "How to Make the Most of the Nominal Group Technique." *National Productivity Review*, Vol. II, No. 2, 1983.

Sink, D. S. "Productivity Action Teams: An Alternative Involvement Strategy to Quality
Circles." Institute of Industrial Engineers 1982 Annual Conference *Proceedings*, Norcross, Ga., 1982 (New Orleans Conference).

QUESTIONS AND APPLICATIONS

1. Participate in a nominal group technique session led by an experienced NGT facilitator. Pay close attention to the process and the facilitator's behaviors. There is no real
substitute for experience, first as a participant and then as a coordinator or facilitator in
real NGT sessions addressing significant tasks. There are many opportunities outside
the classroom to experiment in a realistic setting. Find a situation in which the NGT has
a real possibility and try it out. Potentially good applications are

    (a) various committees and task forces
    (b) church groups
    (c) scout groups; youth groups
    (d) citizen's groups; community associations
    (e) advisory boards
    (f) special-interest groups; hobby groups
    (g) small businesses

2. Read *Group Techniques for Program Planning* by Delbecq, Van de Ven, and Gustafson.

3. Identify an application for the NGT. Design and develop the application. Facilitate
the NGT session.

4. The NGT is a participatory data collection and consensus-seeking device. Discuss
reasons that this social science technique is so critical for American organizations and
groups of managers and employees.

5. Comparing the NGT with quality circles is like comparing apples with oranges.
Discuss and evaluate the reasoning behind this comment.

6. Select a specific organizational system (preferably a specific work group in an organization or a small organization). Design and develop (identify and sequence specific
activities and milestones) an NPMM application for the organizational system. What do
you foresee as major roadblocks or difficulties to implementing the NPMM? What are
the desired outcomes? What are the independent variables? (What factors are being
manipulated?) What are the expected cause-and-effect linkages that are being set up by
the NPMM?

7. Develop and discuss a two-to-five-year strategic plan for a productivity measurement
effort.

8. An output from the NPMM is a disaggregated vector or set of productivity ratios or indexes. The set of productivity measures'are prioritized and essentially weighted by the votes received. How might one collapse the set of measures into one indicator that could tell us what is happening to productivity?

9. If you involve participants in the NPMM, particularly the NGT session, that do not have a disciplined (Chapter 2) view of productivity, experience has proven that regardless of the task statement, you will obtain a mixed bag of performance measures. In other words, if you ask uninformed persons to identify measures of productivity, they usually give you few true productivity measures (ratios or indexes). This suggests that if you really want true productivity measures and not surrogate productivity measures, then you will have to either educate the participants or in stage 2 (analyst intervention) convert the surrogate measures to true productivity measures. Experiment with this concept. Ask people for productivity measures (measures of productivity) and see what you obtain. Think through how you might convert the normative *productivity* measurement methodology to the normative *performance* measurement methodology.

10. How might the NGT be utilized to enhance the quality of a traditional quality circles program?

11. How might the NGT be utilized in a goal-setting and objective-setting process for a department?

12. If decoupled productivity measurement systems were developed for each work group in an organization, develop several plans or approaches that might be utilized to "couple" or integrate these measurement systems into higher-level measurement and evaluation systems.

13. Write NGT task statements for groups with the following objectives:
    (a) Developing and weighting criteria for the evaluation of new systems design
    (b) Setting one-year objectives for a department
    (c) Developing performance criteria for an industrial engineering department
    (d) Identifying the major considerations that should guide the development and evolution toward the "factory of the future"
    (e) Designing a decision support system
    (f) Designing a productivity measurement system

14. Simulated experience as a participant and collaborator is also of value. Form groups of five or six people. Each group member selects a task of interest and concern to the group members and designs an NGT for the task. Each member then serves as leader/facilitator for a shortened version of the NGT. Silent generation is limited to three to five minutes; round robin listing is limited to two or three rounds or three to five minutes; clarification is very brief, limited primarily to combining and clarifying; and the voting and ranking is limited to five cards with all participants helping to record the results.

15. Discuss how the leader/facilitator of an NGT session responds to the following situations:
    (a) Participants talk during silent generation period.
    (b) Some participants jump in out of turn during round robin and clarify or evaluate ideas being presented.
    (c) Some of the ideas that appear in the round robin listing are completely unrelated to the task statement.
    (d) A participant refuses to vote and rank.

(e) During the discussion module, after voting and ranking, a participant addresses the group regarding the true importance of his or her idea despite the fact that it only obtained one vote.

(f) The clarification module bogs down almost to the point of failure because of the group's inability to deal with hierarchical differences in ideas. One set of participants want to lump all ideas into six or seven major catch-all ideas; another set want to do no combining at all.

16. Discuss the design of a change process that might accomplish the steps in the achievement of a fully functioning productivity measurement system (Morris, 1979) outlined below. The change process may or may not involve the use of the structured group processes, such as the nominal group technique.

(a) Identification and evaluation of measures (surrogate, ratios, indexes) appropriate for assessing productivity in a particular organizational system

(b) Identification of specific quantification methods, scales, and so forth that will operationalize the measures

(c) Structuring the measures into a measurement system and determining data sources that are presently available, those that require further development, and so forth. Technical assistance may be supplied at this step.

(d) Creation of necessary and specific measurement instruments and methods.

(e) Executing within the organizational system and organization a carefully designed program of implementation that will result in an operating productivity measurement system available to management as a decision support system.

MULTIFACTOR PRODUCTIVITY MEASUREMENT MODEL (MFPMM)

HIGHLIGHTS

- Introduction
- Historical Background
- Development of the Model
- MFPMM Basics
- Description of the MFPMM
    - Columns 1–6
    - Columns 7–9
    - Columns 10 and 11
    - Columns 14–16
    - Columns 17–19
- MFPMM Simulation Routine: Decision Support System Developments
- Using Quality Costs in Productivity Measurement
- Quality Costs in the Model
- MFPMM Software Support
    - Data Input
    - Basic-to-Current-Period Analysis
    - Sensitivity Analysis
- References
- Questions and Applications

OBJECTIVES

- To introduce the reader to the theoretical developments for this model.
- To teach through use of a simple example the multifactor productivity measurement model.
- To demonstrate and make the reader aware of the potential of this model as a diagnostic and evaluative productivity measurement technique.

- To acquaint the reader with software support available for the model. The software can be utilized for educational enrichment purposes or for operational applications.
- To contrast the normative approach with this accounting-oriented, data-driven model.
- To begin to develop an appreciation on the part of the reader for the range of techniques available and for the special characteristics of these techniques relative to productivity measurement and evaluation system development.

Introduction

As outlined at the beginning of this part, three major, distinct approaches to measuring productivity were to be presented. We have just completed a lengthy discussion, presentation, and illustration of what has been labeled the normative productivity measurement methodology. As we have seen, the NPMM is an action-research, involvement-participative, organizational, development-oriented approach to measuring productivity. It is, because of its participative character, appropriate primarily for smaller units of analysis, such as the work group and department. As such, it can be labeled a decoupled or decentralized approach to productivity measurement. This approach, however could also be utilized for larger units of analysis (plant, function, and firm) if (1) sampled representation, (2) the Delphi technique, or (3) coupling of the various measurement systems were employed. Recall, particularly from the case studies provided, that the NPMM can be expected to develop surrogate "measures" of productivity if a formal productivity basics training program is not first implemented. If the strict definition of productivity is not communicated to participants in the NPMM program, they will likely provide "measures" of productivity that reflect (1) roadblocks to productivity improvement, (2) opportunities for productivity improvement, (3) surrogate measures of productivity (things that they perceive correlate highly with productivity, and/or (4) other types of performance measures (quality, effectiveness, and so forth). This is not necessarily a bad characteristic of the NPMM; it is simply one that has to be recognized. If you are after a true productivity measurement system, you will either have to educate participants prior to Stage 1 or plan for analyst intervention to convert the surrogate "measures" into ratios and/or indexes of productivity. The advantages of the NPMM, as you will recall, are shared commitment and understanding and hence higher probability of successful implementation and positive behavior change.

The second productivity measurement approach to be presented will be titled the multifactor productivity measurement model. In various forms it has been known as the "total-factor productivity model," the APC model (named for the American Productivity Center, which promoted the approach in 1977), or perhaps most generically as a "price-weighted, indexed, and aggregated multifactor productivity measurement model." For the purpose of this book and more specifically this chapter, we will simply call it the multifactor productivity measurement model or, for short, the MFPMM.

What follows in this chapter is a reasonably detailed description and explanation of this approach to productivity measurement. The MFPMM approach, in contrast to the NPMM, is more consultative; significantly less involvement/ participative-oriented; data base/accounting-system-oriented (primary source of data is not people but system documentation); top-down as opposed to bottom-up in character; more restrictive in terms of operationalizations of the definitions of productivity (this approach adheres strictly to definitions provided in Chapter 2 and utilizes only ratios and indexes to measure productivity); diagnostic in a passive, absolute, and objective sense as opposed to an active, relative, and subjective sense; and a highly developed and reasonably self-contained decision support system (the NPMM, in contrast, is in fact a process by which to develop a decision support system). So, as you can see, these approaches to measuring productivity differ significantly.

It is recommended that the reader maintain an objective, nonvalue-laden point of view as this next productivity measurement approach is detailed. Just because these approaches differ from one another does not mean that one is necessarily better than the other. Each approach has its own individual benefits, costs, risks, and so forth. Each has a distinct range of applicability and appropriateness. It is also, of course, important to point out that the approaches are not in any way inconsistent or incompatible with each other. It is possible that an organization might utilize both approaches simultaneously.

What is to follow will be a drastic departure from the presentation to this point. There will be formulas, equations, sample data, tables filled with numbers, and formatted reports. However, if you keep in mind the basics presented in Chapter 2, the model will be fairly easy to follow and should significantly enhance your understanding of productivity, particularly at the firm level of analysis.

Historical Background

The development of this particular approach has a relatively short history. In the 1880s output per person-hour measures were being developed for various industries in the United States. The Bureau of Labor Statistics began its program of industrial productivity measurement in 1940. However, as you know, ratios of output to labor or any other single class of input for that matter is only a partial measure of productivity. Partial measures of productivity are subject to a variety of influences, such as factor substitutions (capital for labor, materials, and so forth), efficiency changes in the entire process, and demand for product and/or service fluctuations. Hiram Davis argued that net savings of real costs per unit of output and productive efficiency in general could not be measured unless output was related to all associated inputs. Because of this argument, David was one of the early pioneers in the development of the concept and measurement of "total-factor productivity" with particular reference to the firm.

As Kendrick points out in his introduction to Davis' 1978 reprint edition of *Productivity Accounting*, the concept of total-factor productivity was not new. It was only after World War II, though, that this concept was statistically implemented. Early measures dealt with specific industries. But in the 1950s a number of economists evidently developed multifactor productivity estimates for the United States' private domestic economy as a whole. "These aggregate measures were based on estimates of real Gross National Product (GNP) first published by the United States Department of Commerce in 1951; employment and hours estimates from the Department of Labor; and newly developed estimates of real stocks of capital, including land, from studies sponsored by the National Bureau of Economic Research" (Kendrick, 1978).

At the same time these macroeconomic productivity measures were being developed, Davis who was then a member of the staff of the Industrial Research Department at the Wharton School of Finance and Commerce of the University of Pennsylvania was developing measures of multifactor productivity for a textile firm. The Davis study, according to Kendrick, was the first attempt to measure total productivity at the micro level. Davis' book, *Productivity Accounting*, published in 1955, explains the methodology he developed to obtain the company estimates that were used to illustrate the model. Davis points out that each industry and firm faces unique problems and that the productivity analyst engaged in productivity measurement must be prepared to use some ingenuity in adapting general principles to specific company situations. This is good advice for the reader to heed with respect to this book also.

Davis' work influenced Kendrick's work, as is apparent in the handbook Kendrick coauthored with Daniel Creamer for the Conference Board in 1965 entitled *Measuring Company Productivity: Handbook with Case Studies*. Davis firmly believed in productivity measurement at the firm level. An interesting technique for analyzing and distributing productivity savings or losses among investors, employees, suppliers, and customers is presented in Chapter V of Davis' book. Several companies in the United States are currently experimenting with this basic model as the measurement basis for such gainsharing plans as Rucker, Improshare, and certain hybrids.

In 1977, the American Productivity Center, founded by C. Jackson Grayson, and a private-sector productivity center arranged with the Wharton School's Industrial Research Unit to reissue *Productivity Accounting*. During the late seventies, staff from the APC, with the assistance of Dr. Kendrick, Basil J. van Loggerenberg, and others developed a short course on productivity measurement. The course was devoted almost entirely to a refined version of the "total-factor productivity measurement model." This course up until about 1980 served as one of the few major sources of disciplined productivity measurement training, education, and development in the United States. The APC is one of a dozen or so productivity and quality of work life centers in the United States today. However, certainly in the areas of top management awareness and the "total or multifactor productivity measurement model," the APC has been the clear leader in terms of dissemination of information.

This has been a reasonably brief and certainly not a historically comprehensive presentation of the total evolution of the multifactor model. However, the presentation hit major milestones. It is estimated that today somewhere between 50 and 100 companies in the United States are experimenting, developing, and/or utilizing this model in a major fashion. There are perhaps another 100 or so firms that have one or more persons at smaller units of analysis trying out the approach. Some of the more notable firms with experience in this approach are Phillips Petroleum Company, Hershey Foods, Inc., Anderson Clayton, Honeywell, and General Foods.

The model can be and is being utilized (1) to obtain an overall, integrated measure of productivity for the firm; (2) to provide an analytical audit of past performance; (3) for budget control of current performance; (4) for common-price financial statements; (5) to assess and evaluate bottom-line impact on profitability as a result of productivity shifts; (6) to track the results of specific productivity improvement efforts, such as quality circles, quality control, incentive systems, and technological innovation; (7) to measure initial distribution of benefits flowing from gains and/or losses in the productivity of the firms; and (8) to assist with setting productivity objectives and general strategic planning with regard to capacity utilization, marketing efforts, cost management, staffing, quality management, pricing strategies, and so forth.

Development of the Model

As mentioned, Davis began the development of his model with a total-factor approach. In his book, he discusses the need for an accounting measure of productivity and presents his concepts on productivity accounting. He begins with what he terms the productivity statement, which appears in Table 5.1. Productivity accounting, as Davis terms it, requires devalued or revalued prices and costs. Constant value sales are utilized as output in the model. Product-mix problems are discussed in some detail. Constant value costs for "all" resources consumed are used as input. Labor and management, materials, supplies and business services, capital goods, and investor input are the major categories of input that Davis quantifies in the "Total-Factor Productivity Measurement Model."

Davis devotes an entire section of his book on revaluation and related problems. Shifts in product price lines, shifts in input cost lines, revaluing new qualities and new products, the base-period problem, capital goods revaluation, management input, investor input, and taxes are all called out and addressed. Uses of the model in business are discussed, including applications for measuring company efficiency, analytical audits, budget control, and common-price

Table 5.1 Productivity Statement for the X Manufacturing Company, 1947

	THOUSANDS OF DOLLARS			
	1939	1947		1947
				IMPLICIT
	IN	IN	IN	PRICE
	1939	1939	1947	INDEXES
	PRICES	PRICES	PRICES	(1939 = 100)
	Output			
Class M (medium unit cost)	3, 384	3,893	12,157	312
Class N (high unit cost)	3,120	5,535	14,438	261
Class O (high unit cost)	1,303	345	974	282
Class P (low unit cost)	162	309	982	318
Total	7,969	10,082	28,551	283
	Input			
Labor and management	2,108	2,092	6,660	318
Materials	2,947	3,284	12,566	383
Supplies and business services	2,295	2,463	4,110	167
Capital good (depreciation)	246[a]	300	560[a]	187
Investor				
Gross	373	400	4,655	1164
Taxes	157	168[b]	2,078	1237
Net	216	232	2,577	1111
Total	7,969	8,539	28,551	334
	Productivity Change			
Output per $ of input	1.00	1.18		
Increase in output per $ of input		.18		
Savings over base year (dollar increase for input of $8,539)		1,540[c]		

[a]Reported depreciation after adjustment to the replacement prices of the year indicated and for year's rate of operation
[b]Taxes at their base-year ratio to gross investor input
[c]In this example, the savings do not equal the difference between 1947 output and input totals revalued to 1939 prices because of rounding of the productivity increase to 18 cents.

SOURCE: H.S. Davis, Productivity Accounting, 1978 (first released in 1955). Reprinted with permission.

accounting. Finally, measuring the distribution of savings and losses is discussed with implications for productivity gainsharing.

Since Davis' book and his research with this model, development of the concepts and approach has tended to focus on simplifying the model and on making it more attractive to a broader spectrum of managers. The National Productivity Institute of South Africa, The American Productivity Center, the Oklahoma Productivity Center, the Virginia Productivity Center, and a variety of private-sector organizations and consulting firms have been the prime movers in this development.

In particular, the multifactor productivity measurement model to be presented in this book is a third-generation development from Davis' original work. Several modifications have been made to the total factor model developed for and by the APC that you should be aware of:

1. Capital has been removed as an input variable in the model. Hence, the model to be presented is accurately labeled multifactor.

2. Variance analysis, which is present in both the Davis and the APC versions of the model, has been removed.

3. A "what-if" simulation routine has been added to the model to allow the analyst/manager to forecast prices, costs, and quantities for future periods and to analyze the effect of these changes.

These modifications simplify the model without sacrificing too much in the way of detail, rigor, and power. The evolution has occurred through a reasonably systematic potential-user evaluation. Over 200 managers and potential users have been exposed to various versions of the model. The design modifications have been based on feedback from those managers. The resulting design has been reviewed by at least one experienced user of the APC version in a large corporation to ensure that the changes were reasonable. In each case, the experienced user felt the design modifications were justified and a reasonable tradeoff between practicality or relevance and rigor.

In the case of capital being left out as an input, the experienced user informed me that he had long since dropped capital out of the analysis in his company. The rationale for this move on his part and mine is as follows. The accountants, the financial analysts, the comptroller, and the engineers with economic analysis should be doing a pretty good job managing capital "productivity." In fact, if there is one resource that is probably "best" managed, it is capital. Aside from perturbations in the system, such as uncertainty and risk associated with the cost of capital, taxes, and cash flows, the techniques for making investment decisions and thus ensuring at least "satisfactory" capital productivity are reasonably well defined. These techniques are not always used consistently or appropriately, but they are available.

The other major problem with capital productivity that these techniques do not address is associated with the selection of an appropriate planning horizon and an appropriate discount factor. The issues of strategic planning and building for the future have become critical in light of Japanese planning strategies. The role that superordinate goals play in determining an appropriate discount factor, a minimum attractive rate of return, and a cost of capital rate is beyond the scope of this book. However, the reader should note that this issue is at the heart of the capital productivity, capital investment, technological investment dilemma in the United States. (See Appendix A for a brief article on new tax laws and capital productivity.)

Variance analysis is another component of the Davis and APC versions that has been deleted from the revised model. Variance analysis attempts to analyze the degree of variability associated directly with volume changes. In most organizations there are fixed and variable costs. Hence, some portion, subclass, or type of each major input is directly variable with volume; another portion is semivariable, and still another remains unchanged or is fixed. The purpose of this line of analysis is to ascertain cause-and-effect relationships for changes in input utilization relative to levels of output. We try to get the model to tell us if an increase in the use of material, for example, is the result of scrap, quality, actual volume changes, design changes, and so forth. This undoubtedly would be very useful information and is the kind of problem that excites cost accountants.

Building this variance analysis into the model, however, adds a level of complexity to the basic productivity analysis that frightens nonaccounting-based managers away from the model. In an attempt to increase the probability that this model can become a useful decision support system for a wider range of managers, we have eliminated this component from the model in this book. For those interested in treatment of capital and the variance analysis, the APC has a course on their model. Davis' book is also an excellent source of additional information.

Finally, the third design modification made to the basic model is the inclusion of a simulation routine. This development is unique to the model presented in this book and has taken place at Oklahoma State University during the past four years. This addition significantly enhances the model's decision support capabilities. As you will see, it creates a whole new dimension to the productivity measurement model.

Davis points out that the measurement of productivity has largely been in the hands of the economic statisticians and the industrial engineers. The inference, of course, is that productivity, in 1955, had been treated at the macro and micro levels or units of analysis but not at the "intermediate" or organizational/firm level. He goes on to define "industrial productivity" as the change in product obtained for the resources expended. In broadest terms, it is the change in results obtained (effectiveness) for the resources expended (denominator of the efficiency quotation). Davis also states that productivity "is always a relative measurement . . . present versus past performance, but it may [also] be the results obtained for the resources expended by one producing unit compared with another unit at the same time, or compared with what the unit could do under some set of assumed circumstances such as ideal operating conditions" (Davis, 1951).

MFPMM Basics

As we have mentioned earlier, productivity measurement can be impeded by product variety and the multiplicity of various resources utilized. Person-hours cannot be combined with tons of steel, dollars of capital equipment, kilowatt-hours, and so forth for a resource total. Nor could a Westinghouse or a General Electric add up the number of motors, refrigerators, electrical components, and so forth to get a measure of total product. The dollar, in the case of the United States, is a convenient common denominator.

Since productivity gains or losses are distributed via the price system (the customer, stockholder, owner, and employee benefit or lose according to shifts in productivity), it seems appropriate to use the yardstick of that system—money—to analyze the distribution. However, the dollar or any other currency is, particularly in the current economic period, a variable standard. Therefore, in order to use the dollar as an aggregating measure, the variability needs to be taken out (Davis, 1955). One major characteristic of the model to be presented is a requirement for and incorporation of a "revaluing," devaluing, or indexing mechanism. In essence, the model "partials out" or holds constant price and cost changes over time. This is accomplished either with the actual revaluing of outputs and/or inputs prior to use in the model or by selecting a base period for the model and "automatically" indexing prices and costs back to that period.

The basic concept of productivity measurement utilizing constant value prices and costs is presented in Table 5.2. As one can see, by revaluing or indexing to base year values, the analysis simply partials out or removes the influence in price and cost changes from the base year or period to the current year or period. What remains is the constant dollar value of output and input resources consumed. When these two values are compared for the base year, we establish a productivity ratio labeled output per dollar of input. When the current year or period productivity ratio is compared to the base year or period, we establish a productivity index. This table and these measures of productivity are consistent with the development presented in Chapter 2.

From a pragmatic business sense, the underlying purpose of productivity measurement and evaluation is to improve business operations and competitive position so as to enhance accomplishment of longer-term goals of survival, profitability, missions, effectiveness, and so forth. "Without productivity objectives, a business does not have direction. Without productivity measurement, it does not have control" (Drucker, 1980). The MFPMM can be utilized to measure productivity change in labor, materials, energy, and even capital, although it is not explicitly treated in this book. It can also be used to measure the effects of these changes separately as well as in aggregate on corresponding change in business profitability or, in the case of public-sector nonprofit firms, in budget maintenance. As van Loggerenberg and Cucchiaro (1982) point out, this "new" technique can be utilized to

1. Monitor historical productivity performance and measure how much, in dollars, profits were affected by productivity growth or decline

2. Evaluate company profit plans to assess and determine the acceptbility and reasonableness or productivity changes in relation to those plans

Table 5.2 Illustrative Calculation of Productivity Change Using Output and Input Data Revalued at Constant Prices
(Output and input totals in millions of dollars)

ITEM	BASE YEAR	GIVEN YEAR REVALUED AT BASE-YEAR PRICES
Case A. Increase in Productivity: Profits Earned Both Years		
Value of output	$200	$275
Cost of input (including profit at base-year rate)	$200	$250
Output per dollar of input	$ 1.00	$ 1.10
Productivity change, given/base year:		
Percentage		+ 10 percent
Per dollar of input		+$ 0.10 percent
Total dollars		+$ 25
Case B. Increase in Productivity: Losses Incurred Both Years		
Value of output	$170	$252
Cost of input	$200	$280
Output per dollar of input	$ 0.85	$ 0.90
Productivity change, given/base year:		
Percentage		+ 5.9 percent
Per dollar of input		+$ 0.05
Total dollars		+$ 14
Case C. Decrease in Productivity: Profits Earned Both Years		
Value of output	$200	$228
Cost of input (including profit at base-year rate)	$200	$240
Output per dollar of input	$ 1.00	$ 0.95
Productivity change, given/base year:		
Percentage		− 5 percent
Per dollar of input		−$ 0.05
Total dollars		−$ 12

SOURCE: H.S. Davis, *Productivity Accounting*, 1955. Reprinted with permission.

3. Measure the extent to which the firm's productivity performance is strengthening or weakening its overall competitive position relative to its peer group(s)

These three uses for the MFPMM in addition to the eight additional uses mentioned earlier represent significant benefits accruable from this model.

An organization's financial performance (one of the seven measures of performance previously mentioned) is a result of interactions of a wide variety of controllable and uncontrollable factors. Managers in organizational systems attempt to improve performance by managing (allocating, utilizing, controlling, delegating, and so forth) resources under their control while being constrained or influenced by the uncontrollable factors. Typical uncontrollable factors are

• economic environment
• industry/market growth or decline
• resource prices (costs), particularly in an inflationary period
• rates of inflation for product prices versus resource costs
• budget allocation
• organizational processes and procedures

   Typical controllable factors are

• technological innovation
• resource substitutions
• training and motivaton of employees
• asset redeployment
• resource quality

   It is interesting to note that a number of variables will influence or determine which specific factors a given manager preceives as controllable or uncontrollable. Such variables as position with the firm, personality type, leadership style, and locus of control will shape the manager's perceptions. It would seem reasonable that a manager's actual behaviors are affected more directly and strongly by perceptions than "reality." Managers today view themselves as being significantly constrained by uncontrollable factors. This is a potentially consequential dilemma with respect to prospects for productivity improvement.

   The MFPMM makes it possible to measure explicitly, in terms of dollars the profit impacts of these uncontrollable as well as controllable factors and to determine and analyze how various management strategies could increase or decrease profitability. Fundamentally, profit change comes about because of a difference between revenues and costs. If revenues increase faster than costs, there would obviously be a positive change in profits (see Figure 5.1). Yet revenues and costs do not always present a complete picture because of underlying complex relationships between controllable and uncontrollable factors. Therefore, as Davis, and Scott (1950) before him, pointed out, "[t]he net profit figure alone is an inadequate basis for judgment as to whether industrial operations are being carried out efficiently and labour and materials utilized effectively; it may merely tell us that a satisfactory balance has been struck between the value received and the value given." With essentially the same basic accounting information used to calculate revenues and costs, however, it is possible to use the MFPMM to gain additional and significantly more detailed insight into what is driving profits.

   Column 1 of Figure 5.2 depicts, as presented in Chapter 2, the basic productivity index relationship, a change in output quantities over a change in resource quantities. In every organizational system, there exists a unique productivity index for each resource. Column 2 depicts what has been called a "price recovery

$$\frac{\text{Total Revenue (TR)}}{\text{Total Cost (TC)}} = \frac{\sum\limits_{i=1}^{n} Q_i^O p_i^O}{\sum\limits_{i=1}^{m} Q_i^I p_i^I} = \begin{array}{l}\text{a measure}\\\text{of profitability}\end{array}$$

where: Q_i^O = quantities of type i output (superscript O)

p_i^O = the unit price for each output type

Q_i^I = Quantities of type i input type (superscript I)

p_i^I = the unit cost for each input type

n = the number of different types of output

m = the number of different types of input

$$\frac{\Delta TR}{\Delta TC} = \frac{\dfrac{\left(\sum\limits_{i=1}^{n} Q_i^O p_i^O\right) \text{period}_2}{\left(\sum\limits_{i=1}^{n} Q_i^O p_i^O\right) \text{period}_1}}{\dfrac{\left(\sum\limits_{i=1}^{n} Q_i^I p_i^I\right) \text{period}_2}{\left(\sum\limits_{i=1}^{n} Q_i^I p_i^I\right) \text{period}_1}} = \begin{array}{l}\text{measure of}\\\text{change in}\\\text{profitability}\end{array}$$

Figure 5.1   Profitability Assessment

Δ in output quantity	$\times$	Δ output price	$=$	Δ revenue
$\parallel$		$\parallel$		$\parallel$
$\dfrac{\sum\limits_{i=1}^{n} [(Q_i^O)\ \text{period}_2]}{\sum\limits_{i=1}^{n} [(Q_i^O)\ \text{period}_1]}$	$\times$	$\dfrac{(p_i^O)\ \text{period}_2]}{(p_i^O)\ \text{period}_1]}$	$=$	$\dfrac{TR\ \text{period}_2}{TR\ \text{period}_1}$

<hr>

$\dfrac{\sum\limits_{i=1}^{m} [(Q_i^I)\ \text{period}_2]}{\sum\limits_{i=1}^{m} [(Q_i^I)\ \text{period}_1]}$	$\times$	$\dfrac{(p_i^I)\ \text{period}_2]}{(p_i^I)\ \text{period}_1]}$		$\dfrac{TC\ \text{period}_2}{TC\ \text{period}_1}$
$\parallel$		$\parallel$		$\parallel$
Δ resource quantity	$\times$	Δ resource costs (prices)	$=$	Δ cost
Column 1 Productivity	$\times$	Column 2 "Price Recovery"	$=$	Column 3 Profitability

Figure 5.2   Productivity, Price Recovery, Profitability Relationship.

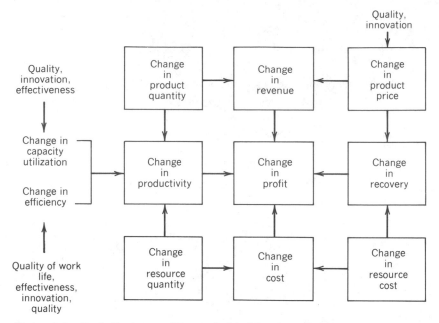

Figure 5.3   Basic Factors and Interrelationships Contributing to Performance
(Adapted from B. J. van Loggerenberg and S. J. Cucchiaro, "Productivity Measurement and the Bottom Line," *National Productivity Review*, Winter 1981–82)

index." The price recovery index is a change in output prices over a change in resource costs (prices). Column 3 reflects the profitability index, a change in revenues over a change in costs. Note that if all other factors are held constant, namely prices and costs, a positive change in the productivity index will cause or translate into a positive change in profits. Similarly, if quantities are held constant and the price recovery index is positive (output prices increase at a faster rate than resource costs), then profits, at least in the short run, will be positive. Figure 5.3, adapted from van Loggerenberg and Cucchiaro, is another representation of these relationships.

The MFPMM reflects an attempt to add to and enhance conventional profit analysis represented by Column 3. The ability to evaluate profitability changes in terms of where they come from and how they were caused is increasingly coming to be viewed as an important control system element. Similar to redesigning the control panel for an aircraft, we are beginning to see management in the United States reevaluate the instruments, dials, knobs, and controls in the control system for organizations.

Description of the MFPMM

Table 5.3 depicts the format for the MFPMM. The easiest way to describe the model is to work through the format with an example, moving from left to right

Table 5.3 Multiple Productivity Measurement Model Format (VPI/VPC Version MFPMM)

	PERIOD 1			PERIOD 2			WEIGHTED CHANGE RATIOS			COST/REVENUE RATIOS		PRODUCTIVITY RATIOS		WEIGHTED PERFORMANCE INDEXES			DOLLAR EFFECTS ON PROFITS		
	QUANTITY, Q_{i1}	PRICE, P_{i1}	VALUE, $(Q_{i1})(P_{i1})$	QUANTITY, Q_{i2}	PRICE, P_{i2}	VALUE, $(Q_{i2})(P_{i2})$	$\dfrac{Q_2P_1}{Q_1P_1}$	$\dfrac{Q_2P_2}{Q_2P_1}$	$\dfrac{Q_2P_2}{Q_1P_1}$	$\dfrac{I_{ij1}}{\Sigma(O_{i1})(P_{i1})}$	$\dfrac{I_{ij2}}{\Sigma(Q_{i2})(P_{i2})}$	PERIOD 1	PERIOD 2	CHANGE IN PRODUCTIVITY $(7/7)$	CHANGE IN PRICE RECOVERY (8×8)	CHANGE IN PROFITABILITY $(9 \cdot 9)$	CHANGE IN PRODUCTIVITY	CHANGE IN PRICE RECOVERY	CHANGE IN PROFITABILITY
	1	2	3	4	5	6	7	8	9	10	11	12	13	14	15	16	17	18	19
OUTPUT(S)																			
$O_{i,\,i=1\ldots n}$																			
$\sum_{i=1}^{n} O_i$																			
INPUTS(S)																			
LABOR																			
I_{iL}																			
$\sum_{i=1}^{n} I_{iL}$																			
MATERIAL																			
I_{iM}																			
$\sum_{i=1}^{n} I_{iM}$																			
ENERGY																			
I_{iE}																			
$\sum_{i=1}^{n} I_{iE}$																			
INVESTMENT																			
I_{iI}																			
$\sum_{i=1}^{n} I_{iI}$																			
SERVICES																			
I_{iS}																			
$\sum_{i=1}^{n} I_{iS}$																			
ETC.																			
TOTAL INPUTS																			
$\sum_{i=1}^{n} I_i$																			
DIFFERENCE																			

or from Column 1 to Column 19. For the purpose of instructional clarity, a simple example involving a fiberglass boat manufacturer is utilized to explain and "teach"the workings of MFPMM.

Columns 1–6

The first six columns of the MFPMM are data input. Column 1 represents quantities of outputs the organizational system produced and/or sold and quantities of input resources consumed in order to produce those outputs for period 1. As mentioned previously, period 1 in this model will be designated as a base period. Selection of a base period is primarily a matter of selecting a representative period in time against which you wish to compare current period performance. It might be a period of time in 1967, which just happens to be the base year utilized in the consumer price index. Or it might be a unique period in time representative of current business conditions. The base period designation can be "standards" or even simply last period. However, note that if one selects the last period as the base period and hence allows the base period to change each time the current period moves ahead, then the built-in indexing mechanism in the model is negated. In such a case, an external indexing mechanism will have to be imposed. This involves utilizing a published index, such as the producers price index or the GNP index. For more detail on indexing, refer to Anderson, Sweeney, and Williams (1981) and American Productivity Center (1978).

Recall also that the organizational system boundaries or unit of analysis for this model are flexible. A productivity process modeling exercise should precede any attempted development of an application of the MFPMM. This will ensure accurate definition of unit of analysis, inputs, outputs, and outcomes. Another parameter to be determined prior to application of this model is the length of the analysis period. Depending on decision-maker needs and interests, data availability, product cycle time, and so forth, the length might be almost any period of time (weekly, monthly, quarterly, semiannually, annually). When determining the length of the period, keep in mind your data collection needs and data matching requirements. The goal is to match outputs produced during a given period to the input resources utilized during that same period in time.

So, Column 1 represents quantities of outputs produced during the base period and quantities of inputs utilized to produce those outputs during the same base period. Table 5.4 depicts data for the base period of our boat company example. Note that in period 1 (base period) the company produced 50 Boat As and 30 Boat Bs, and utilized 320 units (in this case, hours) of management labor, 800 units (hours) of fiberglass labor, 1120 units (hours) of assembly labor, 2200 units of fiberglass, 750 units of wood, 8000 units (in this case, KWHs) of electricity, and 100 units (in this case, MCF) of gas. Note also that the scale or units utilized for outputs and inputs is a decision that can be made by the analyst. In addition, the number and class of categories, types (subcategories), and levels (sub-sub-categories) in inputs and outputs (the rows in the model) is a decision that can

Table 5.4 LINPRIM Boat Company Example (VPI/VPC Version MFPMM): Period 1 (Columns 1–6)

	PERIOD 1			PERIOD 2		
	(1) QUANTITY	(2) PRICE	(3) VALUE	(4) QUANTITY	(5) PRICE	(6) VALUE
BOAT A	50.0	5000.00	250000.00	70.0	5500.00	385000.00
BOAT B	30.0	10000.00	300000.00	35.0	12000.00	420000.00
TOTAL OUTPUTS			550000.00			805000.00
LABOR MANAGEMENT	320.0	20.00	6400.00	304.0	22.00	6688.00
LABOR GLASS	800.0	8.00	6400.00	760.0	9.00	6840.00
LABOR ASSEMBLY	1120.0	6.00	6720.00	1064.0	7.00	7448.00
TOTAL LABOR			19520.00			20976.00
FIBERGLASS	2200.0	50.00	110000.00	3000.0	85.00	255000.00
WOOD	750.0	3.00	2250.00	1000.0	3.00	3000.00
TOTAL MATERIALS			112250.00			258000.00
ELECTRICITY	8000.0	0.10	800.00	8200.0	0.10	820.00
NATURAL GAS	100.0	4.00	400.00	90.0	4.00	360.00
TOTAL ENERGY			1200.00			1180.00
TOTAL INPUTS			132970.00			280156.00

be made by either the analyst, decision maker(s), or other users of the model. For example, one could break out, by level, management labor (president, supervisor, plant manager, and so forth). The model will accommodate at least three levels (class, type, and level) of output and input. Since the model is computerized, it can handle, depending on how it is programmed, almost any number of rows. For example, the VPI/VPC version of the model for a HP3000 system is programmed to accept up to 100 row elements for each category (output, labor, energy, materials). Minicomputer programs of the model, such as on the IBM PC with 126K storage, have capacity for slightly more than 50 total row elements.

Column 2 represents the unit price for outputs and unit cost for inputs during period 1 (base period). From Table 5.4 you can see that Boat *A* sold for $5000, and Boat *B* sold for $10,000; management labor cost $20.00 per unit (hour); fiberglass labor cost $8.00 per unit (hour); assembly labor cost $6.00 per unit (hour); fiberglass cost $50.00 per unit; wood cost $3.00 per unit; electricity cost $.10 per unit (KWH); and gas cost $4.00 per unit (MCF). Note that since the analyst or user of the model can define the unit of measurement to be utilized for each output and input, the unit price and cost is also controllable. For instance, labor cost can reflect base salary, or wage rate plus bonuses or benefit calculations. The only requirement is that the unit cost remain consistent with the units of quantity.

Column 3 reflects the value (quantity × price) for each row element (outputs and inputs). Therefore, column 3 represents revenues for outputs and costs for inputs. This column is calculated automatically by the programmed version of this model. So, from Table 5.4 you can see that this company had revenues of $250,000 from sales of Boat As and $300,000 from sales of Boat Bs for a total revenue figure of $550,000; at the same time, cost for management labor was $6400; fiberglass labor, $6400; assembly labor $6720; fiberglass, $110,000; wood, $2250; electricity, $800; and gas, $400. Again, Column 3 is automatically calculated in the programmed version of this model.

Columns 4–6 are the same as columns 1–3 except that they are data for period 2 or the current period. Again, columns 4 and 5 are the data input requirements and column 6 is simply column 4 × column 5. From Table 5.4 you can see the following:

1. Boat A production went from 50 in the base period to 70 in current period; the price for Boat A went from $5000 in period 1 to $5500 current period.

2. The company utilized 16 less units (hours) of management labor but increased the cost for that category of labor from $20.00 to $22.00.

3. Fiberglass utilization increased by 800 units, and the unit cost rose from $50.00 to $85.00.

Interpretation of other changes should by now be evident and self-explanatory.

These first six columns of the MFPMM, in particular Columns 1, 2, 4, and 5, reflect data input required to "run" the model. Data availability appears not to be a critical roadblock to successful implementation of this model. Experience suggests that the basic data required to run this model are typically available from most accounting or comptroller departments. Many decisions and finer points to the actual development of an application of this model could be discussed now. However, it may be more effective to continue this tutorial on this simple example and reserve discussion of finer points until later in this Chapter.

Columns 7–9

The next three columns in the MFPMM are titled "Weighted Change Ratios." The basic purpose of these columns and, in particular, the formula calculations is to determine:

Column 7: Price-weighted and base period price indexed changes in quantities. Essentially, Column 7 partials out or holds constant the effect of prices and just examines the price-weighted changes in quantities of outputs and inputs. (See Figure 5.4 for the formula for Column 7.)

Column 8: Quantity-weighted and current period indexed changes in unit prices and unit costs. Essentially, Column 8 partials out or holds constant the changes in quantities of outputs and inputs and just examines the changes in unit prices and unit costs from period 1 to period 2. (See Figure 5.4 for the formula for Column 8.)

Column 7: $$\dfrac{\sum\limits_{i=1}^{n}(Q_{i2})(p_{i1})}{\sum\limits_{i=1}^{n}(Q_{i1})(p_{i1})}$$

Column 8: $$\dfrac{\sum\limits_{i=1}^{n}(Q_{i2})(p_{i2})}{\sum\limits_{i=1}^{n}(Q_{i2})(p_{i1})}$$

Column 9: $$\dfrac{\sum\limits_{i=1}^{n}(Q_{i2})(p_{i2})}{\sum\limits_{i=1}^{n}(Q_{i1})(p_{i1})}$$ or Column 7 × Column 8

Figure 5.4   Weighted-Change Ratio Formulas for Outputs and Inputs

Column 9: Examines the simultaneous impact of changes in price and quantity from period 1 to period 2 for each row in the model. (See Figure 5.4 for the formula for Column 9.)

From Column 7 (Table 5.5) it can be seen that

1. In period 2, 40 percent more Boat As were produced than in period 1.

$$\frac{Q_2 P_1^*}{Q_1 P_1} = \frac{70(5000)}{50(5000)} = 1.40$$

2. In period 2, 16.67 percent more Boat Bs were produced than in period 1.

$$\frac{35(10000)}{30(10000)} = 1.1667$$

3. In period 2, 27.27 percent more boats of types A and B were produced.

$$\frac{\Sigma Q_2 P_1^*}{\Sigma Q_1 P_1} = \frac{70(5000) + 35(10000)}{50(5000) + 30(10000)} = 1.2727$$

4. In period 2, 5 percent less labor was utilized than in period 1.

$$\frac{304(20) + 760(8) + 1064(6)}{320(20) + 800(8) + 1120(6)} = 0.95$$

*Shorthand formula notation.

Table 5.5   LINPRIM Boat Company Example (VPI/VPC Version MFPMM): Columns 7–11

	WEIGHTED CHANGE RATIOS			COST/REVENUE RATIOS		PRODUCTIVITY RATIOS	
	(7) QUANTITY	(8) PRICE	(9) VALUE	(10) PERIOD 1	(11) PERIOD 2	(12) PERIOD 1	(13) PERIOD 2
BOAT A	1.4000	1.1000	1.540				
BOAT B	1.1667	1.2000	1.400				
TOTAL OUTPUTS	1.2727	1.1500	1.464				
LABOR MANAGEMENT	0.9500	1.1000	1.045	0.0116	0.0083	85.94	115.13
LABOR GLASS	0.9500	1.1250	1.069	0.0116	0.0085	85.94	115.13
LABOR ASSEMBLY	0.9500	1.1667	1.108	0.0122	0.0093	81.85	109.65
TOTAL LABOR	0.9500	1.1311	1.075	0.0355	0.0261	28.18	37.75
FIBERGLASS	1.3636	1.7000	2.318	0.2000	0.3168	5.00	4.67
WOOD	1.3333	1.0000	1.333	0.0041	0.0037	244.44	233.33
TOTAL MATERIALS	1.3630	1.6863	2.298	0.2041	0.3205	4.90	4.58
ELECTRICITY	1.0250	1.0000	1.025	0.0015	0.0010	687.50	853.66
NATURAL GAS	0.9000	1.0000	0.900	0.0007	0.0004	1375.00	1944.44
TOTAL ENERGY	0.9833	1.0000	0.983	0.0022	0.0015	458.33	593.22
TOTAL INPUTS	1.2990	1.6220	2.107	0.2418	0.3480	4.14	4.05

5. In period 2, 36.36 percent more fiberglass was utilized than in period 1.

$$\frac{Q_2 P_1^*}{Q_1 P_1} = \frac{3000(50)}{2200(50)} = 1.3636$$

6. In period 2, 33.33 percent more wood was utilized than in period 1.

$$\frac{1000(3)}{750(3)} = 1.3333$$

7. In period 2, 36.3 percent more materials were utilized than in period 1.

$$\frac{3000(50) + 1000(3)}{2200(50) + 750(3)} = 1.3630$$

*Shorthand formula notation.

8. Total price-weighted and indexed change in inputs utilization was 29.90 percent.

$$\frac{304(20) + 760(8) + 1064(6) + 3000(50) + 1000(3) + 8200(.10) + 90(4)}{320(20) + 800(8) + 1120(6) + 2200(50) + 750(3) + 8000(.10) + 100(4)}$$

Hence, Column 7 simply tells us the rate of price-weighted quantity change with prices and costs held constant at period 1 levels.

From Column 8 it can be seen that

1. The prices of Boat A went up 10 percent.

$$\frac{Q_2P_2{}^*}{Q_2P_1} = \frac{70(5500)}{70(5000)} = 1.10$$

2. The quantity-weighted average price change for Boats A and B was 15 percent.

$$\frac{\Sigma Q_2P_2{}^*}{\Sigma Q_2P_1} = \frac{70(5500) + 35(12000)}{70(5000) + 35(10000)} = 1.15$$

3. Management labor unit cost increased 10 percent

$$\frac{304(22)}{304(20)} = 1.10$$

4. Quantity-weighted average cost increase for labor was 13.11 percent.

$$\frac{304(22) + 760(9) + 1064(7)}{304(20) + 760(8) + 1064(6)} = 1.1311$$

5. Fiberglass unit cost increased 70 percent

$$\frac{3000(85)}{3000(50)} = 1.70$$

*Shorthand formula notation.

6. Quantity-weighted average cost increase for materials was 68.63 percent.

$$\frac{3000(85) + 1000(3)}{3000(50) + 1000(3)} = 1.6863$$

7. There were no changes in the price of gas or electricity.

$$\frac{8200(.10) + 90(4)}{8200(.10) + 90(4)} = 1.00$$

8. Total quantity-weighted change in input costs was 62.20 percent.

$$\frac{304(22) + 760(9) + 1064(7) + 3000(85) + 1000(3) + 8200(.10) + 90(4)}{304(20) + 760(8) + 1064(6) + 3000(50) + 1000(3) + 8200(.10) + 90(4)}$$

Hence, Column 8 simply indicates the rate of quantity-weighted price and cost change with quantities of outputs and inputs held constant at period 2 levels.
From Column 9 it can be seen that

1. Revenues from Boat A increased 54 percent.

$$\frac{Q_2 P_2^*}{Q_1 P_1} = \frac{70(5500)}{50(5000)} = 1.54$$

2. Combined impact on revenue change from period 1 to period 2 from both Boat A and Boat B was 46.36 percent.

$$\frac{\Sigma Q_2 P_2^*}{\Sigma Q_1 P_1} = \frac{70(5500) + 35(12000)}{50(5000) + 30(10000)} = 1.4636$$

3. Total labor cost increased 7.46 percent from period 1 to period 2.

$$\frac{304(22) + 760(9) + 1064(7)}{320(20) + 800(8) + 1120(6)} = 1.0746$$

*Shorthand formula notation.

4. Total input costs increased 110.69 percent.

$$\frac{304(22) + 760(9) + 1064(7) + 3000(85) + 1000(3) + 8000(.10) + 90(4)}{320(20) + 800(8) + 1120(6) + 2200(50) + 750(3) + 8000(.10) + 100(4)}$$

Hence, Column 9 simply indicates the rate of change of revenues and costs (simultaneous changes in prices, costs, and quantities of outputs and inputs).

Columns 10 and 11

Columns 10 and 11 are labeled "Cost/Revenue Ratios." They indicate the ratio of input row elements for Columns 3 and 6. The formula for these columns appears in Figure 5.5. Note that Column 10 is the cost-to-revenue ratio for period 1 and Column 11 is the cost-to-revenue ratio for period 2.

From Column 10 one can observe that management labor costs (Column 3) represent 1.16 percent of total revenues in period 1 ($6400/$550,000). Similarly, total labor costs represent 3.55 percent of total revenues, fiberglass costs reflect 20 percent of total revenues, and total input costs reflect 24.18 percent of total revenues. Note that since this model is not attempting to be a total factor productivity measurement model, there is no way to tell directly whether the 75.82 percent of remaining revenues is all profits or consumed by other input resource costs not captured in this model. Note also that the information in these two columns will very likely be already available and familiar to most managers. Most managers are knowledgeable about certain cost categories as a percentage of either total costs, total revenues, or some other aggregate budget number.

The purpose of these two columns is not to provide new information but to integrate this information into the MFPMM so as to provide a manager with

$$\text{Column 10: } \frac{I\,ij_1}{\displaystyle\sum_{i=1}^{n}(O_{i1})(p_{i1})} = \frac{\text{Input elements, column 3}}{\text{Total, column 3}}$$

$$\text{Column 11: } \frac{I\,ij_2}{\displaystyle\sum_{i=1}^{n}(O_{i2})(p_{i2})} = \frac{\text{Input elements, column 6}}{\text{Total column 6}}$$

$$\text{Column 12: } \frac{\displaystyle\sum_{i=1}^{n}(O_{i1})(p_{i1})}{(I_{ij1})(p_{ij1})} = \frac{\text{Total, column 3}}{\text{Input elements, column 6}}$$

$$\text{Column 13: } \frac{\displaystyle\sum_{i=1}^{n}(O_{i2})(p_{i1})}{(I_{ij2})(p_{ij1})} = \frac{\text{Base period price weighted total, column 6}}{\text{Base period price weighted input elements, column 6}}$$

Figure 5.5   Cost/Revenue Ratio Formulas

insights as to where leverage exists. If Columns 10 and 11 are rank ordered, the manager can then invoke Pareto's Principle and make productivity improvement decisions, in terms of cost reduction, on the higher priority input resources. From this example one can easily see that a manager's leverage is with fiberglass and, in particular, with fiberglass prices.

From Column 11 it can be observed that labor costs are now (in period 2 or current period) 2.61 percent of revenues, a decrease from 3.55 percent in period 1. Fiberglass costs are now 31.68 percent of revenues, an increase from 20 percent. And total costs are now 34.8 percent of revenues, up from 24.18 percent.

Columns 12 and 13

Columns 12 and 13 are titled "Productivity Ratios." Column 12 reflects the output-to-input ratios for period 1, while column 13 reflects the output-to-input ratios for period 2. This is a relatively new edition to this model and exists only on certain versions of the software for this particular productivity measurement technique. The formulas for these two columns appear in Figure 5.5.

Columns 14–16

Columns 14–16 (Table 5.6) are titled "Weighted Performance Indexes." Column 14 reflects price-weighted productivity indexes. Column 15 represents quantity-

Table 5.6   LINPRIM Boat Company Example (VPI/VPC Version MFPMM): Columns 14–19

	WEIGHTED PERFORMANCE INDEXES			DOLLAR EFFECTS ON PROFITS		
	(14)	(15)	(16)	(17)	(18)	(19)
	CHANGE IN			CHANGE	CHANGE	CHANGE
	PRODUC-TIVITY	PRICE RECVRY	PROFIT-ABILITY	IN PRODUC-TIVITY	IN PRICE RECOVERY	IN PROFIT-ABILITY
BOAT A						
BOAT B						
TOTAL OUTPUTS						
LABOR MANAGEMENT	1.340	1.045	1.401	2065.45	613.82	2679.27
LABOR GLASS	1.340	1.022	1.369	2065.45	461.82	2527.27
LABOR ASSEMBLY	1.340	0.986	1.321	2168.73	218.91	2387.64
TOTAL LABOR	1.340	1.017	1.362	6299.64	1294.54	7594.18
FIBERGLASS	0.933	0.676	0.631	-10000.00	-84000.00	-94000.00
WOOD	0.955	1.150	1.098	-136.36	429.55	293.18
TOTAL MATERIALS	0.934	0.682	0.637	-10136.38	-83570.44	-93706.81
ELECTRICITY	1.242	1.150	1.428	198.18	152.73	350.91
NATURAL GAS	1.414	1.150	1.626	149.09	76.36	225.45
TOTAL ENERGY	1.294	1.150	1.488	347.27	229.09	576.36
TOTAL INPUTS	0.980	0.709	0.695	-3489.45	-82046.81	-85536.27

weighted price recovery indexes. And Column 16 depicts profitability indexes. The formulas for these three columns appear in Figure 5.6. Note that there are no entries for the cells corresponding to the output row elements. This is because Columns 14–16 are now calculating output over input indexes, or changes in performance ratios, from period 1 to period 2. The essence of the MFPMM appears in Columns 12–19.

As discussed in Chapter 2, there are at least four generic types of productivity "measures": (1) partial factor, static ratio; (2) total factor, static ratio; (3) partial factor, dynamic index; and (4) total factor, dynamic index. Recall that a dynamic productivity index is essentially a productivity ratio at one period in time, say, period 2 (current period), over that same productivity ratio at a previous period in time, say, period 1 (base period). Columns 14–16 calculate and depict dynamic *performance* indexes. Column 14 calculates and depicts dynamic *productivity* indexes. Figure 5.7 conceptually depicts what the MFPMM is doing.

In Figure 5.7, formulas and development of static productivity ratios are depicted. We take a snapshot of the organizational system for a given period of time and place some or all of the outputs in the numerator and one, some, or all of the inputs in the denominator. For a decoupled, disaggregated system, such as the NPMM, we do not necessarily need to use indexed prices and costs as a common denominator. For an aggregated system, such as the MFPMM, indexed prices and costs are necessary.

Column 14:

$$\dfrac{\dfrac{\sum\limits_{i=1}^{n}(O_{i2})(p_{i1})}{\sum\limits_{i=1}^{n}(O_{i1})(p_{i1})} = \text{Column 7 for total outputs}}{\dfrac{(Iij_2)(p_{i1})}{(Iij_1)(p_{i1})} = \begin{array}{l}\text{Column 7 for each individual}\\ \text{input}\end{array}} \qquad \text{Productivity}$$

Column 15: Column 14/Column 12
or

$$\dfrac{\dfrac{\sum\limits_{i=1}^{n}(O_{i2})(p_{i2})}{\sum\limits_{i=1}^{n}(O_{i2})(p_{i1})} = \text{Column 8 for total outputs}}{\dfrac{(Iij_2)(p_{i2})}{(Iij_2)(p_{i1})} = \text{Column 8 for each input}} \qquad \text{Price recovery}$$

Column 16:

$$\dfrac{\dfrac{\sum\limits_{i=1}^{n}(O_{i2})(p_{i2})}{\sum\limits_{i=1}^{n}(O_{i1})(p_{i1})} = \text{Column 9 for total outputs}}{\dfrac{(Iij_2)(p_{i2})}{(Iij_1)(p_{i1})} = \text{Column 9 for each input}} \qquad \text{Profitability}$$

Figure 5.6   Weighted Performance Indexes

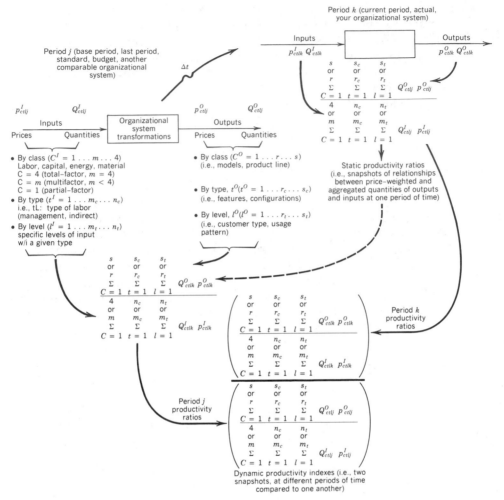

Figure 5.7   Price-Weighted and Aggregated Multifactor Productivity Measurement Model

Figure 5.7 also depicts formulas and development of dynamic productivity indexes. A snapshot of the organizational system's partial, multi-, and perhaps even total static productivity ratio is developed for period k (period 2, current period). An equivalent snapshot of the organizational system's partial, multi-, or perhaps even total static productivity ratio is developed for period j (period 1, base period, budget, standards, another comparable system, and so forth). The productivity ratios for period k (period 2 or current period) are then divided by the productivity ratios for period j (period 1 or base period). The resultant formulation is highlighted in Figure 5.7, and it is this calculation that is depicted in Column 14 of the MFPMM.

From Column 14 the following observations can be made:

1. Labor productivity increased by 34 percent.

$$\frac{\text{Column 7 for total outputs} = 1.2727}{\text{Column 7 for total inputs} = .95} = 1.34$$

(Note that Column 7 is the price-weighted changes in quantities for outputs and inputs. As an exercise, see question 13 at the end of this chapter to convince yourself that

$$\frac{O}{I} = \frac{Q_2^O/Q_1^O}{Q_2^I/Q_1^I} = \frac{Q_2^O/Q_2^I}{Q_1^O/Q_1^I}$$

This tells us that price-weighted change in outputs from period 1 to period 2 went up 27.27 percent while labor input went down 5 percent creating a corresponding gain in productivity of 34 percent.

2. Materials productivity decreased 7 percent.

$$\frac{\text{Column 7 for total outputs} = 1.2727}{\text{Column 7 for total materials} = 1.363} = 0.93$$

3. Total inputs productivity declined by 2 percent. Again, total price-weighted and indexed outputs from this company increased by 27.27 percent, while total price-weighted and indexed input quantities increased by 29.9 percent. Hence, $1.2727/1.299 = 0.98$ and the calculated 2 percent decline in productivity for all inputs measured in this model formulation.

Column 15 depicts rates of change for quantity-weighted and indexed prices over costs. It reflects rate of price increases in relation to the rate of cost increases. In a sense it reflects the degree to which the organizational system was able to increase its price in relation to elemental input costs. It is simply termed price recovery.
From Column 15 it can be observed that

1. Price recovery for management labor was up 5 percent.

$$\frac{\text{Column 8 for total outputs} = 1.15}{\text{Column 8 for management labor} = 1.10} = 1.045$$

This indicates that the organization was able to raise prices approximately 5 percent faster than management unit prices (costs) increased.

2. Price for fiberglass increased approximately 32 percent faster than management was able to raise the prices of boats.

$$\frac{\text{Column 8 for total outputs} = 1.15}{\text{Column 8 for fiberglass input} = 1.7} = 0.676$$

3. On the whole, price recovery fell off by 29 percent.

$$\frac{\text{Column 8 for total outputs} = 1.15}{\text{Column 8 for total inputs} = 1.622} = 0.71$$

Changes in output prices were 71 percent of the changes in input costs. The company was not able or did not (for whatever reason) raise prices fast enough to compensate for increases in costs. (Note: Fiberglass price under-recovery was the major source of the relatively poor price recovery ratio of .71.)

Column 16 indicates profitability indexes, which reflect rates of change for simultaneous changes in price and quantity. The simplest way to think about Column 16 is that it is revenues/costs (a measure of profitability) for period 2 divided by revenues/costs for period 1. Hence, Column 16 is in reality a profitability index.

From Column 16 it can be seen that labor contributed to a 36 percent increase in profitability from period 1 to period 2. That is, revenues went up 46.36 percent from period 1 to period 2 (Column 9 for total outputs), while total labor costs increased by 7.46 percent (Column 9 for total labor) creating a 36 percent (1.4636/ 1.0746 = 1.3619) labor relative increase in profitability from period 1 to period 2. Materials created a period 1 to period 2 relative drain on profitability of 36 percent. Revenues changed at a rate of 46.36 percent, while material costs increased at a rate of 129.84 percent. Note that most of this drain on potential profits, which could have been achieved from the 46.36 percent increase in revenues, was caused by the 131.82 percent increase in fiberglass costs (both increased unit cost and increased quantity usage).

Overall, Column 16 depicts a 31 percent decline in potential profitability. This company was 31 percent less profitable in period 2 than it was in period 1. The company may well have made profits, but it could have made 31 percent more profits had certain price under-recovery situations not occurred. It should by now be clear that a number in Column 14, 15, or 16 that is greater than 1.00 reflects a positive change and a number less than 1.00 reflects a negative change. Therefore, our overall evaluation of this particular organization's productivity, price recovery, and profitability performance on a period 1 to period 2 basis is not favorable. In particular, management or an analyst could be concerned about fiberglass cost recovery.

Columns 17–19

Columns 17–19 reflect the dollar equivalence of corresponding cells in Columns 14–16. In other words, these columns indicate what impact an increase in productivity (Column 17) or price recovery (Column 18) has on profits. The total impact on profits from productivity and price recovery is indicated in Column 19. The formulas for these columns appear in Figure 5.8. From these columns we see the following.

1. *Column 17:* Management labor productivity contributed $2065.45 to profits from period 1 to period 2.

$$(1.2727 - .95)\$6400 = \$2065$$

Column 18: The model does not directly calculate Column 18, effect of price recovery on profits. Column 18 values are calculated by subtracting Column 17 values from Column 19 values. In other words, Column 17 + Column 18 = Column 19.

Column 19: Management labor contributed positively to profits between period 1 and period 2 to the tune of $2679. About $2065 came from productivity gains and $613 came from price recovery gains.

$$(1.4636 - 1.045)\$6400 = \$2679$$

Column 17:

$$\left[\begin{array}{c}(lij_1)(P_{i1})\\ \text{or}\\ \text{Column 3}\\ \text{for each}\\ \text{corresponding}\\ \text{input}\end{array}\right]\left[\left(\frac{\sum_{i=1}^{n}(Q_{i2})(p_{i1})}{\sum_{i=1}^{n}(Q_{i1})(p_{i1})}\begin{array}{c}\text{Column 7}\\ \text{or for total}\\ \text{outputs}\end{array}\right) - \left(\frac{(lij_2)(p_{i1})}{(lij_1)(p_{i1})}\begin{array}{c}\text{Column 7}\\ \text{or for each}\\ \text{input}\end{array}\right)\right]$$

Column 18: Column 19 − Column 17

Column 19:

$$\left[\begin{array}{c}(lij_i)(p_{i1})\\ \text{or}\\ \text{Column 3}\\ \text{for each}\\ \text{corresponding}\\ \text{input}\end{array}\right]\left[\left(\frac{\sum_{i=1}^{n}(Q_{i2})(p_{i2})}{\sum_{i=1}^{n}(Q_{i1})(p_{i1})}\begin{array}{c}\text{Column 9}\\ \text{or for total}\\ \text{outputs}\end{array}\right) - \left(\frac{(lij_2)(p_{i2})}{(lij_1)(p_{i1})}\begin{array}{c}\text{Column 9}\\ \text{or for each}\\ \text{input}\end{array}\right)\right]$$

Figure 5.8 Weighted Performance Indexes, Individual Effects on Profits

2. *Total materials Column 19:* Low productivity in materials utilization created a drain on profits from period 1 to period 2 of $-\$10,136$. About $\$10,000$ of this decline came from low fiberglass productivity alone.

$$(1.2727 - 1.363)\$112,250 = -\$10,136$$

Total materials Column 19: Very poor price recovery on fiberglass and low productivity created a $-\$93,706$ drain on profits for this company from period 1 to period 2.

$$(1.4636 - 2.2984)\$112,250 = -\$93,706$$

This reflects the drain on profits caused by an inability to recover rising costs from period 1 to period 2. As one can see, the biggest source of lowered profits from period 1 to period 2 is this category.

3. Overall, this boat manufacturing company was $\$85,536$ less profitable in period 2 than in period 1 had nothing changed in the company. About $\$82,047$ of this decline in profits is attributable to relatively poor price recovery. And, as indicated, very poor fiberglass price recovery is the major source of this total decline in profits.

This completes the description and example for the MFPMM. There are obviously many fine details, application and implementation issues, and refinements that could be discussed. Some of these points will be dealt with in this section. However, at this point, the reader should have a good grasp of the basic character of the model. It is a relatively simple model and yet it has tremendous potential as an integrative decision support system. There are applications at the end of this chapter that can be utilized to develop more skill and a deeper understanding of how to interpret program output. Those desiring to purchase the model software can experiment with the model quite painlessly. You might even wish to collect data from a specific example of your own and run the program. Like any decision support system, the model itself is a critical but rather minor component of an application. Integrating the model into an existing control system, collecting the data, getting management to accept and feel comfortable with the system, and selling the system based on benefit-to-cost projects are all activities that actually play a more critical role in successful implementation of such a system.

In an attempt to improve the decision support capabilities of the model, staff at the Oklahoma Productivity Center and now at the Virginia Productivity Center have developed a simple simulation routine to allow management to project the impact of productivity improvement interventions on profits. This development is the focus of the discussion in the next section.

MFPMM Simulation Routine: Decision Support System Developments

Imagine the following setting in relation to the boat manufacturing company just presented. The president of the firm, his managers of purchasing, marketing, production, personnel, industrial engineering, quality control, and finance (comptroller), and a staff industrial engineer who is the company's productivity analyst are in a monthly planning meeting to discuss the performance of the company this past month. The productivity analyst has just presented a briefing summarizing the report just described to you in the last section. The productivity analyst has learned from past meetings with this group of upper-level managers that their tolerance for long complex briefings and reports is low. So, the analyst has worked hard to develop a simple yet effective set of graphics that succinctly summarize and depict the key data from the MFPMM. The analyst has learned that some of this group of managers feel more comfortable with raw data, tables, and figures while others prefer to see charts, graphs, and other pictorial-type representations of the MFPMM report. A few samples from the analyst's briefing materials are presented on the next several pages.

The analyst has assumed the role of facilitator for this session. Each manager is provided with a briefing package prior to this performance/productivity improvement planning session. This package includes the managers' own individualized summary graphics of the most recent MFPMM run, graphics such as the ones depicted in Figure 5.9, and the actual output in an appendix (if requested). In preparation for this monthly session, each manager reviews the MFPMM results in addition to any other performance measurement control

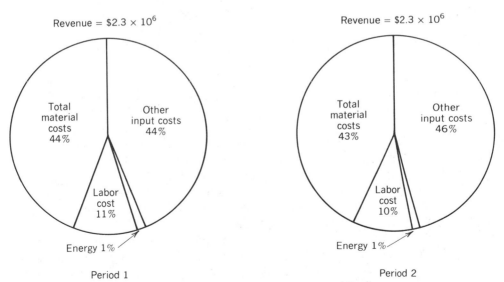

Figure 5.9a   Input Costs as a Percentage of Revenue

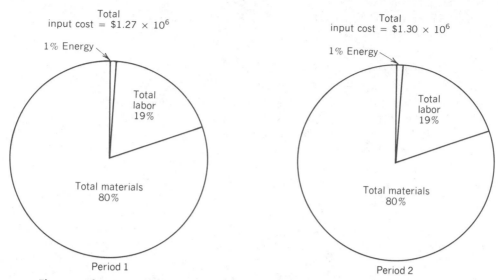

Figure 5.9b   Cost of Classes of Input as a Percentage of Cost of Measured Input

system data the managers individually have access to (production, inventory, quality, scheduling, personnel type reports/information systems, and so forth). Each manager is also expected to make projections for input and/or output changes he or she foresees taking place either independently of intervention on his or her part or with appropriate action on the management team's part.

Each manager then comes to this monthly performance/productivity improvement planning session with some or potentially all of the form indicated in Figure 5.10 completed. In the meeting itself, the productivity analyst starts out by making a short, general review briefing of the most recent MFPMM report. The president of the firm is allowed five minutes to express his overall perception of the company's performance from a strategic standpoint. Each manager is then given no more than five minutes to state his or her assessment of the company's performance in the previous month. Data other than that provided by the MFPMM are often presented, and graphics in the form of overheads or handouts are frequently utilized.

At the end of this briefing and review session, the analyst loads the MFPMM software onto the company's business computer. The MFPMM simulation routine is called up, and the forecast portion of this planning session begins. The MFPMM simulation routine allows management to develop "what if" scenarios with the model. The only data input required are three point estimates (pessimistic, most likely, optimistic) for specific input and/or output values. One, some, or all of the prices, costs, output quantities, or input quantities can be changed. The analyst and managers have the option of comparing period 2 (current or immediate past period) with period 3 (forecasted, projected period, next month), or period 1 (base period) with period 3. Recall that if the group decides to compare periods 2 and 3, then the model's built-in base period in-

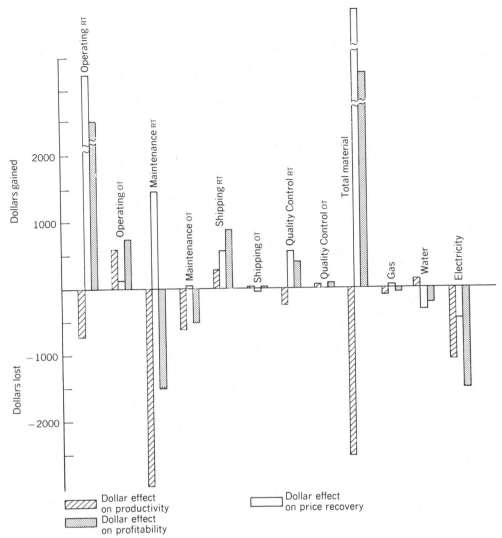

Figure 5.9c   Dollar Effect on Profitability, Productivity, and Price Recovery

dexing process is negated. In this case, the analyst would have to externally or manually index all prices and costs to some constant value. If the group chooses to compare periods 3 and 1, then the MFPMM automatically removes the effect of inflation from the productivity analysis.

The next step in the simulation subroutine and planning process is to indicate a desired value for Column 19 for total inputs. The question is, what would the management group like the total effect on profits of their efforts to be at the end of the next month? The analyst asks them to indicate a desired value for the change in profitability for the following month. Recall that this value was

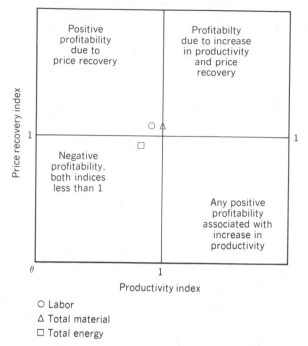

Figure 5.9d   Productivity, Price Recovery, and Profitability

−$85,536 in period 2, or for the immediate past month. The management group agrees that given current economic and business conditions, they would be pleased if they could cause the number to go to $0.00. The analyst enters this decision into the computer.

The next step in the simulation subroutine and for this monthly session is to enter the three-point estimates for specific expected value changes for input quantities and costs, as well as for output quantities (in this case, sold boats) and prices. First, the projected changes for output quantities and prices are entered. There is much discussion among the managers of production, marketing, purchasing, industrial engineering, and quality control as to what the pessimistic, most likely, and optimistic values should be. Additional data from various sources are referred to for support of estimates. The analyst's job is one of keeping the session moving and striving for consensus. It is not critical that a unified estimate be arrived at. Several different scenarios can be developed, each based on a different set of assumptions. The simulation routine can be run for several scenarios and results can then be compared.

Next, labor projections are made in terms of staffing, workload scheduling, and pay determinations. Again, three-point estimates are arrived at among relevant managers and entered by the analyst. The same process of making estimates is completed for materials, energy, and capital if it was included. A given scenario for this example is depicted in Tables 5.7a–d. This table presents the

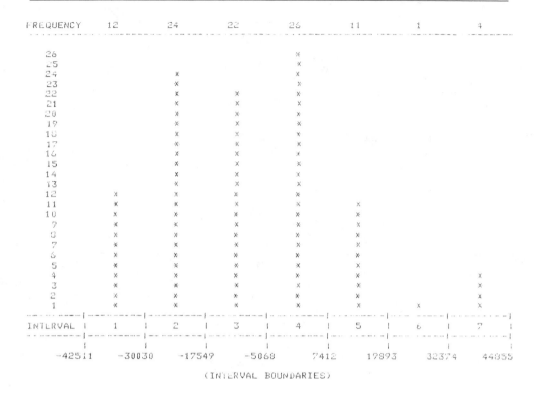

PROBABILITY (FROM 100 SIMULATION RUNS) OF EQUALLING OR
EXCEEDING YOUR DESIRED VALUE OF   0 FOR TOTAL INPUTS
EFFECT ON CHANGE IN PROFITABILITY = .30

(e)

**Figure 5.9e   Histogram Plot of Column 19 Values: Simulated Impact of Pro-
ductivity Management Interventions on Profits**

entire simulation output comparing projected period 3 and base period 1. You
may wish to compare this output with the output comparing periods 1 and 2
depicted in Tables 5.4–5.6. The simulation subroutine calculates expected values
for all variables that were changed. Note when management changes or makes
a three-point estimate for a given variable, that variable becomes stochastic or
probabilistic in character. The triangular distribution is utilized to derive the
expected values (Sullivan and Orr, 1982; Buck, 1982; and Pritsker, 1979). See
Columns 4 and 5 of the MFPMM for these values from the scenario developed
by this management group. If no changes are made for a given quantity or price/
cost, the model assumes that the current period value (period 2, immediate past
month, in this case) remains the same for period 3, or the next/projected period.
Once all variables that are to be changed for a given scenario have been entered,
the program can be run. On minicomputers, such as the TRS-II, IBM-PC, and

*CATEGORY NUMBER	CATEGORY NAME	ESTIMATE SEQUENCE ↓±	QUANTITY	PRICE
Input			*Future Period:* ___	
		P = pessimistic		
		M = most likely		
		O = optimistic		
		P		
		M		
		O		
● ● ● ●		● ● ● ●		

*CATEGORY NUMBER	CATEGORY NAME	ESTIMATE SEQUENCE ↑±	QUANTITY	PRICE
Output			*Future Period:* ___	
		P		
		M		
		O		
		P		
		M		
		O		
		● ● ●	● ● ●	● ● ●

Figure 5.10   Simulation Change Sheet

Apple, the simulation routine may take up to 15 minutes (a good time for a coffee break). On small business computers, such as the HP3000, or larger systems the response time depends primarily on printer speed.

The analyst's skills in interpreting the MFPMM now play a big role in the effectiveness of this session. Results of the MFPMM will appear on the terminal (each manager might have his or her own), a hard copy can be generated, and with video out and the right equipment the results can be presented on a big screen TV or on a projection screen. The first and most obvious element to examine is Column 19 for total inputs. The management group can see how close they came to meeting their target of −$0.00 change in profits from period 1 to period 3. Additionally, a histogram depicting the results of 100 simulation runs is generated and the managers can obtain a better feel for the characteristics of the probability distribution for the specific scenario they have developed. The

Table 5.7a Example Simulation Scenario for LINPRIM Boat Company Example (VPI/ VPC Version MFPMM)

```
YOUR CURRENT VALUES FOR QUANTITY ARE AS FOLLOWS:
NO.    CATEGORY NAME                QUANTITY
                            PESS.      M.LIKELY     OPTM.
  1    BOAT A                50.        65.         70.
  2    BOAT B                35.        45.         55.
  3    TOTAL OUTPUTS          0.         0.          0.
  4    LABOR MANAGEMENT     304.       304.        304.
  5    LABOR GLASS          760.       760.        760.
  6    LABOR ASSEMBLY      1064.      1064.       1064.
  7    TOTAL LABOR            0.         0.          0.
  8    FIBERGLASS          3000.      3000.       3000.
  9    WOOD                1000.      1000.       1000.
 10    TOTAL MATERIALS        0.         0.          0.
 11    ELECTRICITY         8200.      8200.       8200.
 12    NATURAL GAS           90.        90.         90.
 13    TOTAL ENERGY           0.         0.          0.
 14    TOTAL INPUTS           0.         0.          0.

YOUR CURRENT VALUES FOR PRICE ARE AS FOLLOWS:
NO.    CATEGORY NAME                 PRICE
                            PESS.      M.LIKELY     OPTM.
  1    BOAT A            5500.00      6000.00     6500.00
  2    BOAT B           12000.00     12000.00    12000.00
  3    TOTAL OUTPUTS         .00          .00         .00
  4    LABOR MANAGEMENT    22.00        22.00       22.00
  5    LABOR GLASS          9.00         9.00        9.00
  6    LABOR ASSEMBLY       7.00         7.00        7.00
  7    TOTAL LABOR           .00          .00         .00
  8    FIBERGLASS          75.00        70.00       55.00
  9    WOOD                 3.00         3.00        3.00
 10    TOTAL MATERIALS       .00          .00         .00
 11    ELECTRICITY           .10          .10         .10
 12    NATURAL GAS          4.00         4.00        4.00
 13    TOTAL ENERGY          .00          .00         .00
 14    TOTAL INPUTS          .00          .00         .00
```

(a)

This part of the program performs a "what if" game on the model. It can give you all available options based on any desired value in Col. 17 for total *inputs*. In order to effectively use this program, you have to make estimates regarding quantity and price for the next period as we go along in this part. If you want to continue in this part of the program answer "Yes"; otherwise, "No" ? ?Yes. Do you want to compare periods 2 & 3? Yes/No: ?No. Do you want to compare periods 1 & 3? Yes/No: ?Yes. Enter desired value in Col. 17 for total inputs: ?0.

Now, you are required to indicate a three point estimate, namely: pessimistic, most likely, and optimistic values for quantity and price of each category you choose to vary for the next period. For variables, these estimates are not made. It is assumed that the present values of quantity and price are same for the next period, too.

histogram plots the column 19 values associated with each of the 100 simulated trials. Other graphics, such as this histogram, could be developed and immediately prepared in this planning session for review. (see Figure 5.9e)

Productivity, price recovery, and profitability index trend charts (performance trend charts) can be updated with forecasted scenario output and presented for review (see Figure 5.11). This evaluation of scenarios can go on as long as management desires. Typically, this part of the monthly performance/productivity improvement planning session would last approximately one hour.

The next step in this planning session is to begin to develop specific action plans relative to the most preferred scenario. For this step, the management group must agree on the desired scenario they wish to work toward and then thinking through specific actions that will need to be taken by specific individuals

Table 5.7b   Forecasted Scenario Output for LINPRIM Boat Example (VPI/VPC Version)

```
MULTI-FACTOR PRODUCTIVITY MEASUREMENT                    1 OF 3
VPI/VPC VERSION
```

	PERIOD 1			PERIOD 2		
	(1) QUANTITY	(2) PRICE	(3) VALUE	(4) QUANTITY	(5) PRICE	(6) VALUE
BOAT A	50.1	5000.00	250000.00	61.1	6000.00	366000.00
BOAT B	30.1	10000.00	300000.00	45.1	12000.00	540000.00
TOTAL OUTPUTS			550000.00			906000.00
LABOR MANAGEMENT	320.1	20.00	6400.00	304.1	22.00	6688.00
LABOR GLASS	800.1	8.00	6400.00	768.1	8.00	6540.00
LABOR ASSEMBLY	1120.1	6.00	6720.00	1064.1	7.00	7448.00
TOTAL LABOR			19520.00			20976.00
FIBERGLASS	2200.1	50.00	110000.00	3000.1	66.671	200000.00
WOOD	750.1	3.00	2250.00	1000.1	3.00	3000.00
TOTAL MATERIALS			112250.00			203000.00
ELECTRICITY	8000.1	.10	800.00	8200.1	.10	820.00
NATURAL GAS	90.1	4.00	360.00	90.1	4.00	360.00
TOTAL ENERGY			1160.00			1180.00
TOTAL INPUTS			132930.00			225156.00

or groups in order to cause the scenario to become a reality. Again, as has been mentioned, some changes that are projected to occur are controllable, others are not. Obviously, the action plans that are specifically developed will be proactive about those things that can be controlled and reactive to negative trends and their impacts. An example of an action plan assignment (responsibility and accountability) format for this case scenario is depicted in Figure 5.12.

In summary, it may be valuable to step back and examine this performance/productivity planning process in generic perspective. The productivity management process model presented in Chapter 2, Figure 2.1, will facilitate this. In this planning meeting just discussed, the management team has reviewed the output from their performance measurement decision support systems. As presented, the MFPMM played an integral role in this performance measurement process. The management group evaluated the data from the MFPMM and other performance measurement systems. They developed and evaluated projected scenarios of company performance utilizing the MFPMM simulation subroutine. They developed specific action plans necessary to cause the most desirable scenario to be achieved. The commitment to and quality of follow-through on these action plans by the various managers and functions involved will eventually determine the degree to which performance and productivity will be controlled and improved. Planning was necessary and was involved in order to

Table 5.7c LINPRIM Boat Example, Columns 7–13 (MFPMM) (VPI/VPC Version)

	WEIGHTED CHANGE RATIOS			COST/REVENUE RATIOS		PRODUCTIVITY RATIOS	
	(7) QUANTITY	(8) PRICE	(9) VALUE	(10) PERIOD 1	(11) PERIOD 3	(12) PERIOD 1	(13) PERIOD 3
BOAT A	1.2333	1.2000	1.480				
BOAT B	1.5000	1.2000	1.800				
TOTAL OUTPUTS	1.3788	1.2000	1.655				
LABOR MANAGEMENT	0.9500	1.1000	1.045	0.0116	0.0073	85.94	124.73
LABOR GLASS	0.9500	1.1250	1.069	0.0116	0.0075	85.94	124.73
LABOR ASSEMBLY	0.9500	1.1667	1.108	0.0122	0.0082	81.85	118.79
TOTAL LABOR	0.9500	1.1311	1.075	0.0355	0.0231	28.18	40.89
FIBERGLASS	1.3636	1.3333	1.818	0.2000	0.2198	5.00	5.06
WOOD	1.3333	1.0000	1.333	0.0041	0.0033	244.44	252.78
TOTAL MATERIALS	1.3630	1.3268	1.808	0.2041	0.2231	4.90	4.96
ELECTRICITY	1.0250	1.0000	1.025	0.0015	0.0009	687.50	924.80
NATURAL GAS	0.9000	1.0000	0.900	0.0007	0.0004	1375.00	2106.48
TOTAL ENERGY	0.9833	1.0000	0.983	0.0022	0.0013	458.33	642.66
TOTAL INPUTS	1.2990	1.3036	1.693	0.2418	0.2474	4.14	4.39

(c)

Table 5.7d LINPRIM Boat Example, Columns 14–19 (MFPMM) (VPI/VPC Version)

	WEIGHTED PERFORMANCE INDEXES			DOLLAR EFFECTS ON PROFITS		
	(14) CHANGE IN PRODUCTIVITY	(15) CHANGE IN PRICE RECVRY	(16) CHANGE IN PROFITABILITY	(17) CHANGE IN PRODUCTIVITY	(18) CHANGE IN PRICE RECOVERY	(19) CHANGE IN PROFITABILITY
BOAT A						
BOAT B						
TOTAL OUTPUTS						
LABOR MANAGEMENT	1.451	1.091	1.583	2744.24	1156.85	3901.09
LABOR GLASS	1.451	1.067	1.548	2744.24	1004.85	3749.09
LABOR ASSEMBLY	1.451	1.029	1.493	2881.46	789.09	3670.55
TOTAL LABOR	1.451	1.061	1.540	8369.94	2950.79	11320.73
FIBERGLASS	1.011	0.900	0.910	1666.67	-19666.67	-18000.00
WOOD	1.034	1.200	1.241	102.27	620.45	722.73
TOTAL MATERIALS	1.012	0.904	0.915	1768.95	-19046.24	-17277.28
ELECTRICITY	1.345	1.200	1.614	283.03	220.61	503.64
NATURAL GAS	1.532	1.200	1.838	191.52	110.30	301.82
TOTAL ENERGY	1.402	1.200	1.683	474.55	330.91	805.45
TOTAL INPUTS	1.061	0.921	0.977	10613.44	-15764.53	-5151.09

(d)

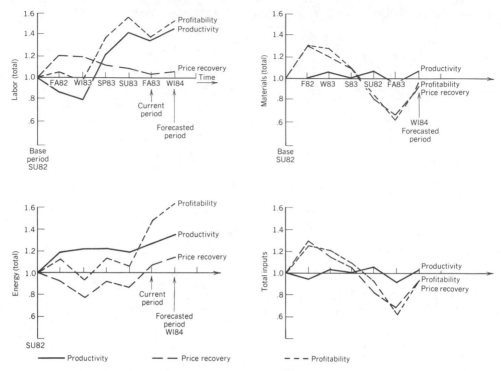

Figure 5.11 Performance Trend Charts (Columns 14–16 plotted over time)
━━━ **Productivity** ━━ **Price Recovery** ------- **Profitability**

develop action plans that would have a high probability of controlling and improving performance and productivity. So from Figure 5.13, which was first presented in Chapter 2, it can now perhaps be more easily observed and understood how the productivity management process actually can and should take place in an organization.

Using Quality Costs in Productivity Measurement

Sink and Keats (1983) presented concepts regarding the integration of quality costs into the MFPMM. Although not fully developed, use of this model to evaluate the impact of productivity improvement interventions such as the following seems highly promising:

1. Quality management programs

2. Quality circles

3. Capital investments

4. Gainsharing plans

5. Shifts in quality emphasis from appraisal to prevention

PRODUCTIVITY IMPROVEMENT	IDEA FOR: PERFORMANCE IMPROVEMENT	PERIOD 2 VALUE	PERIOD "3" TARGETED CHANGE	SIMULATED POTENTIAL IMPACT	RESPONSIBILITY & ACCOUNTABILITY
Hold labor constant • wages • hours	Sales of: Boat A	70	61	• ↑ Rev. to $906,000 from $805,000	Marketing, Sales, Production
	Boat B	35	45	•	
	↑ Price of: Boat A	5500	6000		
	Boat B	12000	12000		
	Reduce Fiberglass Cost (unit)	35	67	• $8369 Prod. • $2808 P.R. • $11,178 Total	Human Resources, Production, All management, Industrial Engineering
				• $60,000 reduction in price recovery loss from period 2 to 3	Purchasing, Industrial Engineering
Materials Management		(.93) -7% prod. loss Col. 12 1.29	(1.01) 1% prod. gain Col. 12 1.36 36% gain in prod.	• $1768	Manufacturing, Industrial Engineering
Energy Management				• $419	Industrial Engineering

THE MAJOR, PLANNED FOR PRODUCTIVITY IMPROVEMENT STRATEGIES

SECONDARY STRATEGIES / MANAGEMENT BASICS

Figure 5.12 MFPMM Decision Support System—Monthly Productivity Improvement Planning Session Action Plan Assignment

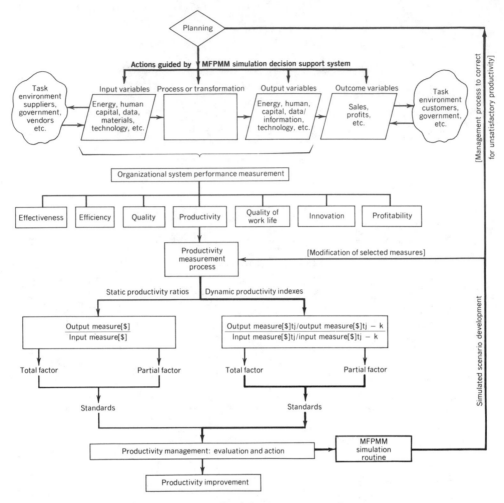

Figure 5.13   Productivity Management Process

The next section presents basic concepts on how to separate quality costs in the MFPMM in an attempt to evaluate the performance of quality improvement efforts.

Quality Costs in the Model

Table 5.8 presents, for selected levels (as obtained from the *Quality Control Handbook*) within each of the four quality cost categories the effects of these levels on the output and inputs of the multifactor model. The effect is indicated as either being directly observable (D) or indirectly observable (I) in the model. D/I entries indicate that, depending on the record-keeping practices of the company, the effect could be either directly or indirectly

Table 5.8 Directly (D) and Indirectly (I) Observable Effects of the Multifactor Model

	INTERNAL							EXTERNAL				APPRAISAL					PREVENTION					
	Scrap	Rework	Retest	Downtime	Yield losses	Disposition	Complaint adj.	Returned Mat'l.	Warranty chgs.	Allowances	Incom. insp.	Insp. & test	Accur. of equip.	Mat'l. & services	Stock evaluation	Qual. planning	New products	Training	Process control	Data acq. & anal.	Qual. report	Improvement
Output	I	I	I	I	I	I	D															
Labor	D	D/I		I		I	I	D/I	D/I	D	D	D/I		D/I	D/I	I	D/I	D/I	D/I	D/I	D/I	D/I
Material	I	D/I		I			D/I	D/I					D			D		D				D
Energy	I	I		I			I	I			I		I					I				
Capital										D	D						D		D	D		

observable. Blank entries in the matrix do not necessarily indicate no effects in the model—
only those levels that have a substantial effect have been marked.

With respect to output, internal failures will obviously have a strong effect. In a period
with large amounts of scrap and rework, for example, output will be smaller than what it
would have been had scrap and rework been under control. Returned items (external
failure costs) can be subtracted from units produced either in the time period when these
units were made, or in the time period when the returns occurred.

Space considerations will not allow comment on each of the input effects and thus only
a few effects will be illustrated. Any activity that will reduce scrap and rework will also
decrease labor, material, and energy inputs. Repairs and adjustments under warranty will
result in labor, material, and energy changes that may or may not be directly measured.
Capital expenditures are required for all quality equipment. These expenditures include
depreciation, repairs, space charges, and so forth. Quality materials as appraisal costs
would include x-ray film, materials used in printed reports and charts, as well as any
disposables used in assessing the condition of the product. All labor used in prevention
will be reflected in increased or decreased inputs although such labor may not be tracked
according to such levels as planning, training, and reporting.

To depict use of the multifactor model with quality costs, an example is presented in
Table 5.9. The example is extremely simplified for illustrative purposes. In practice, one
would find hundreds of input variables and considerably more quality inputs than are
presented here.

The quantities shown in Columns 1 and 4 for output represent units sold in a period
for Product X. Period 1 is the base period and period 2 is any subsequent period chosen
for comparison purposes. For input Columns 1 and 4 are person-hours of labor expended,
units of material consumed, and 1000 or 10000 BTUs of electricity and gas, respectively.
The prices of Columns 2 and 5 are the selling price of Product X (output) and unit costs
(all inputs). The values of Columns 3 and 6 are the products of price and quantity. Column
7 represents the price-weighted productivity change ratio, P_1Q_2/P_1Q_1. Column 8 is the
quantity-weighted price recovery ratio, P_2Q_2/P_1Q_2. Column 9 is the ratio P_2Q_2/P_1Q_1. The
cost to revenue ratios of Columns 10 and 11 represent input values (Columns 3 and 6)
divided by the output values of $1,691,200 and $2,166,000, respectively.

Column 12, change in productivity, is the result of dividing the total output quantity
change ratio from Column 7 (1.2983) by the corresponding input entities from Column
7. Column 13 is the result of dividing the total output price change ratio in Column 8 by
the corresponding input price change ratios from Column 8. Column 14 is the product of
Columns 12 and 13. It can also be thought of as between period rate of change profitability
margins calculated by allowing quantities and prices to vary simultaneously.

Comparing period 1 and period 2 with respect to quality labor, inspection and testing
labor has been reduced while other labor hours have increased. The price of labor has
increased except for management salaries. Quality materials usage has increased from
period 1 to period 2 as have capital expenditures associated with quality equipment
(inspection, testing, computer).

The price-weighted change ratios of Columns 7 through 9 are nearly all greater than 1,
reflecting increased quantities, prices, and values.

The cost/revenue ratios of Columns 10 and 11 track the same type of information that
is measured by quality cost studies. That is the model measures quality labor material and
capital expenditures by each type as a fraction of sales (total revenue) for period 1 and
period 2.

All of the productivity ratios of Column 12 are greater than 1 indicating that the output

Table 5.9 Multifactor Productivity Model Example

	PERIOD 1			PERIOD 2			PRICE WEIGHTED CHANGE RATIOS			COST/REVENUE RATIOS		CHANGE IN			EFFECTS ON PROFITS		
	QUANTITY (1)	PRICE (2)	VALUE (3)	QUANTITY (4)	PRICE (5)	VALUE (6)	QUANTITY (7)	PRICE (8)	VALUE (9)	PERIOD 1 (10)	PERIOD 2 (11)	PRODUC-TIVITY (12)	PRICE RECVY. (13)	PROFIT-ABILITY (14)	PRODUCTIVITY (15)	PRICE RECOVERY (16)	PROFITABILITY (17)
Product X	3020.	560.00	1691200.00	3800.	570.00	2166000.00	1.2583	1.0179	1.2807								
Total Outputs			1691200.00			2166000.00	1.2583	1.0179	1.2807								
Mgmt. Non-qual.	5230.	20.00	104600.00	5360.	20.00	107200.00	1.0249	1.0000	1.0249	.0618	.0495	1.23	1.02	1.25	24415.91	2350.28	26766.19
Mftg. Non-qual.	52600.	8.00	420800.00	53100.	8.80	467279.94	1.0095	1.1300	1.1105	.2488	.2157	1.25	.93	1.15	104683.50	-33024.94	71658.56
Non-qual. Labor			525400.00			574480.00	1.0126	1.0798	1.0934	.3107	.2652	1.24	.94	1.17	129099.38	-30674.63	98424.75
Mgmt. Quality	1076.	18.00	19368.00	1240.	18.00	22320.00	1.1524	1.0000	1.1524	.0115	.0103	1.09	1.02	1.11	2050.33	435.18	2485.52
Rework Labor	2074.	9.00	18666.00	2120.	9.90	20988.00	1.0222	1.1300	1.1244	.0110	.0097	1.23	.93	1.14	4407.02	-1488.59	2918.43
Incom. Inspection	1080.	10.00	10800.00	1150.	11.00	12650.00	1.0648	1.1300	1.1713	.0064	.0058	1.18	.93	1.09	2089.40	-907.33	1182.07
Inspec. and Test	1522.	9.00	13698.00	1300.	9.90	12870.00	.8541	1.1300	.9396	.0081	.0059	1.47	.93	1.36	5535.89	-862.21	4673.68
Quality Labor			62532.00			68828.00	1.0331	1.0554	1.1007	.0370	.0318	1.22	.96	1.16	14082.64	-2822.94	11259.70
Total Labor			587932.00			643308.00	1.0147	1.0783	1.0942	.3476	.2970	1.24	.94	1.17	143182.00	-33497.63	109684.38
Raw Mtl.	10000.	25.00	250000.00	12200.	26.80	326960.00	1.2200	1.0720	1.3078	.1478	.1510	1.03	.95	.98	9569.56	-16342.69	-6773.13
Mtls. Qual.	200.	2.00	400.00	230.	2.05	471.50	1.1500	1.0250	1.1788	.0002	.0002	1.09	.99	1.09	43.31	-2.51	40.80
Total Material			250400.00			327431.50	1.2199	1.0719	1.3076	.1481	.1512	1.03	.95	.98	9612.88	-16345.25	-6732.38
Elec. MBtu	2700000.	.01	27000.00	3100000.	.01	34100.00	1.1481	1.1000	1.2630	.0160	.0157	1.10	.93	1.01	2973.51	-2493.33	480.18
Gas 10MBtu	980000.	.04	39200.00	1100000.	.04	46200.00	1.1224	1.0500	1.1786	.0232	.0213	1.12	.97	1.09	5324.50	-1319.20	4005.30
Total Energy			66200.00			80300.00	1.1329	1.0707	1.2130	.0391	.0371	1.11	.95	1.06	8298.02	-3812.53	4485.48
Cap. Non-qual.	1000000.	.20	200000.00	1050000.	.20	210000.00	1.0500	1.0000	1.0500	.1183	.0970	1.20	1.02	1.22	41655.63	4493.84	46149.47
Inspec. Equip. Inc.	50000.	.20	10000.00	50300.	.20	10060.00	1.0060	1.0000	1.0060	.0059	.0046	1.25	1.02	1.27	2522.78	224.69	2747.47
Inspec. Equip.	25000.	.20	5000.00	25300.	.20	5060.00	1.0120	1.0000	1.0120	.0030	.0023	1.24	1.02	1.27	1231.39	112.35	1343.74
Test Equip.	100000.	.20	20000.00	101000.	.20	20200.00	1.0100	1.0000	1.0100	.0118	.0093	1.25	1.02	1.27	4965.56	449.39	5414.95
Computer Qual.	40000.	.20	8000.00	40500.	.20	8100.00	1.0125	1.0000	1.0125	.0047	.0037	1.24	1.02	1.26	1966.22	179.75	2145.98
Subt. Qual. Equip.			43000.00			43420.00	1.0098	1.0000	1.0098	.0254	.0200	1.25	1.02	1.27	10685.96	966.18	11652.14
Total Cap.			243000.00			253420.00	1.0429	1.0000	1.0429	.1437	.1170	1.21	1.02	1.23	52341.56	5460.06	57801.63
Total Inputs			1147532.00			1304459.50	1.0723	1.0501	1.1368	.6785	.6022	1.17	.96	1.13	213434.50	-48195.25	165239.25

change ratio of column 7 (1.2583) was higher than any of the input change ratios in the same column. Some of the price recovery changes of Column 13 are below 1 and others are greater than 1. Those below 1 indicate that their costs in period 2 relative to period 1 were higher than the corresponding revenue ratio (i.e., costs increased more as a percent of their base period costs than did revenue as a percent of base period revenue). This can be thought of as a loss in price recovery. On the other hand, Column 13 ratios greater than 1 are indicative of smaller cost increases relative to revenue increase (a gain in price recovery).

The total effects on profits are shown in Columns 15, 16, and 17. These effects represent the dollar value impact on profits as a result of changes in productivity, price recovery, and profitability. All of the entries in column 15 are positive. They have been calculated by subtracting the appropriate input quantity price-weighted change ratios of Column 7 from the output quantity price-weighted change ratio (1.2583) and multiplying by the corresponding value of input from Column 3. Each is a productivity gain since the output change ratio is larger than the input change ratio. Quality labor accounted for a $14,083 increase in profits. The price recovery figures in column 16 are obtained by subtracting Column 15 from Column 17. Column 17, dollar change in profitability, is obtained by subtracting the appropriate input price-weighted value change ratio in Column 9 from the output price-weighted value change ratio (1.2807) and multiplying by the corresponding value of input from Column 3. Negative values in Column 16 indicate that profits were adversely affected due to input costs increasing at a higher rate than the price of Product X. Column 17 reflects the simultaneous changes in productivity and price recovery.

The model may also be used for control purposes as well as for comparing outcomes with quality goals. If quality goals, standards, or desired outcomes are substituted for actual data in period 1, then period 2 could be used to record the actual quantities and costs incurred in a period of interest. The model will generate productivity, price recovery, and profitability ratios and changes, which may be used to verify the extent which quality goals have been achieved.

MFPMM Software Support

The MFPMM is a highly interactive, online system possessing several appealing software features that make it a valuable decision support system. Complete documentation in the form of a user's guide that describes the model and its features is available to assist the new user with implementation of the MFPMM. The software features will be briefly described here in essentially the same sequence they would appear during execution.

Data Input

Quantity and price for each output and input of the entity being analyzed are required to run the model. This information can be entered into the model by reading it from an existing data file or inputting it interactively. If the data are input interactively, the user has the option of running the model only once with the data or sending the data to a file to be saved for future use. Either way the data are entered, only a short response is required from the user. (Sometimes a "Y" for yes or an "N" for no is all that is necessary.)

After the data are entered, the model displays the data as they exist to the user for verification. The data are segmented into small sections for ease of viewing on a CRT. As each section is displayed, the user has the opportunity to change any of the data for that particular execution. If the data that were changed existed in a file, the file will not be changed.

Base-to-Current-Period Analysis

Once quantity and price data have been entered and verified, three tableaus are displayed providing the user with a dynamic productivity report. After the tableaus are displayed, the user has the opportunity to receive a hard copy of the tableaus by simply answering "Y" for yes. The user will then be given instructions pertaining to the sensitivity analysis stage of the model and asked if he or she wants to continue. Answering "Y" will enable the user to proceed; answering "N" will terminate the session.

Sensitivity Analysis

Probably the most appealing feature of the VPI/VPC model is the simulation routine, which allows a sensitivity analysis of projected data. This feature essentially allows the user to play a "what-if" game with the model. The user has the opportunity to compare a projected period to the base period or the current period, but first he or she must enter a desired future value for "total inputs' effect on change in profitability" (last row in Column 19). For example, if the user were comparing a projected period to the current period and he or she did not want his or her level of profitability to change, he or she would enter a desired value of "0."

In order to perform the sensitivity analysis of projected data, pessimistic, most likely, and optimistic estimates must be entered for each category the user wishes to vary in the projected period. Any combination of output and input quantities and prices can be projected.

After all quantity and price projections have been made, a histogram plot and probability statement are displayed. The histogram depicts the results of Monte Carlo simulation, which has generated 100 random outcomes. The 100 data points on the histogram represent the 100 simulated values for total inputs' effect on change in profitability based on projections. The probability statement tells the user how many of the 100 simulated values were greater than or equal to the desired value. The histogram and probability statement give the user a very good indication as to whether or not the projected scenario will result in the desired change in profitability without even looking at the tableaus.

After the histogram plot and probability statement are displayed, the user has the opportunity to create another scenario (make more projections or change some of the previous projections). For each scenario created, a histogram plot and probability statement can be generated. Again, the user has the opportunity to get a hard copy if desired.

When the user has completed this part of the session, the three tableaus will be displayed again with the results reflecting projected period values. During both the base-to-current period analysis and the projected-period analysis, the user has the option of seeing all three tableaus or just the total inputs line of the third tableau (Columns 14–19). Throughout execution of the model, the user is given opportunities to have hard copies printed of what is on the CRT if a printer is available.

The model as it exists is extremely "user-friendly," but the model is continually being developed and refined. Also, added graphics capabilities are being investigated as a means of improving the clarity with which results can be displayed and understood. Industry pilot studies with the model are underway, and it is anticipated that the experience gained through participation in these pilot studies will contribute to further MFPMM enhancements. More information regarding software support can be obtained by contacting the author.

REFERENCES

Anderson, D. R., Sweeney, D. J., and Williams, T. A. *Introduction to Statistics: An Applications Approach*. St. Paul, Minnesota: West Publishing Co., 1981.

Buck, J. R. "Risk Analysis Method Can Help Make Firms' Investments Less of a Gamble," *Industrial Engineering*, November 1982.

Davis, H. *Productivity Accounting*, The Wharton School Industrial Research Unit, University of Pennsylvania, Philadelphia, Pennsylvania, 1955 (reprint, 1978).

Davis, H. S. "The Meaning and Measurement of Productivity." *Industrial Productivity*, Industrial Relations Research Association, Madison, Wisconsin, 1951.

Drucker, P. F. *Managing in Turbulent Times*. New York: Harper-Row, 1980.

Hellriegel, D., and J. W. Slocum. "Managerial Problem-Solving Styles." *Business Horizons*, December 1975.

"How to Measure Productivity at the Firm Level," American Productivity Center, Short Course Notebook and Reference Manual, Houston, Texas, 1978.

Kendrick, J. W., and D. Creamer. *Measuring Company Productivity: Handbook with Case Studies*, Studies in Business Economics, No. 89, New York, Conference Board, 1965.

Kendrick, J. W., and E. S. Grossman, *Productivity in the United States*. Baltimore: Johns Hopkins Press, 1980.

Myers, I. B. *Introduction to Type—The Meyers Briggs Type Indicator*. Palo Alto, Calif.: Consulting Psychologists Press, 1980.

Pritsker, A. A. B., and C. D. Pegden. *Introduction to Simulation and SLAM*. West Lafayette, In.: Systems Publishing Company, 1979.

Robey, D., and W. Taggart. "Measuring Managers' Minds: An Assessment of Style in Human Information Processing." *Academy of Management Review*, Vol. 6, No. 3, 1981.

Scott, J. A. *The Measurement of Industrial Efficiency*, Sir Isaac Pitman and Sons, Ltd., London, 1950.

Sullivan, W. G., and R. G. Orr. "Monte Carlo Simulation Analyzes Alternatives in Uncertain Economy." *Industrial Engineering*, November, 1982.

van Loggerenberg, B. J., and S. J. Cucchiaro. "Productivity Measurement and the Bottom Line." *National Productivity Review*, Winter 1981–82.

QUESTIONS AND APPLICATIONS

(Software programs are available to facilitate instruction of this model and to use in some of these classroom-type exercises. University discounts are available. For more information, write or call: the Virginia Productivity Center, VPI and State University, Blacksburg, VA 24061; (703) 961-4568.)

1. Write a program to execute the basic MFPMM formulas.

2. Utilizing any spread sheet you have access to (for example, Visicalc, Supercalc, 1-2-3, or Multiplan), develop an MFPMM formulation.

3. The data in Tables 5.10 and 5.11 were obtained from a motel operation. Using these data, develop an MFPMM analysis. Your analysis should include not only output, but also analysis and interpretation, executive summaries, and graphics for management.

4. If you were hired as a management systems analyst for this motel operation, how would you integrate the MFPMM into the overall organizational performance measurement, evaluation, control, and improvement system? Before answering this question,

Table 5.10 Hotel Data Example

	PERIOD 1			PERIOD 2		
	QUANTITY (1)	PRICE (2)	VALUE (3)	QUANTITY (4)	PRICE (5)	VALUE (6)
Single	629.	17.00	10693.00	660.	18.00	11880.00
Double	86.	24.00	2064.00	105.	25.00	2625.00
Triple	??.	26.00	572.00	18.	28.00	504.00
Meeting Room	4.	40.00	160.00	5.	50.00	250.00
Restaurant	1.	1500.00	1500.00	1.	2000.00	2000.00
Carpet Store	1.	300.00	300.00	1.	400.00	400.00
Total Outputs			15289.00			17659.00
Personnel	3.	1000.00	3000.00	2.	1200.00	2400.00
Management Labor			3000.00			2400.00
Desk Clerk	400.	3.50	1400.00	360.	4.00	1440.00
Housekeeper	325.	3.00	975.00	280.	3.60	1008.00
Maintenance	165.	4.00	660.00	150.	4.50	675.00
Laundrymen	110.	3.00	330.00	125.	3.50	437.50
Hourly Labor			3365.00			3560.50
Total Labor			6365.00			5960.50
Cleaning Supp.	1.	584.00	584.00	1.	473.00	473.00
Office Supp.	1.	100.00	100.00	1.	128.00	128.00
Maintenance Supp.	1.	452.00	452.00	1.	361.00	361.00
Total Materials			1136.00			962.00
Natural Gas	260.	2.55	663.00	242.	3.70	895.40
Electricity	15950.	.05	797.50	15012.	.08	1200.96
Water & Sewage	362.	.70	253.40	370.	.90	333.00
Total Energy			1713.90			2429.36
Total Inputs			9214.90			9351.86

Table 5.11 Hotel Example: Period 3/Forecasted Period Projections for Quantity and Price

		QUANTITY		
NO.	CATEGORY NAME	PESSIMISTIC	LIKELY	OPTIMISTIC
1	Single	650.	675.	700.
2	Double	90.	100.	140.
3	Triple	20.	24.	30.
4	Meeting Room	4.	6.	8.
5	Restaurant	1.	1.	1.
6	Carpet Store	1.	1.	1.
7	Total Outputs	0.	0.	0.
8	Personnel	3.	2.	1.
9	Management Labor	0.	0.	0.
10	Desk Clerk	450.	375.	350.
11	Housekeeper	360.	310.	285.
12	Maintenance	200.	160.	140.
13	Laundrymen	140.	110.	95.
14	Hourly Labor	0.	0.	0.
15	Total Labor	0.	0.	0.
16	Cleaning Supp.	1.	1.	1.
17	Office Supp.	1.	1.	1.
18	Maintenance Supp.	1.	1.	1.
19	Total Materials	0.	0.	0.
20	Natural Gas	300.	260.	235.
21	Electricity	16500.	15000.	13800.
22	Water & Sewage	400.	360.	310.
23	Total Energy	0.	0.	0.
24	Total Inputs	0.	0.	0.

		PRICE		
NO.	CATEGORY NAME	PESSIMISTIC	LIKELY	OPTIMISTIC
1	Single	18.00	20.00	22.00
2	Double	24.00	26.00	28.00
3	Triple	26.00	28.00	30.00
4	Meeting Room	40.00	55.00	60.00
5	Restaurant	2000.00	2000.00	2000.00
6	Carpet Store	450.00	450.00	450.00
7	Total Outputs	.00	.00	.00
8	Personnel	1500.00	1400.00	1000.00
9	Management Labor	.00	.00	.00
10	Desk Clerk	4.15	4.15	4.15
11	Housekeeper	3.75	3.75	3.75
12	Maintenance	4.75	4.75	4.75
13	Laundrymen	3.75	3.75	3.75
14	Hourly Labor	.00	.00	.00
15	Total Labor	.00	.00	.00
16	Cleaning Supp.	550.00	450.00	400.00
17	Office Supp.	190.00	125.00	90.00

Table 5.11 (*Continued*)

		PRICE		
NO.	CATEGORY NAME	PESSIMISTIC	LIKELY	OPTIMISTIC
18	Maintenance Supp.	500.00	410.00	380.00
19	Total Materials	.00	.00	.00
20	Natural Gas	4.20	4.20	4.20
21	Electricity	.09	.09	.09
22	Water & Sewage	1.00	1.00	1.00
23	Total Energy	.00	.00	.00
24	Total Inputs	.00	.00	.00

you might want to either visit with a manager of a motel or carefully think through what performance means to a small motel.

5. Much work has been done recently on psychological type, problem-solving style, and managerial style in relation to preferences for decision support system types. Investigate some of this work and discuss implications of this area of research on development of decision support systems for productivity measurement (see Hellriegel and Slocum, 1975; Myers, 1980; Robey and Taggart, 1981.)

6. Using the data on the boat company example developed in this chapter, execute the following exercise. Select persons to play the roles of president, marketing manager, quality manager, procurement/purchasing manager, industrial engineering manager (this person will also play the role of productivity analyst), human resource director, production/plant manager, and comptroller/accountant. Each of these persons is to come to the exercise familiar with the current-to-base period output from the MFPMM. They should also have thought through and completed the projected period forms (Figure 5.10). Note each person, playing his or her specified role, will provide projected data only for relevant and appropriate aspects of the company. The session, run or led by the industrial engineering manager, should be a computer and management group interactive session. Where available (in large classes or management training sessions), a terminal with video-out capabilities should be linked to a big-screen TV or overhead projection system so as to allow the rest of the class to view the computer simulation run. In a semistructured meeting, the top management group, under the direction of the productivity analyst, discusses and develops two or three scenarios of what might happen in the company from the current period (this meeting) to the projected period (next month, quarter, year). Given scenario data are input and the simulation routine is generated. Hard copies are generated, and either in this session or in subsequent sessions the alternative scenarios are analyzed and interpreted.

Each manager is also required to develop specific productivity improvement action plans (see Figure 5.12) to accompany each potential scenario. Each manager must identify the specific program or plan, describe it, discuss project planning, determine responsibilities and accountabilities, and discuss cost/benefit implications (perhaps using model results).

7. Discuss how the MFPMM could be improved as a productivity measurement decision support system.

8. Design graphics that you would utilize in a management briefing. Realize that you will not likely be giving management the output from the MFPMM. Some sort of executive summary will have to be tailor-made for each manager. What would such a summary

look like for:
  (a) The president of the company?
  (b) The industrial engineering manager?
  (c) The human resource manager?
  (d) The quality manager?
  (e) The comptroller?
  (f) The board of directors?
  (g) First-line supervision?
  (h) The union executive board?
  (i) Employees?

9. Explain and discuss the implications of base-period price weighting and current-period quantity weighting in the MFPMM.

10. Explain and discuss why Columns 16 and 13 are not calculated directly.

11. Perform a graphic change in price and in quantity analysis to depict what Columns 12–14 and 15–16 are actually doing. For example

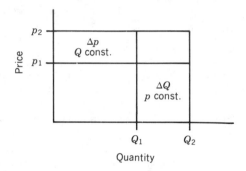

12. Develop a "sales pitch" you might use to get management to at least experiment with the MFPMM. What are major points of resistance you expect to encounter when trying to get this model accepted?

13. Prove that

$$\frac{Q_2^0/Q_1^0}{Q_2^1/Q_1^1} = \frac{Q_2^0/Q_2^1}{Q_1^0/Q_1^1}$$

CHAPTER

6

MULTICRITERIA PERFORMANCE/ PRODUCTIVITY MEASUREMENT TECHNIQUE (MCP/PMT)

HIGHLIGHTS

- Introduction
- Background
- Description of the MCP/PMT
- MCP/PMT Procedure
- References
- Questions and Applications

OBJECTIVES

- To introduce the reader to yet another approach to measuring and evaluating productivity and performance.
- To expose the reader briefly to the historical developments underlying this technique.
- Through the use of examples, to instruct the reader how to develop a productivity or performance measurement and evaluation system utilizing this technique.

Introduction

As outlined at the beginning of Part III, there are three major distinct techniques for measuring productivity—the NPMM and the MFPMM have been presented and discussed, although a review may be in order. Figure 6.1 reflects a spectrum of units of analysis for which measurement and, in particular, some form of productivity measurement may be of interest. Note that aggregated

189

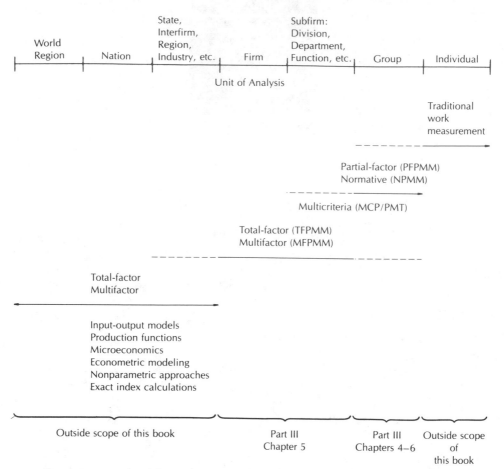

Figure 6.1 Productivity Measurement Approaches, Models, and Techniques for Various Units of Analysis

total-productivity, total-factor productivity, and/or multifactor productivity measurement approaches, models, and/or techniques are appropriate and often useful for more macroscopic units of analysis (nation, region, industry, firm), while multifactor and partial-factor productivity measurement approaches, models, and/or techniques are appropriate and often useful at lower units of analysis (firm, plant, function, group).

An industrial engineer trained in work measurement and improvement is armed with techniques that are quite appropriate and useful at the individual work-center level; however, this body of techniques loses much of its utility once it is applied to larger units of analysis. Most measurement systems for larger units of analysis are either functional or resource control systems, such as quality control, inventory control, production/process control, management by objectives, and attitude surveys, or they are financial in character, such as cost accounting, budgets, and corporate finance. Hence, a gap in the development of performance measurement systems seems to exist. It would appear that the techniques presented to this point have the potential to fill some of these measurement gaps. Keep in mind that the objective is not to replace existing measurement and control systems or to design one system that will do everything for everybody. The objective is to develop a comprehensive, integrated, effective, and efficient performance measurement, evaluation, and control system that incorporates productivity as well as the other six performance criteria.

The two techniques presented thus far fill this measurement gap in different ways and at different levels. The normative productivity measurement methodology (NPMM) is a process by which measures (surrogate), ratios, and/or indexes of productivity can be participatively identified and developed into a measurement, evaluation, control, planning, and improvement system. As discussed, this approach may initially sacrifice a little quality and rigor in the measurement system and may take a little longer; however, its real strength is that it builds acceptance of and commitment to the measurement, evaluation, control, and improvement process. The multifactor productivity measurement model (MFPMM) is actually a decision support system model that operates with organizational system data on prices/costs and quantities of output (goods and/ or services) and input resources. Both approaches have merit and applications for which they are very appropriate.

The NPMM appears to be best suited for smaller units of analysis, such as the group level, and in situations where decoupled productivity measurement systems are appropriate or desired. Again, the participative character of the NPMM facilitates linking the measurement and evaluation activities to the planning, control, and improvement activities. The MFPMM, on the other hand, appears to be best suited for intermediate units of analysis, such as division, plant, or firm levels. The model provides an integrated and, therefore, coupled measurement system and perspective. It is much less a process in which persons in the organization can become involved and much more a decision support system for management of the particular organizational system.

When the terms "total productivity," "total-factor productivity," and "multifactor productivity measurement" are used, one should think in terms of coupled, integrated, typically price- and cost-weighted, indexed, and aggregated models, such as the MFPMM. The term partial-factor productivity, however, implies productivity ratios that are decoupled and that reflect some or all of the output from a given organizational system over some portion of one category or factor of input from that same organizational system for some given period of time. As you can by now imagine and as was discussed during the presentation of the NPMM, there are a tremendous number of partial-factor productivity ratios for any organizational system. (Table 6.1 provides examples of a variety of partial-factor "productivity" ratios.)

As mentioned at the close of the last chapter and as illustrated in Table 6.1, there are an extremely large number of partial-factor ratios of productivity that could be looked at for most organizational systems. In order to keep this section of the chapter in perspective, a quick review is in order.

Productivity, as strictly defined, is a ratio of outputs (goods and/or services; quantities) produced from a given organizational system during a given period

Table 6.1 Examples of Partial-Factor "Productivity" Ratios

EXAMPLES OF OVERALL RATIOS (Mali)

1. *Business and industry*

   (a) $\dfrac{\text{Sales}}{\text{Employees}}$

   (b) $\dfrac{\text{Space utilized}}{\text{Space available}}$

   (c) $\dfrac{\text{Market share now}}{\text{Market share in base year}}$

   (d) $\dfrac{\text{Sales lost}}{\text{Customer complaints}}$

   (e) $\dfrac{\text{Profit}}{\text{Equity capital}}$

   (f) $\dfrac{\text{Actual price paid}}{\text{Market price}}$

2. *Government*

   (a) $\dfrac{\text{Benefits}}{\text{Costs}}$

   (b) $\dfrac{\text{Legislation authorized}}{\text{Legislation proposed}}$

   (c) $\dfrac{\text{Quits}}{\text{Employees}}$

   (d) $\dfrac{\text{Budget performance}}{\text{Authorized budget}}$

   (e) $\dfrac{\text{Prices now}}{\text{Prices at base year}}$

   (f) $\dfrac{\text{Gains from legislative enactments}}{\text{Cost of enactments}}$

3. *Education*

   (a) $\dfrac{\text{Enrollment}}{\text{Faculty}}$

   (b) $\dfrac{\text{Class count} \times \text{credit hours}}{\text{Direct costs}}$

   (c) $\dfrac{\text{Personnel costs}}{\text{Employees}}$

   (d) $\dfrac{\text{Income/expense}}{\text{Faculty}}$

   (e) $\dfrac{\text{Research projects completed}}{\text{Costs of projects}}$

   (f) $\dfrac{\text{Tuition}}{\text{Administrative staff}}$

4. *Health and human services*

   (a) $\dfrac{\text{Cost of patient care}}{\text{Number admitted}}$

   (b) $\dfrac{\text{Treatment plans implemented}}{\text{Total treatment plans}}$

   (c) $\dfrac{\text{Client caseload}}{\text{Professional staff}}$

   (d) $\dfrac{\text{Beds occupied}}{\text{Beds available}}$

   (e) $\dfrac{\text{Revenues}}{\text{Patients}}$

   (f) $\dfrac{\text{Training costs}}{\text{Employees}}$

EXAMPLES OF OBJECTIVE RATIOS (Mali)

1. *Business and industry*

   (a) $\dfrac{\text{Projects completed}}{\text{Projects planned}}$

   (b) $\dfrac{\text{Progress in labor negotiations}}{\text{Expected schedule}}$

   (c) $\dfrac{\text{Marketing products adopted}}{\text{Feasible ideas}}$

   (d) $\dfrac{\text{Work packages}}{\text{Expected work packages}}$

   (e) $\dfrac{\text{Sales level}}{\text{Expected inventory}}$

   (f) $\dfrac{\text{Quits}}{\text{Desired level of quits}}$

2. *Government*

   (a) $\dfrac{\text{Highways built}}{\text{Highways needed}}$

   (b) $\dfrac{\text{Actual contributed value}}{\text{Expected contributed value}}$

   (c) $\dfrac{\text{Settlement of claims}}{\text{Total claims}}$

   (d) $\dfrac{\text{Convictions}}{\text{Arrests}}$

   (e) $\dfrac{\text{Benefits}}{\text{Expected benefits}}$

   (f) $\dfrac{\text{Contracts renegotiated}}{\text{Needed renegotiated}}$

Table 6.1 (*Continued*)

EXAMPLES OF OBJECTIVE RATIOS (Mali)

3. *Education*

(a) $\dfrac{\text{Benefits from research projects}}{\text{Expected benefits}}$

(b) $\dfrac{\text{Skills prevailing}}{\text{Skills needed}}$

(c) $\dfrac{\text{Behavioral outcomes}}{\text{Behavioral outcome desired}}$

(d) $\dfrac{\text{Handicapped children trained}}{\text{Total to be trained}}$

(e) $\dfrac{\text{Minorities completing program}}{\text{Expected completions}}$

(f) $\dfrac{\text{Faculty ratings in current year}}{\text{Expected ratings}}$

4. *Health and human services*

(a) $\dfrac{\text{Steps completed in treatment}}{\text{Total steps}}$

(b) $\dfrac{\text{Treatment plans}}{\text{Budget allocation}}$

(c) $\dfrac{\text{Prescriptions filled}}{\text{Expected prescriptions filled}}$

(d) $\dfrac{\text{Preventive medical programs}}{\text{Total desired}}$

(e) $\dfrac{\text{Client caseloads}}{\text{Expected total}}$

(f) $\dfrac{\text{Patients admitted}}{\text{Patients needed}}$

EXAMPLES OF COST RATIOS (Mali)

1. *Business and industry*

(a) $\dfrac{\text{Sales}}{\text{Operating costs}}$

(b) $\dfrac{\text{Borrowed capital}}{\text{Borrowing costs}}$

(c) $\dfrac{\text{Inventory}}{\text{Advertising costs}}$

(d) $\dfrac{\text{Rejects}}{\text{Costs}}$

(e) $\dfrac{\text{Turnover}}{\text{Costs}}$

(f) $\dfrac{\text{Rework}}{\text{Costs}}$

2. *Government*

(a) $\dfrac{\text{Transactions}}{\text{DP costs}}$

(b) $\dfrac{\text{Renegotiated contracts}}{\text{Costs of renegotiations}}$

(c) $\dfrac{\text{Recruits selected}}{\text{Costs}}$

(d) $\dfrac{\text{Mail processed}}{\text{Payroll cost}}$

(e) $\dfrac{\text{Benefits from proposal}}{\text{Cost of proposal}}$

(f) $\dfrac{\text{Legislative enactments}}{\text{Cost of enactments}}$

3. *Education*

(a) $\dfrac{\text{Tuition generated}}{\text{Cost of generation}}$

(b) $\dfrac{\text{Dropouts}}{\text{Cost of enrollment}}$

(c) $\dfrac{\text{Benefits of research projects}}{\text{Cost of projects}}$

(d) $\dfrac{\text{Students graduating}}{\text{Annual costs}}$

(e) $\dfrac{\text{Budget value}}{\text{Allocated budget}}$

(f) $\dfrac{\text{Meals served}}{\text{Cost of cafeteria operation}}$

4. *Health and human services*

(a) $\dfrac{\text{Trainees completing programs}}{\text{Training costs}}$

(b) $\dfrac{\text{Design of therapeutic treatment}}{\text{Cost of design}}$

(c) $\dfrac{\text{Research reports}}{\text{Allocated budget}}$

(d) $\dfrac{\text{Clients caseloads}}{\text{Cost of interviews}}$

(e) $\dfrac{\text{Beds occupied}}{\text{Cost of bed occupancy}}$

(f) $\dfrac{\text{Patients admitted}}{\text{Cost of admission}}$

Table 6.1 (*Continued*)

1. *Business and industry*

   (a) $\dfrac{\text{Machines operating}}{\text{Setup time}}$

   (b) $\dfrac{\text{Value of returned goods}}{\text{Purchases}}$

   (c) $\dfrac{\text{Grievances settled}}{\text{Grievances investigated}}$

   (d) $\dfrac{\text{Workload assignments}}{\text{Engineering staff}}$

   (e) $\dfrac{\text{Actual labor per unit}}{\text{Scheduled labor per unit}}$

   (f) $\dfrac{\text{Accepted products}}{\text{Products produced}}$

2. *Government*

   (a) $\dfrac{\text{Benefits from a project}}{\text{Total task required}}$

   (b) $\dfrac{\text{Settlement of unfair labor charges}}{\text{Investigation of charges}}$

   (c) $\dfrac{\text{Compliance of board orders}}{\text{Investigation of on-compliance}}$

   (d) $\dfrac{\text{Board adjudications}}{\text{Total hearings}}$

   (e) $\dfrac{\text{Buying costs}}{\text{Purchases}}$

   (f) $\dfrac{\text{Value added}}{\text{Contract changes}}$

3. *Education*

   (a) $\dfrac{\text{Research projects completed}}{\text{Procedure used}}$

   (b) $\dfrac{\text{Achievement attainment}}{\text{Standardized test}}$

   (c) $\dfrac{\text{Graduates}}{\text{Standardized curriculum}}$

   (d) $\dfrac{\text{Graduates reading in 50th percentile}}{\text{Standardized reading test}}$

   (e) $\dfrac{\text{Implemented recommendations}}{\text{Committees}}$

   (f) $\dfrac{\text{Graduates}}{\text{Curriculums}}$

4. *Health and human services*

   (a) $\dfrac{\text{Client caseload}}{\text{Standard caseload}}$

   (b) $\dfrac{\text{Prescriptions filled}}{\text{Standard procedure}}$

   (c) $\dfrac{\text{Patients admitted}}{\text{Standard admissions}}$

   (d) $\dfrac{\text{Absenteeism}}{\text{Industry standard}}$

   (e) $\dfrac{\text{Skills displayed in a situation}}{\text{Skills trained for in a procedure}}$

   (f) $\dfrac{\text{Rework backlog}}{\text{Rework procedure}}$

1. *Business and industry*

   (a) $\dfrac{\text{Production}}{\text{Working days}}$

   (b) $\dfrac{\text{Actual machine hours per unit}}{\text{Scheduled machine hours per unit}}$

   (c) $\dfrac{\text{Reject work}}{\text{Standard hours to produce}}$

   (d) $\dfrac{\text{Inventory buildup}}{\text{Average daily purchases}}$

   (e) $\dfrac{\text{Overtime hours}}{\text{Total hours}}$

   (f) $\dfrac{\text{Rework}}{\text{Time for rework}}$

2. *Government*

   (a) $\dfrac{\text{Working time}}{\text{Total time}}$

   (b) $\dfrac{\text{Person-days lost}}{\text{Person-days worked}}$

   (c) $\dfrac{\text{Gains from legislative enactments}}{\text{Time period of the gain}}$

   (d) $\dfrac{\text{Service to noncrime calls}}{\text{Time devoted to noncrime calls}}$

   (e) $\dfrac{\text{Benefits from project}}{\text{Time for renegotiation}}$

   (f) $\dfrac{\text{Renegotiated contracts}}{\text{Time for renegotiation}}$

Table 6.1 (*Continued*)

3. *Education*

(a) $\dfrac{\text{Teaching days in a schedule}}{\text{Teaching days lost}}$

(b) $\dfrac{\text{Research projects completed}}{\text{Time required}}$

(c) $\dfrac{\text{Faculty plans submitted}}{\text{Time required}}$

(d) $\dfrac{\text{Minorities in program}}{\text{Standard time required}}$

(e) $\dfrac{\text{Skills level attained}}{\text{Standard time required}}$

(f) $\dfrac{\text{Benefits from project}}{\text{Total hours}}$

4. *Health and human services*

(a) $\dfrac{\text{Meals served}}{\text{Standard time}}$

(b) $\dfrac{\text{Implementation of new therapeutic treatment}}{\text{Person-hours to complement}}$

(c) $\dfrac{\text{Prescriptions filled}}{\text{Average person-hours}}$

(d) $\dfrac{\text{Patients admitted}}{\text{Person-hours to admit}}$

(e) $\dfrac{\text{Client caseloads}}{\text{Person-hours to complete}}$

(f) $\dfrac{\text{Sickness treatment}}{\text{Standard time}}$

MATERIALS HANDLING (FUNCTIONAL) PRODUCTIVITY RATIOS (White, 1979)

Resource Utilization Measures

Labor	MHL Ratio =	$\dfrac{\text{Personnel assigned to materials handling duties}}{\text{Total Operating Workforce}}$
	DLMH Ratio =	$\dfrac{\text{Materials handling time spent by direct labor}}{\text{Total direct labor time}}$
Equipment	Production Equipment Utilization =	$\dfrac{\text{Actual output}}{\text{Theoretical output}}$
	Handling Equipment Utilization =	$\dfrac{\text{Weight moved/hour}}{\text{Theoretical capacity}}$
Space	Storage Space Utilization =	$\dfrac{\text{Storage space occupied by material}}{\text{Total storage space}}$
	Aisle Space Percentage =	$\dfrac{\text{Space occupied by aisles}}{\text{Total space}}$
Energy	EUI =	$\dfrac{\text{BTUs consumed/day}}{\text{Cubic space}}$

Management Control Measures

Materials	Inventory Turnover Ratio =	$\dfrac{\text{Annual sales}}{\text{Average annual inventory investment}}$
	Inventory Fill Ratio =	$\dfrac{\text{Line item demands filled/day}}{\text{Line item demands/day}}$

Table 6.1 (Continued)

MATERIALS HANDLING (FUNCTIONAL) PRODUCTIVITY RATIOS (White, 1979)

		Management Control Measures
Movement, Flow	Movement/Operation Ratio $=$	Total number of moves / Total number of productive operations
	Average Distance/ Move Ratio $=$	Total distance traveled/day / Total number of moves/day
Loss	Damaged Loads Ratio $=$	Number of damaged loads / Number of loads
	Inventory Shrinkage Ratio $=$	Inventory investment verified / Inventory investment expected
		Operating Efficiency
Receiving and Shipping	RP Ratio	Pounds received/day / Labor hours/day
	SP Ratio	Pounds shipped/day / Labor hours/day
Storage and Retrieval	OP Ratio	Equivalent lines or orders picked/day / Labor hours required/day
	TPI	Throughput achieved/day / Throughput capacity/day
Manufacturing	Manufacturing Cycle Efficiency (MCE)	Total time spent on machines / Total time spent in production system
	Job Lateness (JL) Ration	Number of jobs completed or in process that are late/week / Number of jobs completed/week

of time to inputs (energy, labor, capital, materials; quantities) produced from that same organizational system during the same period of time. In Chapter 2 it was indicated that various types of productivity "measures" can be developed.

1. Static
    (a) Partial-factor ratios (Table 6.1)
    (b) Multifactor ratios
    (c) Total-factor ratios

2. Dynamic
    (a) Partial-factor indexes (MFPMM)
    (b) Multifactor indexes (MFPMM)
    (c) Total-factor indexes (TFPMM)

3. Surrogate "productivity" measures (i.e. performance measures), ratios, and/
or indexes (NPMM without productivity basics training to participants).

Note that under the category "Dynamic Productivity Measures," the MFPMM
creates multifactor productivity indexes "automatically" (Column 17) (see Figure
5.6). If one were to capture all the output data and all the inputs in the model,
then essentially the MFPMM becomes a "Total-Factor Productivity Measurement
Model" (TFPMM) and category 2c would be covered. As was indicated in Chapter
2, some authors differentiate between total- and partial-factor productivity and
total and partial productivity. This distinction is based on whether gross output
is placed in the numerator (total and partial productivity) or whether just value-
added output (gross output—materials, intermediate goods and services consumed in production, and so forth) is placed in the numerator (total- and partial-
factor productivity). This text does not make this distinction.

The NPMM has been and can be quite useful and successful for developing
productivity measures of the type in categories 1a, 2a, and 3 of the preceding
list. Examples have been provided for developing these categories of productivity
"measures."

So, it can be seen that the only categories of productivity "measures" not
discussed are 1b, static multifactor ratios, and 1c, static total-factor ratios. Productivity process modeling and/or the NFPMM can assist in the identification,
prioritization, and development of partial-factor ratios and indexes. (Recall that
a productivity index is simply a productivity ratio at one point in time over that
same productivity ratio at a previous point in time. Therefore, if you have a
ratio, obtaining the index is simply a matter of collecting the data.) As mentioned
previously, development of partial-factor ratios and indexes will result in a
disaggregated and likely decoupled productivity measurement system. One might
find it desirable to aggregate these productivity ratios into a single indicator of
productivity performance. Of course, the MFPMM does this implicitly; however,
as you have observed, there are certain costs associated with development of
the MFPMM and certainly situations in which that approach and model is not
appropriate. In cases where the NPMM or similar partial-factor development
strategy is appropriate, the question has often arisen as to how to pull together
or integrate partial-factor productivity ratios. In the past ten years, there have
been developments toward this end. This is the focus of this chapter.

Background

During the Ohio State studies mentioned in Chapter 4, the question of how to
aggregate unlike productivity "measures" was asked frequently. One of the
graduate researchers, William T. Stewart, addressed this issue in his dissertation
effort entitled "The Facilitation of Productivity Measurement and Improvement
in Manufacturing Organizations" (Stewart, 1978). His approach was to develop

a prioritized set of productivity "measures" (mostly surrogate) utilizing the NGT. He then developed a utility curve for each of the priority (top eight to ten) "measures." A ranking and rating process was executed so as to weight the relative importance of each productivity "measure" to organizational system (in this case, plant or firm) performance. The utility curve was utilized to "transform" actual performance against each specific "measure" into a common 0 to 1.0 performance score. This performance score was then multiplied by the relative weight for each measure to obtain a performance value for that measure. The various performance values for each of the top priority measures were then added together to obtain a productivity performance index. Stewart developed this procedure based primarily on the works of Morris (1977), The Ohio State Productivity Research Group (1977), and Keeney and Raiffa (1976). Those interested in pursuing this background development further are urged to refer to those references.

Since those early developments in 1976–78 at Ohio State, several other efforts have been made in this general area. In 1980, Stewart successfully applied this approach in the Common Carrier industry, as well as in several specific organizations. The American Productivity Center has also done limited applications of this approach in the past several years. In 1981, while a graduate associate in the Oklahoma Productivity Center at Oklahoma State University, William Viana applied a hybrid design of this procedure in a fairly large, diversified manufacturing firm (gate valves, rockbits, wellheads, butterfly valves, ball valves, and so forth) in Brazil. His successful application is presented in his thesis entitled "Productivity Measurement and Improvement Strategy Program" (Viana, 1982). More recently, Riggs and Felix, in their book entitled *Productivity by Objectives: Results-Oriented Solutions to the Productivity Puzzle* (1983), present an approach called the "objectives matrix," which is analogous to Stewart's. There are likely other similar developments in this general area, although this list appears to be the most visible and perhaps the most significant.

Again, the question at hand is how to bring together, integrate, and aggregate partial-factor ratios of productivity in such a way that we can measure productivity performance with one number (that is, convert a partial-factor, decoupled, disaggregated productivity measurement system into a multifactor, coupled, aggregated productivity measurement system). If a technique, approach, or methodology enabling one to accomplish this were available, then a productivity analyst, a manager, or an industrial engineer would have at his or her disposal a full complement of techniques for measuring productivity in a broad range of situations and applications. That individual would have the NPMM, the MFPMM, and now a technique for developing static, multi-, or perhaps even total-factor productivity ratios. And, of course, if this technique can create these types of ratios, then the ratios could easily be converted into indexes. For the sake of at least internal consistency within this text, the generic title of multicriteria performance/productivity measurement technique (MCP/PMT) will be given to this general approach.

Description of the MCP/PMT

A convenient way to keep this methodology in perspective with the other techniques is to view it as an integral component of the NPMM. The NPMM is presented again for your convenience in Figure 6.2. Milestone t_6, stage 2, Productivity Analyst Intervention, is precisely where the MCPMM can integrate with the NPMM. What will be presented is a specific technique for refining the partial-factor ratios into an integrated and aggregated productivity measurement system.

Assume that you have just completed stage 1 of the NPMM and have a list of performance measures such as those depicted in Table 6.2. In other words, a group of employees, supervisors, or managers have identified and prioritized a list of performance measures for a given organizational system. Note that some are productivity ratios while others are surrogate "productivity measures." We can assume, therefore, either that the task statement for the NGT session (Stage 1) asked specifically for performance measures; or, if it asked for productivity ratios, specifically, the group had not been trained in productivity basics. The next question to be addressed is how to take the output from the NGT session and mold it into a usable productivity measurement system. What is presented next is a simplified version of Stewart's approach that was developed at Ohio State, Oklahoma State, and at Oregon State (in the form of the objectives matrix in Figure 6.3).

MCP/PMT Procedure

Once the criteria against which productivity (or, in this case, performance) is to be evaluated have been identified and prioritized, a mechanism for aggregating or collapsing this vector of criteria needs to be developed. Typically, each criterion is quantifiable or measurable. If one is not, surrogate measurement, behaviorally anchored rating scales, and so forth can be developed. A performance scale (what Stewart called a utility scale) needs to be developed for each priority criterion. These performance scales can range over any interval; however, 0 to 1.0, 0 to 10, or 0 to 100 are the usual, with 0 to 1.0 being the most common. For the purpose of this presentation, we will utilize 0 to 10, as do Riggs and Felix (1983) in their objectives matrix procedure. Note that the major purpose of a performance or utility scale is to have a common denominator for unlike criteria. In the MFPMM, dollars provide the basis for a common denominator.

In our performance scale, level 0 would represent the lowest level of performance possible for that given criterion, level 5 would represent an arbitrary "acceptable performance level," and level 10 wold represent the perception of best performance possible or excellence. Therefore, levels 0, 5, and 10 should be clearly defined and accepted by the group involved in and responsible for using and performing against this measurement system (Archer, 1970). Figure

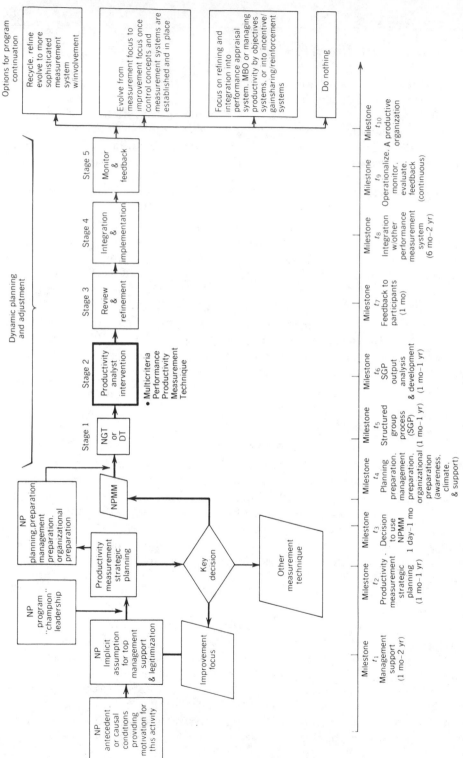

Figure 6.2 Design, Development, and Implementation of a Productivity Measurement System Utilizing the Normative Productivity Measurement Methodology

Table 6.2 Sample List of Productivity Measures for a Computer Center

MEASURE	PRIORITY
Reports/projects completed and accepted Constant value budget dollars	1
Customer satisfaction	2
Quality of decision support from systems developed	3
Meeting user flexibility requirements	4
Existence of and use of work scheduling/project management techniques	5
Projects completed on time Total projects completed	6
Number of requests for reworking/redoing a project	7
Existence of and quality of strategic planning for facilities, equipment, management processes, and operational systems	8

6.4 depicts performance scales for several of the priority criteria listed in Table 6.2. The process by which the actual performance curves are developed will likely require analyst assistance.

If the process is to be participative and, therefore, normative, the analyst will need to attempt to get the group members to reach consensus on the range of potential or actual performance (x-axis); the shape of the performance curve or transformation function; and the benchmarks (levels 0, 5, 10) themselves. Procedures for assessing and developing subjective utility curves have been developed and are presented in several sources (Swalm, 1966; Morris, 1977; Keeney and Raiffa, 1976; Stewart, 1978). The goal is, of course, to have not only accepted performance scales but also valid ones. Literature on goal setting and motivation suggests that the criteria should be

1. Consistent and congruent with group and organizational mission goals and objectives

2. Within the control of the group itself

3. Comprehensive and, as much as possible, mutually exclusive

4. Explicit and as objective as possible

5. Challenging, not too easy, not too difficult

6. Measurable. There should be reasonable visibility of the cause-and-effect relationships between group activities and each performance criterion variability.

The literature and research also suggest that the benchmarks should be

1. Challenging but achievable: Level 5 should represent a level of performance that is challenging, given present levels, but that the group feels is accomplishable.

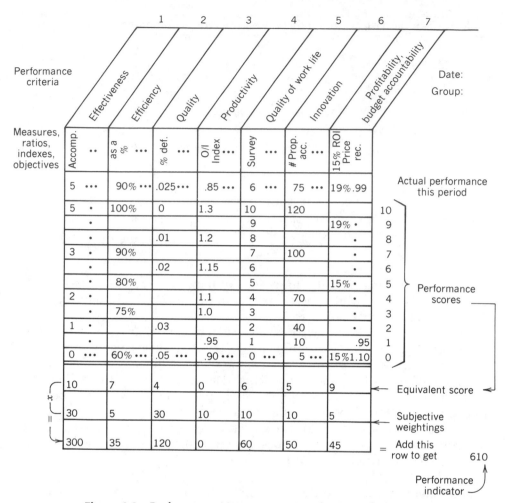

Figure 6.3   Performance Measurement Matrix General Format

2. Behaviorally or quantitatively anchored: Each level should be explicit enough to create a clear objective perception of its needed impact on the individuals, groups, and/or organizational systems' behaviors.

3. Flexible: Levels 0, 5, and 10 can vary over time. It is possible and probable that factors will change causing these "standards" to change accordingly. Essentially, the shape of the performance curve will be shifting.

The next step in the MCPMM procedure is to rate the relative importance of the performance criteria. Note that the criteria have already been ranked in stage 1 of the NPMM utilizing the NGT. In essence, the NGT ranking generated an ordinal scale and the goal is now to generate an interval scale so that relative

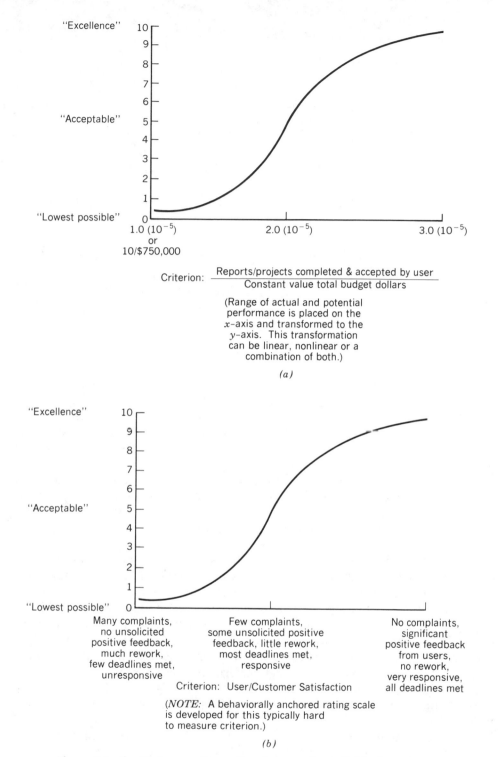

"Excellence" 10
9
8
7
6
"Acceptable" 5
4
3
2
1
"Lowest possible" 0

1.0 (10⁻⁵)   2.0 (10⁻⁵)   3.0 (10⁻⁵)
or
10/$750,000

Criterion: $\dfrac{\text{Reports/projects completed \& accepted by user}}{\text{Constant value total budget dollars}}$

(Range of actual and potential
performance is placed on the
x-axis and transformed to the
y-axis. This transformation
can be linear, nonlinear or a
combination of both.)

(a)

"Excellence" 10
9
8
7
6
"Acceptable" 5
4
3
2
1
"Lowest possible" 0

Many complaints,   Few complaints,   No complaints,
no unsolicited   some unsolicited positive   significant
positive feedback,   feedback, little rework,   positive feedback
much rework,   most deadlines met,   from users,
few deadlines met,   responsive   no rework,
unresponsive   very responsive,
                                              all deadlines met
Criterion: User/Customer Satisfaction

(NOTE: A behaviorally anchored rating scale
is developed for this typically hard
to measure criterion.)

(b)

Figure 6.4a–b   Preference Curve Development for Individual Attributes

203

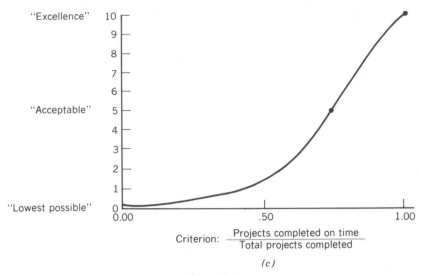

Criterion: $\dfrac{\text{Projects completed on time}}{\text{Total projects completed}}$

(c)

Figure 6.4c

importance can be assessed. This step, as with previous ones, can be executed by the analyst, by mangement, or in a participative fashion involving members of the organizational system itself. Weighting factors will reflect the relative contribution from each performance criterion in the organizational system's over-all performance. In the example being utilized to illustrate this methodology, the weighting factors will reflect the relative importance of each of the eight criteria on computer center performance. (Note that if the focus of this meas-urement system had been just on productivity, the weighting factors would reflect the relative importance of each of the eight criteria on productivity "meas-ures," ratios, or indexes on overall computer center productivity.) This rating process is analogous to performing a stepwise regression analysis on the vari-ables affecting organizational system performance.

It is important during this step to avoid thinking too much of the short term. The focus is not necessarily on how we presently perform best or on how we can most easily score with upper management, but on what criteria the organ-izational system has to concentrate in order to perform and in order to facilitate interface organizational system performance. This step is also critical because we often get what we measure. The MCPMM will end up being highly visible to members of the organizational system if implemented as a part of the NPMM. It will likely result in a very influential performance measurement system. This is why the list of "should be's" presented for the criteria is so critical; for example, if you want a certain outcome but don't measure it in the system, you may not get it.

The actual rating procedure is illustrated in Table 6.3. The first step is to arbitrarily assign 100 points to the top priority criterion. Next, the analyst, man-ager, or group assesses the relative importance of the next criterion. In this case, customer satisfaction was seen as being equally important as the top priority

Table 6.3 Ranking and Rating Procedure

CRITERION	RANK/PRIORITY	RATING	
Reports/projects completed and accepted / Constant value budget dollars	1	100	100/730 = 0.137
Customer satisfaction	2	100	100/730 = 0.137
Quality of decision support from systems developed	3	100	100/730 = 0.137
Meeting user flexibility requirements	4	90	90/730 = 0.123
Existence of and use of work scheduling/ project management	5	90	90/730 = 0.123
Projects completed on time / Total projects completed	6	85	85/730 = 0.116
Number of requests for reworking/ redoing a project	7	85	85/730 = 0.116
Existence of and quality of strategic planning for facilities, equipment, staffing, management processes, and operational systems	8	80	80/730 = 0.101
		730	730/730 = 1.00

criterion; therefore, it was given 100 points. This paired, relative-importance comparison continues on down through the list until all criteria have been assigned points. There has been much theory developed for, and some research on, these types of processes. The concern is over the validity of the results. A pragmatic view suggests that face validity as perceived by the analyst, manager, and/or members in the organizational system is the best test. However, interested readers are urged to consult the following references: Keeney and Raiffa, 1976; Morris, 1977; Stewart, 1978, 1980; and other such references on decision analysis, multi-attribute decision analysis, and utility theory.

Once the rating points have been assigned, these points are added up. For the example being developed, this total came to 730 points. The relative weighting, then, on a scale from 0 to 1.0 or from 0 to 100 if we multiply by 100 to convert to percentage, is obtained by dividing the points for each criterion by the total points. So, the results are that 100 points or percentage points are divided up among criteria in such a way as to indicate the relative importance of each criterion in relation to the others. Out of 100 possible points, if the computer center performed perfectly in the customer satisfaction category, it would receive 13.7 points. The implications are many. However, a simple and direct one is that the computer center should devote 13.7 percent of its effort or attention to each of the top three criteria.

The next step in this process is to integrate the performance graphs (scales and curves) with the criteria weightings. This will allow the development of a mechanism by which to evaluate the actual performance of the computer center with one performance indicator. Figure 6.5 conceptually depicts this step. Note

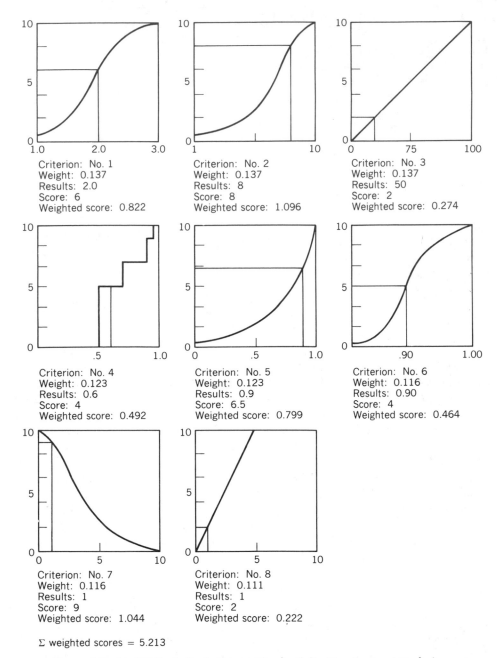

Criterion: No. 1
Weight: 0.137
Results: 2.0
Score: 6
Weighted score: 0.822

Criterion: No. 2
Weight: 0.137
Results: 8
Score: 8
Weighted score: 1.096

Criterion: No. 3
Weight: 0.137
Results: 50
Score: 2
Weighted score: 0.274

Criterion: No. 4
Weight: 0.123
Results: 0.6
Score: 4
Weighted score: 0.492

Criterion: No. 5
Weight: 0.123
Results: 0.9
Score: 6.5
Weighted score: 0.799

Criterion: No. 6
Weight: 0.116
Results: 0.90
Score: 4
Weighted score: 0.464

Criterion: No. 7
Weight: 0.116
Results: 1
Score: 9
Weighted score: 1.044

Criterion: No. 8
Weight: 0.111
Results: 1
Score: 2
Weighted score: 0.222

Σ weighted scores = 5.213

Figure 6.5   Multicriteria Performance/Productivity Measurement Technique

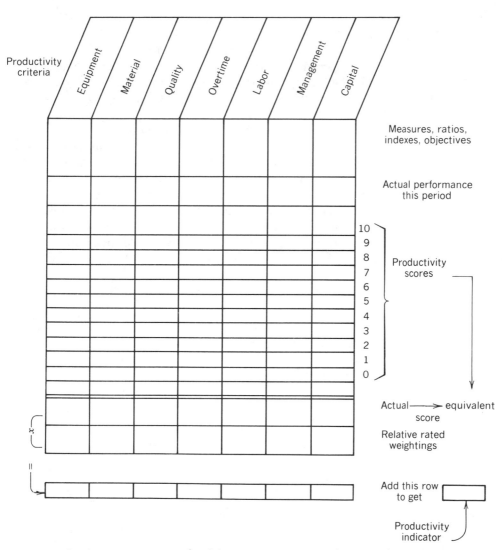

Figure 6.6   NPMM Productivity Measurement Matrix General Format

that actual performance as measured against the scales represented on the x-axis is transformed into a performance score (0 to 10) on the y-axis. These performance scores are then multipled by the criteria weighting factors to obtain the weighted scores. Note that these weighted scores all have common units while the x-axes in this example all reflect different units. So, we have a common denominator making it possible to aggregate the partial-factor measures and ratios into one performance indicator.

Keep in mind that this computer center example is developing a performance measurement system, not a productivity measurement system. There exists only one pure productivity measure: criterion number 1. This procedure is equally applicable if all eight of the criteria were productivity ratios and/or indexes.

The final computational step in this procedure is to add together all the weighted scores. The value for the computer center example is 5.213. Note that a maximum score would be 10.00. The individual performance scores in addition to the total weighted score or overall performance indicator can be tracked over time and utilized to develop necessary performance improvement programs. In essence, this methodology can be used in a fashion similar to the way the MFPMM is utilized to provide input for the productivity improvement planning process.

This MCP/PMT can, if developed with care and patience, result in a highly visible and effective performance and productivity measurement and evaluation system. A general matrix format for this same methodology is presented in Figure 6.6 (Riggs and Felix, 1983). Incorporated into the NPMM process, this methodology can become an extremely powerful and effective mechanism for linking improvement planning and execution to productivity measurement and evaluation.

REFERENCES

Archer, B. L. *Technological Innovation: A Methodology.* Royal College of Art, London (unpublished working paper/manuscript), 1970.

Keeney, R. L., and H. Raiffa, *Decisions with Multiple Objectives: Preferences and Value Trade-offs.* New York: John Wiley and Sons, 1976.

Keeney, R. L., and C. W. Kirkwood. "Group Decision Making Using Cardinal Social Welfare Functions." *Management Science*, Vol. 22, 1975.

Morris, W. T. *Decision Analysis.* Columbus, Ohio: Grid, 1977.

Riggs, J. L., and G. H. Felix. *Productivity by Objectives: Results-Oriented Solutions to the Productivity Puzzle.* Englewood Cliffs, N.J.: Prentice-Hall, 1983.

Stewart, W. T. "A Yardstick for Measuring Productivity." *Industrial Engineering*, Vol. 10, No. 2, 1978.

Stewart, W. T. "The Facilitation of Productivity Measurement and Improvement in Manufacturing Organizations," unpublished Ph.D. Dissertation, The Ohio State University, 1978.

Swalm, R. O. "Utility Theory—Insights into Risk Taking." *Harvard Business Review*, Vol. 44, 1966.

Viana, W. S. B. "Productivity Measurement and Improvement Strategy Program," unpublished M.S. Thesis, Oklahoma State University, 1982.

QUESTIONS AND APPLICATIONS

1. Write a program to execute the MCP/PMT procedure.

2. Outline the steps in the generic rank and rate procedure.

3. Discuss how the MCP/PMT for a productivity measurement application can be broadened to accommodate a performance measurement application.

4. Develop a MCP/PMT application for one or more of the following small business or small organizational systems examples:

  (a) A florist shop
  (b) A clerical pool
  (c) An industrial engineering department
  (d) A computer/data center
  (e) A motel
  (f) A restaurant
  (g) A research and development center
  (h) The boat company example presented in the last chapter
  (i) A faculty/department at a university
  (j) A quality control department
  (k) The maintenance group for a manufacturing plant
  (l) General administration for a high school

5. Discuss your view of how the MCP/PMT can integrate with the NPMM. Do you see both techniques as complementary? If so, in what ways? What are the advantages and disadvantages of using the MCP/PMT to "collapse" or aggregate the vector of productivity or performance measures? What are the theoretical assumptions or premises made with the MCP/PMT?

6. Using question 6 from Chapter 5 and the boat company example, develop an MCP/PMT rather than the MFPMM. Break the class into groups with management roles assigned. Have each management team develop an MCPMM for this boat company. Have each group present and discuss its resulting measurement system.

7. Develop graphics and specific decision support systems from the MCP/PMT for:

  (a) Top management
  (b) Middle management
  (c) Operations management
  (d) First-line supervisors
  (e) Employees

8. Compare and contrast the MFPMM and the MCP/PMT. Give the advantages, disadvantages, cost-benefit ratio, and ease of implementation of each.

9. Design and develop improvements and refinements in the MCP/PMT.

CHAPTER

7

NPMM, MFPMM, AND MCP/PMT SUMMARY AND SURROGATE PRODUCTIVITY MEASUREMENT APPROACHES

HIGHLIGHTS

- A Summary of the NPMM, MFPMM, and MCP/PMT
- Hybrid and Surrogate Measurement and Evaluation Approaches
- Productivity Measurement Using a Product-Oriented Total-Productivity Model
- Productivity (Performance) Audits and Checklists
- References
- Questions and Applications

OBJECTIVES

- To briefly expose the reader to a number of additional approaches to measuring and evaluating performance and productivity.
- To call attention to the fact that there are very likely hundreds of hybrid, in-house, specifically designed approaches and techniques for measuring and evaluating productivity.

A Summary of the NPMM, MFPMM, and MCP/PMT

Thus far, three unique approaches to measuring and evaluating productivity/ performance have been presented. Each has its own specific costs, benefits, advantages, disadvantages, and areas or situations for specific application. Further, although these techniques are unique, they are also somewhat interdependent, as it is highly possible to implement all three as an integrated productivity measurement and evaluation system. For example, the NPMM could be utilized to develop a commitment to productivity measurement, evaluation, control, and improvement on a decoupled, decentralized basis in each work group in an organization. The MCP/PMT could be utilized within the NPMM as Stage 2, analyst intervention, to aggregate the priority productivity "measures" into one productivity indicator. And the MFPMM could be utilized at the division, plant, or organization level to provide a coupled, centralized productivity measurement and evaluation system to support management decision making at a higher level of aggregation.

These three techniques tend to fill certain performance measurement gaps. This chapter is devoted to briefly discussing certain productivity measurement approaches that could be termed hybrids of these three techniques, surrogate approaches.

Hybrid and Surrogate Measurement and Evaluation Approaches

Many hybrid approaches do not really measure productivity but one of the other seven criteria for organizational systems' performance. Despite this fact, it is important to discuss these approaches briefly so that the reader is exposed to a variety of perspectives on how to measure productivity.

Productivity Measurement Using a Product-Oriented Total-Productivity Model

Sumanth (1979, 1984) has developed a total-productivity model (TPM) that is based on work done by Kendrick and Creamer (1965), Craig and Harris (1973), and Hines (1976). This model relates total output to all inputs. It is essentially equivalent to the MFPMM presented earlier except that all outputs and all inputs are to be captured. Hence, the MFPMM presented becomes a total-factor productivity measurement model (TFPMM). (Note that Sumanth differentiates between total-factor productivity, net output/labor + capital and total productivity, and gross output/labor + capital + materials + energy. We have not made this differentiation, so Sumanth's total productivity model would be equivalent to this text's concept of TFPMM.)

Sumanth's approach or model incorporates a product orientation. Therefore, the unit of analysis in his model is the product. In a paper presented at the 1980 Spring Annual IIE Conference, he presented advantages, disadvantages, formulations, and an example for the model. Much attention was devoted to data categories and collection since the model attempts to capture all outputs and inputs. The basic formulation of total productivity of a firm (TPF) is given as

$$TPF = \frac{OF}{IF} = \frac{{}_i O_i}{{}_i I_i} = \frac{{}_i O_i}{{}_{ij} I_{ij}}$$

where OF = total tangible output
           of a firm
      O_i = tangible output
           corresponding to product i
      IF = total tangible input
           to the firm
      I_i = total tangible input
           corresponding to
           product i ($i = 1, \ldots ,N$);
           = ${}_j I_{ij}$;
      j = H,M,C,E,X
      H = human input
      M = material input
      C = capital input
      E = energy input
      X = other expense input
      N = total number of products
           manufactured

The actual computations in Sumanth's model are quite similar to the MFPMM calculations. Productivity indexes are calculated by product and then a weighted total is obtained. Essentially, a separate MFPMM-type analysis is made for each product. Figure 7.1 depicts this. All MFPMM results by product are then combined to provide a firm-level evaluation of total productivity.

Sumanth also develops a break-even concept of total productivity. He relates profits to productivity in an analogous fashion to what the MFPMM does when it calculates effects of profits. If the MFPMM incorporated all outputs and inputs, then we could label it TFPMM. Additionally, the value for total input for Columns 10 and 11 would actually represent profitability margin (total costs/total revenues). In this case, Columns 17–19 take on added meaning since they now reflect actual changes in profits. The simulation routine could be utilized to calculate break-even points for profits based on changes only in productivity, changes only in price recovery, or changes in both productivity and price recovery. Sumanth has developed formulas to calculate the break-even values for total productivity.

In summary, the development of the product-level total productivity measurement model is an interesting and probably useful concept for certain applications. The development is a hybrid of the MFPMM concept and has its "roots" in Hiram Davis' efforts back in 1955. The concepts being developed by Sumanth can be incorporated in the MFPMM by appropriate modifications suggested.

Productivity (Performance) Audits and Checklists

Clearly not as much work has been done in the area of developing productivity audits as is needed. Webster defines "audit" as a regular examination and checking of accounts or financial records. In the business world, the accounting audit or financial audit has received the most attention. However, in the past ten years, increasing attention has been given to a broader variety of audits—for example, the energy audit (Turner, 1982); the materials handling audit (White, 1979); the organizational effectiveness review audit checklist (*R&D Productivity*, Hughes Aircraft Company, 1978); a self-evaluation checklist (TRW Productivity College, 1982); a productivity measurement audit (Bain, 1982); a productivity audit (Mali, 1978); a checklist of 20 questions on productivity (Patton, 1982); and an action guide productivity improvement for profit program (Apple).

The major problem, it would seem, impeding further progress in the area of productivity audit design and development is definitional in character. Few, if any, of the persons developing these audits begin with a solid, systematic, and comprehensive definition of the term "productivity." As a result, many, if not most, audits actually examine performance and not productivity. In addition, many of the audits look for indications of the existence of certain management behaviors, productivity improvement techniques, or types of work processes or general transformation processes instead of examining productivity measures. Part of the confusion stems from an apparent lack of focus. The persons designing the audit fail to develop questions that will pinpoint their goals. For example, is the audit supposed to examine

1. The productivity management process?

2. Specific productivity "measures"?

3. The existence of and effectiveness of certain productivity improvement techniques?

4. The nature and performance characteristics of processes that convert inputs to outputs?

This lack of direction results in confusion of means and ends.

It is impossible to present the contents of even the small list of productivity audits and checklists referenced in this section. It may be useful, though, to categorize or classify each of the approaches. Figure 7.2 presents a matrix that simply identifies the specific focuses of each representative audit. This sample of audits and checklists, of course, is not comprehensive; however, it is a fairly representative set of productivity audits and checklists. Several of the more significant developments will be reviewed briefly.

BY PRODUCT
MULTI FACTOR PRODUCTIVITY MEASUREMENT
OSU/OPC VERSION MPMM-2

RESOURCES CONSUMED ON PRODUCT A

RESOURCES CONSUMED ON PRODUCT B

RESOURCES CONSUMED ON PRODUCT C

OUTPUTS — Product A
INPUTS — LABOR — MATERIAL — ENERGY — CAPITAL — SERVICES — TOTAL

OUTPUTS — Product B
INPUTS — LABOR — MATERIAL — ENERGY — CAPITAL — SERVICES — TOTAL

OUTPUTS — Product C
INPUTS — LABOR — MATERIAL — ENERGY — CAPITAL — SERVICES — TOTAL

PERIOD 1 · PERIOD 2
COST REVENUE RATIOS
PRODUCTIVITY RATIOS
WEIGHTED CHANGE RATIOS
WEIGHTED PERFORMANCE INDEXES
TOTAL EFFECTS ON PROFITS
CHANGE IN PRODUCTIVITY · CHANGE IN PRICE RECOVERY · CHANGE IN PROFITABILITY

QUANTITY · PRICE · VALUE

Figure 7.1 Multifactor Productivity Measurement Model by Product

Figure 7.2 Focus of Various Checklists and Audits

AUDIT, CHECKLIST, OR OTHER PROCEDURE	MEASUREMENT	PRODUCTIVITY EVALUATION	CONTROL	IMPROVEMENT	MANAGEMENT	EFFECTIVENESS	EFFICIENCY	QUALITY	QUALITY OF WORK LIFE	INNOVATION	PROFITABILITY	TRANSFORMATION PROCESS CHARACTERISTICS	GENERAL MANAGEMENT PROCESS
ACTION GUIDE: PRODUCTIVITY IMPROVEMENT FOR PROFIT PROGRAM, APPLE	✓			✓		✓	✓	✓				✓	
CHECKLIST OF 20 QUESTIONS ON PRODUCTIVITY, PATTON	✓	✓	✓		✓	✓	✓		✓				✓
PRODUCTIVITY AUDIT, MALI	✓	✓	✓		✓	✓	✓		✓			✓	✓
PRODUCTIVITY FUNCTIONAL SELF-EVALUATION, TRW PRODUCTIVITY COLLEGE	✓				✓	✓	✓	✓	✓			✓	
YALE MANAGEMENT GUIDE TO PRODUCTIVITY, MATERIALS HANDLING AUDIT, WHITE	✓												
R & D PRODUCTIVITY, ORGANIZATION EFFECTIVENESS REVIEW, HUGHES AIRCRAFT	✓		✓	✓		✓				✓			✓
ENERGY AUDIT, TURNER		✓	✓	✓	✓	✓	✓						
PRODUCTIVITY MEASUREMENT AUDIT, BAIN	✓			✓	✓	✓	✓				✓		
ORGANIZATIONAL CLIMATE, MGMT. ATTITUDE AUDIT, BAIN	✓					✓			✓				✓
CHECKLIST FOR RECOGNIZING METHODS IMPROVEMENT OPPORTUNITIES, BAIN						✓						✓	

FOCUS:

THE PRODUCTIVITY AUDIT

Mali devotes an entire chapter in his book to the productivity audit. He compares productivity auditing to financial, program, operations, compliance, social, and general management auditing. He observes that the auditing process, in general, has the following common characteristics: (1) predetermined standards; (2) evaluation by an independent auditor; (3) a tailoring to the specific organization; and (4) a deductive and comparative analysis for measurement and evaluation. Mali then goes on to develop and present a model of evaluating productivity. This model has five phases:

1. Determine purpose to be achieved by the audit

2. Select standards as criteria for measurement

3. Use measures and compare with standards

4. Correct for significant deviations and variations

5. Compile results into a written report.

The general model is presented in Figure 7.3.

Note the similarity between this model and the productivity management process presented in Figure 2.1. The audit component is highlighted in Mali's model for evaluating productivity; however, the planning, measurement, evaluation, control, and improvement components are also there. What can be observed is that productivity auditing is just a form of productivity measurement. In fact, the MFPMM could easily be utilized as a productivity audit assuming data availability. The NPMM could be viewed as a normative and participative internal productivity audit.

Mali's auditing procedure involves setting standards as criteria for measurement. He presents a list of standards that could be used in a productivity audit. They are is essence statements regarding management processes within ten general performance categories. Mali points out that the list is not necessarily comprehensive. This list of ten performance categories and specific standards statements is

1. Productivity Actions
   (a) All key individuals manage and have made a commitment to do their work according to well-defined, properly stated, and precisely measurable productivity objectives and goals. These objectives are written and open for review.
   (b) Productivity objectives are consistent with mission statements and other objectives of the organization.
   (c) Productivity objectives are set at levels similar to or greater than previous levels.
   (d) Productivity effort exhibits indicators that distinguishes it from other efforts needed and practiced in the organization.

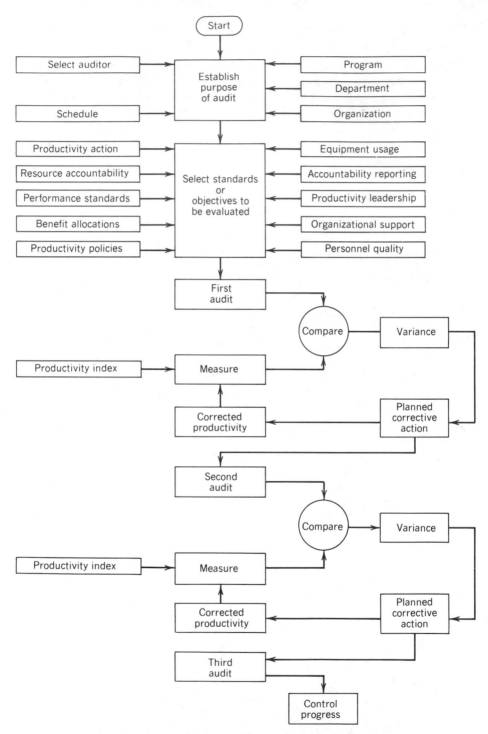

Figure 7.3 Model for Evaluating Productivity (Mali)

(e) Productivity progress toward meeting performance, cost, and schedule milestones can be assessed for early detection and timely correction of problems.

2. Resource Accountability
   (a) All resources are aggregated and accounted for in budgets located at well-organized cost centers that are responsible for achievement of plans and objectives.
   (b) Time is viewed as a critical resource, and managing it is clearly visible with priority systems, schedules, and avoidance of work on trivia.
   (c) Cost centers account through a budgeting information system for resource utilization and waste for the whole organization.
   (d) Resources are at workers' disposal at the time needed to facilitate work.
   (e) Cost centers are closely aligned with functional authority of the organization structure. That is, cost centers are tied to the decision-making process of budget formulation and adoption.

3. Performance Standards
   (a) Individuals are aware of trained-for, expected performance before they start work.
   (b) Performance standards are clearly defined, attainable, accurate, and measurable.
   (c) Individuals are aware of and have controls on the amount of resources they are to use in completing performance standards.
   (d) Performance measures are made and aggregated within cost centers.
   (e) Productivity is measured with resource and performance results at cost centers.

4. Benefit Allocations
   (a) Costs of benefits are precisely identified and assigned in cost centers.
   (b) Reward and benefit system is based on performance and productivity data. Increased benefits are given only with increased productivity.
   (c) Benefits to be achieved are clearly stated for each program, project, or plan.
   (d) Planned benefit allocations are accomplished by planned productivity improvement.
   (e) Productivity is a criterion during negotiations and tradeoffs.

5. Productivity Policies
   (a) A productivity mission statement is issued, actively pursued, and given high priority by top management.
   (b) Managers must submit formal productivity plans to be integrated into an overall productivity plan.
   (c) Productivity results are formally evaluated once a year.
   (d) Gains resulting from increased productivity are shared with those responsible for the increase.
   (e) Productivity improvement is practiced within all cost centers or departments of the organization.

6. Equipment Usage and Technology
   (a) Equipment purchase is justified by data or productivity improvement or by the savings it institutes.
   (b) Of available equipment options, planned usage of equipment is shown to be best for productivity.
   (c) Technological innovations and the use of technological aids are visible in practice.
   (d) Equipment purchase, use, and expenditures are controlled and reported through cost centers.
   (e) Short-range tradeoff benefits should not outweigh long-range benefit expectations.

7. Accountability Reporting
   (a) A system exists for reporting to accountability sources total variances from an established plan of productivity improvement.
   (b) Required productivity is delegated and can be traced to a specific person in the organization.
   (c) Observed productivity is in fact the same as reported productivity.
   (d) Status reporting system is self-correcting through feedback to workers.
   (e) A formal annual report on productivity progress is submitted to top management.

8. Productivity Leadership
   (a) Leaders operate and manage in an action-research manner for productivity improvements. They "tune-up" the organization, searching for ways to improve. There is a high "esprit de corps."
   (b) Leaders operate a system to seek and implement work improvement innovations from workers.
   (c) Leaders allow high "clashes" in productivity ideas but low "clashes" in interpersonal relations. Competition is encouraged for results but is not encouraged among personalities.
   (d) Leaders delegate work responsibilities accompanied by accountability.
   (e) Leaders conduct and make use of periodic productivity audits.

9. Organizational Support
   (a) Power structure is recognizable and close to the formal plan of organization. All unnecessary or marginal functions are eliminated.
   (b) Decision making is delegated to the most critical point of productivity action or cost centers—where work is accomplished and the impact is greatest.
   (c) Balanced effort from all significant units prevails.
   (d) Employees are allowed to participate in decisions affecting their own and related jobs.
   (e) Research and development are a formal and deliberate effort in the organization.

10. Personnel Quality
    (a) Individuals are open to and express a positive attitude toward productivity improvement.
    (b) Self-renewal, obsolescence, and low performance are personal concerns.
    (c) Orientation, training, and coaching for abilities and skills for productivity are provided.
    (d) System of personnel backups prevails to continue level of productivity when turnover occurs.
    (e) Productivity skills are a primary criterion in the recruiting and hiring of personnel (Mali, 1978).

Mali suggests that these "standards" (ten major categories) be ranked and then weighted via a rating process similar to the one described in the section on the MCP/PMT. The criteria that are being audited could actually be evaluated utilizing the procedure described in that section. This would provide a nice mechanism for quantifying performance on the audit.

Mali (1978) suggests that after the audit is completed and results have been evaluated, a report should be developed. He suggests the following outline for that report:

1. Summary (Single-Page)
    (a) Purpose and objectives of the audit
    (b) Results obtained
    (c) Recommendations for improvement
    (d) Brief description of qualification of auditor

2. Contents
    (a) Introduction and background
    (b) Purpose and objectives of the audit
    (c) Why the audit was conducted—facts, current practices, and the evaluation
    (d) Procedure for conducting the audit
    (e) Analysis and measurements of facts and information
    (f) Results obtained
    (g) Conclusions and recommendations

3. Appendix
    (a) Tables or calculations
    (b) Graphs or diagrams
    (c) Supporting documents
    (d) Qualification of auditor

The important points that Mali presents regarding a productivity audit are

1. The difference between a productivity audit and other types of performance audits

2. Productivity auditing as a process. The uniqueness of criteria, standards, and general audit design itself to the specific organizational system

3. The establishment of the purpose, focus, scope, objectives, desired outcomes, and so forth for the audit as a first step to the design process

The productivity audit is something more than productivity measurement. It is a periodic attempt to take stock of and critique the effectiveness, efficiency, and quality of the productivity management effort. It should be a discrete event, perhaps occurring once a year or once every two years, whereas productivity measurement is clearly a continuous process. What one looks for in the audit itself is likely to depend highly on the specific organizational system being audited. Some general criteria and factors to look for can and have been provided; however, the real quality audit will be one that is carefully thought through and designed for each specific organizational system.

THE PRODUCTIVITY MEASUREMENT AUDIT

Bain presents this audit as a starting point for creating a productivity measurement system for a given organizational system. The audit or questionnaire was designed to provide assistance in evaluating existing measurements as well as to gain insight into what has happened in the past in the area of productivity measurement. This audit is presented in Table 7.1.

This particular audit would be very useful during the productivity measurement strategy formulation stage discussed in Chapter 3. It would also be very useful at milestone t_3 (productivity measurement strategic planning) of the NPMM. Bain's book is outstanding, and the chapter on Design Aids from which this audit comes is particularly well developed.

YALE MANAGEMENT GUIDE TO PRODUCTIVITY

White has prepared an outstanding functional (materials handling) performance audit for Eaton Corporation (1979). Its focus is functional, and therefore it is similar to, although more highly developed than, Apple's Productivity Improvement for Profit (PIP) program designed for the Institute of Industrial Engineers. White suggests that there are productivity measures developed for resource utilization (people, equipment, space, energy); management control (materials, movement, loss); and operating efficiency (receiving, storage, manufacturing). However, based on strict definitions developed in Chapter 2, very few, as you will see, are actually real productivity "measures". Most of the measures are surrogate productivity measures at best and more aptly termed various types of performance measures. The reason for presenting a review of this document and development is that it reflects the type and quality of work that needs to be done in other areas of productivity measurement and auditing. The management guide, although just focusing on materials handling, represents a model that could and probably should be followed in other functional areas as well as in the more general, interfunctional area of productivity measurement and auditing.

Table 7.1 Productivity Measurement Audit

QUESTIONS TO BE ANSWERED	INDICATE ANSWER HERE: COLUMN 1	COLUMN 2	IF "YES," PROCEED TO QUESTION:	IF "NO," PROCEED TO QUESTION:
1. Is productivity now being measured in any of your organization's activities?	_____ Yes	_____ No	**4.**	**2.**
2. Was productivity ever previously measured within the organization?	_____ Yes	_____ No	**3.**	**5.**
3. If productivity measurements were implemented but later discontinued, indicate the reasons for discontinuance below:				
Benefits derived failed to justify cost?	_____ No	_____ Yes		
Measurements consumed too much time?	_____ No	_____ Yes		
Measurements generated conflict between management and employees?	_____ No	_____ Yes		
Measurements generated conflict between employees?	_____ No	_____ Yes		
Measurements lacked the support of higher management?	_____ No	_____ Yes		
Other reasons (specify): _____ _____ _____		_____ Yes	**5.**	**5.**

Table 7.1 Productivity Measurement Audit (*Continued*)

QUESTIONS TO BE ANSWERED	INDICATE ANSWER HERE: COLUMN 1	COLUMN 2	IF "YES," PROCEED TO QUESTION:	IF "NO," PROCEED TO QUESTION:
4. To which of the following can existing productivity measurements be related?				
Profit?	Yes	No		
Service?	Yes	No		
Quality?	Yes	No		
Indirect plant labor?	Yes	No		
Direct plant labor?	Yes	No		
Office labor?	Yes	No		
All employees?	Yes	No		
Equipment?	Yes	No		
Materials?	Yes	No		
Facility?	Yes	No		
Sales?	Yes	No		
Credit and collections?	Yes	No		
Scheduling?	Yes	No		
Inventory management?	Yes	No		
Purchasing?	Yes	No		
Receivables?	Yes	No		
Transportation?	Yes	No		
Other (specify):				
_____	Yes			
_____	Yes		**5.**	**5.**

Table 7.1 Productivity Measurement Audit (Continued)

QUESTIONS TO BE ANSWERED	INDICATE ANSWER HERE:		IF "YES," PROCEED TO QUESTION:	IF "NO," PROCEED TO QUESTION:
	COLUMN 1	COLUMN 2		
5. Do engineered work standards exist within the organization?	_____ Yes	_____ No	6.	7.
6. Indicate in which areas engineered work standards are currently being applied: Plant?	_____ Yes	_____ No		
Office?	_____ Yes	_____ No		
Other (specify:) _____	_____ Yes		7.	7.
7. Do specific expectations other than work standards exist within the organization?	_____ Yes	_____ No	8.	14.
8. Indicate in which areas specific work standards are currently being applied: Plant?	_____ Yes	_____ No		
Office?	_____ Yes	_____ No		
Other (specify): _____	_____ Yes		9.	9.
9. Is individual compensation based, in part, on either productivity or profitability: Among plant personnel?	_____ Yes	_____ No		
Among office personnel?	_____ Yes	_____ No		
Among managers?	_____ Yes	_____ No		
Other (specify): _____	_____ Yes		10.	10.

Table 7.1 Productivity Measurement Audit (Continued)

QUESTIONS TO BE ANSWERED	INDICATE ANSWER HERE: COLUMN 1	COLUMN 2	IF "YES," PROCEED TO QUESTION:	IF "NO," PROCEED TO QUESTION:
10. Are people within the organization ever responsible for measuring elements of their own productivity?	_____ No	_____ Yes		**11.**
If yes, which ones? Inputs?		_____ Yes		
Outputs?		_____ Yes	**11.**	
11. Are all employees routinely counseled when their productivity slips below an acceptable level?	_____ Yes	_____ No	**12.**	**12.**
12. Does the operating head of the organization review productivity measurements at least once each month?	_____ Yes	_____ No		**13.**
If yes, does he or she review such measurements at least once each week?	_____ Yes		**13.**	
13. Do existing measurements generally meet the following criteria: *Validity:* Do they actually gauge real changes in productivity?	_____ Yes	_____ No		
Completeness: Do they take into consideration all significant components of total inputs and total outputs?	_____ Yes	_____ No		
Comparability: Do they enable the accurate comparison of productivity between periods?	_____ Yes	_____ No		

Table 7.1 Productivity Measurement Audit (Continued)

QUESTIONS TO BE ANSWERED	INDICATE ANSWER HERE:		IF "YES," PROCEED TO QUESTION:	IF "NO," PROCEED TO QUESTION:
	COLUMN 1	COLUMN 2		
Inclusiveness: Are activities other than production/ manufacturing/order filling being measured?	Yes	No		
Timeliness: Are measurements reported soon enough for appropriate managerial action to be taken when required?	Yes	No		
Cost effectiveness: Are measurements performed in a cost-effective and noninterruptive manner?	Yes	No	14.	14.
14. Does your organization tend to look upon resources committed to measurement as being wasted?	No	Yes	15.	15.
15. Are all resources dedicated to productivity measurement periodically quantified and evaluated on a cost-benefit basis?	Yes	No	16.	16.
16. With what frequency are productivity measurements now being reported to the responsible manager(s): Every hour?	Yes	No		
Every day?	Yes	No		
Every week?	Yes	No		
Every month?	Yes	No		17.
Other (specify):			17.	
_____	Yes			

227

Table 7.1 Productivity Measurement Audit (Continued)

QUESTIONS TO BE ANSWERED	INDICATE ANSWER HERE:		IF "YES," PROCEED TO QUESTION:	IF "NO," PROCEED TO QUESTION:
	COLUMN 1	COLUMN 2		
17. Is the company's financial statement always prepared and distributed within seven working days after the end of the month?	_____ Yes	_____ No	18.	18.
18. With only rare exception, do period expenses match period sales as reported on the financial statement?	_____ Yes	_____ No	19.	19.
19. Is the magnitude of all variances—actual expenses to budgeted expenses or to a standard—expressed in dollars on the financial statement for both the month and for the year to date?	_____ Yes	_____ No	20.	Complete
20. Is detail related to unfavorable variances usually provided with the financial statement?	_____ Yes	_____ No		Complete

After completing the above questionnaire:

Add the total number of responses in column 1, and enter here → _____
  A

Add the total number of responses in column 2, and enter here → _____
  B

Add the grand total number of responses in both columns (column 1 + column 2) and enter here → _____
  C

Divide the total number of responses in column 1 (A above) by the grand total number of responses (C above), and enter here → _____
  D

SOURCE: Bain, D.F. *The Productivity Prescription*, New York: McGraw-Hill, 1982. Reprinted with permission.

If **C** is greater than 40 and **D** is greater than 75 percent, it indicates that managers within the organization are generally interested in measuring productivity. The greater the value of **D** beyond 75 percent, the more meaningful are existing measurements.

If, on the other hand, **C** is less than 40 and **D** is less than 75 percent, then the good news is that you will not have to worry about changing a number of current measurements because current measurements are few in number. Besides being few in number, existing measurements are most likely not very meaningful. More questions have possibly been generated by the questionnaire than have been answered. Leading the list of such questions is, Why, specifically, has little effort been expended to date to install measurements and improve productivity? (A following questionnaire will help you in answering that question.)

The specific measures developed by White along with the "formula" for the measures (note that most are ratios) are presented in Figure 7.4. Example calculations for each ratio are provided in the management guide. In addition, possible areas for productivity improvement relating to each set of measures are discussed in the guide. Worksheets, such as the one depicted in Table 7.2, are provided for convenience in calculating each measure (ratio). These measures for the materials handling function could easily be used as a discrete performance productivity audit mechanism or as a continuous performance/productivity measurement process.

The three perspectives just presented reflect a reasonably comprehensive sample of the types of developments that exist in the general area of "productivity"

Aisle Space Percentage (ASP)—proportion of space devoted to aisles.

$$ASP = \frac{\text{cube space occupied by aisles}}{\text{total cube space}}$$

Average Distance/Move Ratio (AD/M Ratio)—measures the ratio of the length of moves required to the total number of moves made.

AD/M Ratio

$$= \frac{\text{total distance traveled/day}}{\text{total number of moves/day}}$$

Busy Percentage (BP)—the percentage of time handling equipment is busy.

$$BP = \frac{\text{busy observations}}{\text{total observations}}$$

Damaged Loads Ratio (DLR)—measures loss due to poor handling.

$$DL\ Ratio = \frac{\text{number of damaged loads}}{\text{number of loads}}$$

Direct Labor Materials Handling (DLMH) Ratio—measures the proportion of the direct labor time spent in performing materials handling.

$$DLMH\ Ratio = \frac{\text{materials handling time spent by direct labor}}{\text{total direct labor time}}$$

Energy Utilization Index (EUI)—compares the Btu's consumed per day with the total cube space.

$$EUI = \frac{\text{Btu's consumed/day}}{\text{cubic space}}$$

Handling Equipment Utilization (HEU)—determined by considering the weight of a load, the number of loads moved and the time spent in moving the load.

$$HEU = \frac{\text{weight moved/hour}}{\text{theoretical capacity}}$$

Inventory Fill Ratio—provides a measure of out-of-stock conditions.

$$IF\ Ratio = \frac{\text{line item demands filled/day}}{\text{line items demands/day}}$$

Inventory Location Error Ratio—monitors the accuracy of the stock location system.

Figure 7.4 "Productivity" Measures for Materials Handling

Inventory Shrinkage Ratio (ISR)—measures loss of material due to pilferage, error in record keeping, error in material shipments to customer and misplacement of inventory in storage.

IS Ratio

$$= \frac{\text{inventory investment verified}}{\text{inventory investment expected}}$$

Inventory Replenishment Ratio—compares actual stock replenishment lead times against desired lead times.

Inventory Turnover Ratio—measures the return obtained from inventory investments and provides an indication of the movement of materials.

$$\text{IT Ratio} = \frac{\text{annual labor}}{\text{average annual inventory investment}}$$

Job Lateness Ratio (JLR)—measures the due date performance of manufacturing.

$$\text{JL Ratio} = \frac{\text{jobs that are late/week}}{\text{number of jobs completed/week}}$$

Manufacturing Cycle Efficiency (MCE)—compares the time materials spend being processed to the total time spent in the manufacturing department.

$$\text{MCE} = \frac{\text{total time spent on machining}}{\text{total time spent in production system}}$$

Materials Handling Labor Ratio (MHL Ratio)—measures the proportion of the total labor force devoted to materials handling duties.

$$\text{MHL Ratio} = \frac{\text{personnel assigned to materials handling duties}}{\text{total operating workforce}}$$

Movement/Operation Ratio (M/O Ratio)—measures the relative efficiency of the overall materials handling plan.

$$\text{M/O Ratio} = \frac{\text{total number of moves}}{\text{total number of productive operations}}$$

Obsolete Inventory Ratio—measures the percentage of storage space consumed by obsolete materials.

Orderpicking Productivity Ratio (OPR)—measures the productivity of an order-picking activity.

$$\text{OP Ratio} = \frac{\text{equivalent lines or orders picked/day}}{\text{labor hours required/day}}$$

Perpetual Inventory Error Ratio—measures the accuracy of the inventory transactions reporting system.

Production Equipment Utilization (PEU)—provides an indication of the extent to which production equipment is being utilized to its fullest potential.

$$\text{PEU} = \frac{\text{actual output}}{\text{theoretical output}}$$

Receiving Productivity Ratio (RPR)—measures daily output of receiving functions.

$$\text{RP Ratio} = \frac{\text{pounds received/day}}{\text{labor hours/day}}$$

Shipping Productivity Ratio (SPR)—measures daily output of shipping functions.

$$\text{SP Ratio} = \frac{\text{pounds shipped/day}}{\text{labor hours/day}}$$

Storage Slot Occupancy Ratio—measures the percentage of storage slots or bin slots that are occupied.

$$\text{SSO Ratio} = \frac{\text{storage slots occupied}}{\text{total storage slots}}$$

Storage Space Utilization Ratio (SSU Ratio)—proportion of storage space occupied by materials.

$$\text{SSU Ratio} = \frac{\text{materials cube stored}}{\text{total cube space required}}$$

Storage Turnover Ratio—measures the movement or turnover of materials in storage. It is the ratios of loads retrieved to loads stored.

Throughput Performance Index (TPI)—compares the actual throughput with the theoretical throughput capacity for the storage and retrieval system.

$$\text{TPI} = \frac{\text{throughput achieved/day}}{\text{throughput capacity/day}}$$

Figure 7.4   (*Continued*)

Table 7.2 Worksheets for Calculating Ratios
Worksheet No. 1—Sheet 1 FOR COMPUTATION OF YOUR MATERIALS HANDLING LABOR RATIO

$$\frac{\text{Materials Handling}}{\text{Labor Ratio}} = \frac{\text{Materials handling labor}}{\text{All labor}}$$

OUR MATERIALS HANDLING LABOR CONSISTS OF LABOR ASSIGNED SOLELY TO MATERIALS HANDLING DUTIES	NUMBER OF PERSONS	PAYROLL DOLLARS PER YEAR
1. Materials handling equipment operators		
(a) Nonpowered hand trucks		
(b) Powered industrial trucks		
(i) Walkie platform and pallet trucks		
(ii) Walkie stackers		
(iii) Rider pallet trucks		
(iv) Counterbalanced lift trucks		
(v) Narrow aisle reach and straddle trucks		
(vi) Narrow aisle side-reach and sideloader trucks		
(vii) Very narrow aisle turret trucks		
(viii) Orderpicker trucks		
(ix) Man-aboard, aisle captive storage/retrieval machines		
(x) Straddle and van carriers		
(xi) Mobile cranes a. Operators		
b. Riggers		
(xii) Tractor shovels, bulldozers, etc.		
(xiii) _____		
(c) Cranes, stacker cranes, hoists & other overhead equipment (i) Operators		
(ii) Riggers		
(d) Palletizer, depalletizer, shrinkwrap, stretchwrap, and strapping operators		
(e) Conveyor loaders and unloaders		
(f) Intraplant motor trucks		
(g) Intraplant personnel and burden carriers		
(h) Intraplant railroad		
(i) Elevators		
(j) Manual handling labor		
Total		

231

Table 7.2 (*Continued*)
Worksheet No. 1—Sheet 2 FOR COMPUTATION OF YOUR MATERIALS
HANDLING LABOR RATIO (*Continued*)

2. Workforce for activities essentially devoted to:	NUMBER OF PERSONS	PAYROLL DOLLARS PER YEAR
(a) Receiving dock		
(b) Shipping dock		
(c) Raw materials storage		
(d) Work-in-process materials storage		
(e) Finished goods storage		
(f) Distribution warehousing		
(g) Scrap & salvage operations		
(h) Waste materials removal		
Total		

3. Activities partially devoted to, or supporting, the materials handling function	NUMBER OF PERSONNEL	ANNUAL PAYROLL $	% CHARGE-ABLE MH	NET ANNUAL $ FOR MH
(a) Tool room & supplies issue				
(b) Parts preparation				
(c) Maintenance of MH equipment				
(d) Production control, particularly dispatching & expediting				
(e) Packaging operations				
(f) Packing operations				
(g) Inventory control records				
(h) Inbound materials inspection				
(i) Outbound materials inspection				
(j) Traffic				
(k) Training				
Total				

Annual Materials Handling Labor Cost:

1. MH equipment labor _____
   Payroll $
2. MH activities labor _____
   Payroll $
3. Activities partially _____
   chargeable   Payroll $

Item A: Total MH labor _____

Item B: Total annual payroll _____
(Ordinarily computed by
taking operating payroll
and omitting general
administrative and sales
payroll)

$$\text{MH Labor Ratio} = \frac{\text{MH Labor}}{\text{Total Payroll}} = \frac{\text{Item A}}{\text{Item B}} = \frac{\$ \underline{\quad\quad}}{\$ \underline{\quad\quad}} = \underline{\quad\quad} \%$$

Table 7.2 (*Continued*)
Worksheet No. 2—Sheet 1   FOR COMPUTATION OF YOUR MOVEMENT/
OPERATION RATIO

NOTE: There are two ways to obtain these data. *One.* For a quick impression of your handling plan, just walk through the shop, tallying moves and operations. Then compute the ratio.

Two. For a more detailed job and information for corrective action, use the form below. Fill in data on "moves" and "operations" in sequence, and add explanatory detail as desired. Totals of Columns 1 and 2 will give data for the M/O ratio. Remaining columns provide guidance for corrective action. Add comments, queries and suggestions as needed.

$$\frac{Movement/}{Operation\ Ratio} = \frac{Number\ of\ moves}{Number\ of\ production\ operations}$$

Product _____   Observer _____

Location _____To _____Date _____

1	2	3	4	5		6		
	MOVES OR DELAYS (x)			HANDLING TECHNIQUE USED		NEEDS FURTHER INVESTIGATING		
	DISTANCE		DESCRIPTION,					
PRODUCTIVE OPERATION	NO.	FT.	OR REASON FOR ACTIVITY	MANUAL (√)	MECHANIZED (√)	METHODS (√)	LAYOUT (√)	EQUIPMENT (√)

233

Table 7.2 (*Continued*)
Worksheet No. 2—Sheet 2   FOR COMPUTATION OF YOUR MOVEMENT/
OPERATION RATIO (*Continued*)

1	2	3	4	5		6		
	MOVES OR DELAYS (X) DISTANCE			HANDLING TECHNIQUE USED		NEEDS FURTHER INVESTIGATING		
PRODUCTIVE OPERATION	NO.	FT.	DESCRIPTION, OR REASON FOR ACTIVITY	MANUAL ($\sqrt{}$)	MECHANIZED ($\sqrt{}$)	METHODS ($\sqrt{}$)	LAYOUT ($\sqrt{}$)	EQUIPMENT ($\sqrt{}$)

Table 7.2 (*Continued*)
Worksheet No. 3—Sheet 1   MATERIALS HANDLING AUDIT CHECK SHEET

Dept. _____ Building _____ Plant _____

Date _____ Surveyed by _____

CONDITIONS INDICATING POSSIBLE PRODUCTIVITY IMPROVEMENT OPPORTUNITIES	CONDITION EXISTS HERE (√)	TO CORRECT THIS, WE NEED:				
		SUPERVISOR ATTENTION (√)	MANAGEMENT ATTENTION (√)	ANALYTICAL STUDY (√)	CAPITAL INVESTMENT (√)	OTHER (for comments)
1. Delays in material moving						
2. Excessive material on hand						
3. Production equipment idle for material shortage						
4. Long hauls						
5. Cross traffic						
6. Manual handling						
7. Outmoded handling equipment						
8. Inadequate handling equipment						
9. Insufficient handling equipment						
10. Unbalanced sequence of operations						
11. Idle handling equipment						
12. Obstacles to materials flow						
13. Materials piled directly on floors						
14. Poor work place layout for materials						
15. Disorderly storage						
16. Cluttered aisles						
17. Cluttered work space						
18. Crowded dock space						
19. Motor truck and railroad car tieup						

Table 7.2 (*Continued*)
Worksheet No. 3—Sheet 2 MATERIALS HANDLING AUDIT CHECK SHEET (*Continued*)

CONDITIONS INDICATING POSSIBLE PRODUCTIVITY IMPROVEMENT OPPORTUNITIES	CONDITION EXISTS HERE (√)	TO CORRECT THIS, WE NEED:				
		SUPERVISOR ATTENTION (√)	MANAGEMENT ATTENTION (√)	ANALYTICAL STUDY (√)	CAPITAL INVESTMENT (√)	OTHER (for comments)
20. Manual loading techniques						
21. Excessive wasted "cube" in storage						
22. Excessive aisles						
23. Operations unduly scattered						
24. Poor locations of service areas						
25. Lack of in-plant container standard-ization						
26. Lack of unit load technique						
27. Excessive MH equipment mainte-nance cost						
28. Rehandling						
29. Handling done by direct labor						
30. Operators traveling for supplies, mate-rials						
31. Supplies moved by poor techniques						
32. High indirect payroll						
33. Materials waiting for papers						
34. Excessive demurrage						
35. Unexplained delays						
36. Idle labor						
37. Inspection not prop-erly located						
38. Excessive scrap						
39. Hazardous lifting by hand						
40. Misdirected mate-rials						

236

Table 7.2 (*Continued*)
Worksheet No. 3—Sheet 3 MATERIALS HANDLING AUDIT CHECK SHEET (*Continued*)

CONDITIONS INDICATING POSSIBLE PRODUCTIVITY IMPROVEMENT OPPORTUNITIES	CONDITION EXISTS HERE (✓)	TO CORRECT THIS, WE NEED:				
		SUPERVISOR ATTENTION (✓)	MANAGEMENT ATTENTION (✓)	ANALYTICAL STUDY (✓)	CAPITAL INVESTMENT (✓)	OTHER (for comments)
41. Clumsy, dangerous "home made" handling rigs						
42. Lack of standardization on handling equipment						
43. Long travel distances for materials, equipment, and personnel						
44. Backtracking of materials						
45. Non-standard process routing						
46. Opportunity for group technology layout						
47. Opportunity for product layout						
48. Opportunity for process layout						
49. No real time dispatching of equipment						
50. No modular MH system						
51. No modular work stations						
52. Automatic identification system not used						
53. No one-way aisles						
54. MH equipment running empty						
55. Different things treated same						
56. Excessive trash removal						
57. Centralized storage						

Table 7.2 (*Continued*)
Worksheet No. 3—Sheet 4   MATERIALS HANDLING AUDIT CHECK SHEET
(*Continued*)

CONDITIONS INDICATING POSSIBLE PRODUCTIVITY IMPROVEMENT OPPORTUNITIES	CONDITION EXISTS HERE (√)	TO CORRECT THIS, WE NEED:				
		SUPERVISOR ATTENTION (√)	MANAGEMENT ATTENTION (√)	ANALYTICAL STUDY (√)	CAPITAL INVESTMENT (√)	OTHER (for comments)
58. Decentralized storage						
59. No incentive system for MH labor						
60. Low usage of automated MH equipment						
61. Variable path equipment used for fixed path handling						
62. System not capable of expansion and/or change						
63. Low usage of industrial robots						
64. No parts preparation performed prior to manufacturing						
65. No pre-kitting of work						
66. Lack of automated loading/unloading of trailers						
67. Poor MH at the work station						
68. Lack of industrial truck attachments and below hook lifters						
69. Equipment capacity not matched to load requirement						
70. Manual palletizing/ depalletizing						
71. Lack of equipment for unitizing and stabilizing loads						
72. Lack of a long range MH plan						

238

Table 7.2 (*Continued*)
Worksheet No. 3—Sheet 5   MATERIALS HANDLING AUDIT CHECK SHEET
(*Continued*)

CONDITIONS INDICATING POSSIBLE PRODUCTIVITY IMPROVEMENT OPPORTUNITIES	CONDITION EXISTS HERE (√)	TO CORRECT THIS, WE NEED:				
		SUPERVISOR ATTENTION (√)	MANAGEMENT ATTENTION (√)	ANALYTICAL STUDY (√)	CAPITAL INVESTMENT (√)	OTHER (for comments)
73. No short interval scheduling of MH equipment						
74. Lack of narrow aisle and very narrow aisle storage equipment						
75. Low bay storage areas						
76. Poor utilization of overhead space						
77. Single sized pallet rack openings						
78. No palletless handling of unit loads						
79. Storage by part number sequence						
80. Randomized storage						
81. Dedicated storage						
82. Crushed loads in block stacking						
83. No ABC storage classification						
84. Obsolete and inactive materials						
85. Floor stacked material, in receiving, QC, and shipping						
86. Aisles and storage locations not clearly marked						
87. Manual stock locator system						
88. Lack of standardization in part numbers						
89. Cycle counting for physical inventory						

239

Table 7.2 (*Continued*)

Worksheet No. 3—Sheet 6 MATERIALS HANDLING AUDIT CHECK SHEET (*Continued*)

CONDITIONS INDICATING POSSIBLE PRODUCTIVITY IMPROVEMENT OPPORTUNITIES	CONDITION EXISTS HERE (√)	TO CORRECT THIS, WE NEED:				
		SUPERVISOR ATTENTION (√)	MANAGEMENT ATTENTION (√)	ANALYTICAL STUDY (√)	CAPITAL INVESTMENT (√)	OTHER (for comments)
90. No formal audit program in use						
91. No guards to protect racks and columns						
92. Guided aisles without guide rail entry						
93. Loads overhanging pallet						
94. Excessive floor, rack, and structured loading						
95. Equipment operating at excessive speed						
96. Front-to-back rack members not provided						
97. MH equipment does not fit through doors						
98. No sprinklers and smoke detectors						
99. Hazardous and flammable materials not segregated and identified						
100. Lack of ventilation in battery charging area						
101. Entrances and exits not secured						
102. Waste and trash containers located near docks						
103. Inadequate number of fire extinguishers						
104. No contingency plan for fire loss						
105. Sagging load beams and bent trusses on racks						

Table 7.2 (*Continued*)
Worksheet No. 3—Sheet 7   MATERIALS HANDLING AUDIT CHECK SHEET
(*Continued*)

CONDITIONS INDICATING POSSIBLE PRODUCTIVITY IMPROVEMENT OPPORTUNITIES	CONDITION EXISTS HERE (√)	TO CORRECT THIS, WE NEED:				
		SUPERVISOR ATTENTION (√)	MANAGEMENT ATTENTION (√)	ANALYTICAL STUDY (√)	CAPITAL INVESTMENT (√)	OTHER (for comments)
106. No formal training for MH equipment operators						
107. No preventive maintenance program						
108. No equipment replacement program						
109. No dock levelers						
110. Unscheduled arrival of out-bound and in-bound carriers						
111. Decentralized receiving and shipping						
112. In-bound materials not unitized						
113. Inadequate number of dock doors						
114. Receiving numbers not pre-assigned						
115. Picking lists not printed in picking sequence						
116. Orders picked one-at-a-time						
117. Aisle lengths unplanned						
118. Excessive honeycombing in storage						
119. Poor quality pallets, not standardized						
120. Manual sortation in order accumulation						
121. Poor work-in-process control						
122. Energy inefficient lighting						
123. Lights, heaters, and fans poorly located						

Table 7.2 (*Continued*)

Worksheet No. 3—Sheet 8 MATERIALS HANDLING AUDIT CHECK SHEET (*Continued*)

CONDITIONS INDICATING POSSIBLE PRODUCTIVITY IMPROVEMENT OPPORTUNITIES	CONDITION EXISTS HERE ($\checkmark$)	TO CORRECT THIS, WE NEED:				
		SUPERVISOR ATTENTION ($\checkmark$)	MANAGEMENT ATTENTION ($\checkmark$)	ANALYTICAL STUDY ($\checkmark$)	CAPITAL INVESTMENT ($\checkmark$)	OTHER (for comments)
124. No dock enclosures						
125. Excessive heating, ventilation, and air conditioning for materials stored						
126. Poorly insulated walls and roof						
127. Poorly designed enclosures for environmentally controlled areas						
128. Lack of scheduled energy use to reduce peak loads						
129. Unclean floors						
130. Charging too frequently						
131. Other						

auditing. As can be seen, the developments are not strictly focused upon productivity; however, each clearly relates to productivity performance.

To help the reader apply the concepts presented in this chapter, there are several practical exercises following the chapter references, one involves working with an organizational system performance control system audit. Another involves design of a specific audit. These exercises can serve as a mechanism for forming closure on the concepts presented in this chapter.

REFERENCES

Apple, J. M. *Action Guide: Productivity Improvement for Profit Program*, Institute of Industrial Engineers, Norcross, Ga.

Bain, D. *The Productivity Prescription: The Manager's Guide to Improving Productivity and Profits*. New York: McGraw-Hill, 1982.

Craig, C. R., and C. R. Harris. "Total Productivity Measurement at the Firm Level." *Sloan Management Review*, Vol. 14, No. 3, 1973.

Hines, W. W. "Guidelines for Implementing Productivity Measurement." *Industrial Engineering*, Vol. 8, No. 6, 1976.

Kendrick, J. W. (in collaboration with the American Productivity Center). *Improving Company Productivity: Handbook with Case Studies*. Baltimore: The Johns Hopkins University Press, 1984.

Kendrick, J. W., and D. Creamer. *Measuring Company Productivity: Handbook in the Case Studies*. Studies in Business Economics, No. 89, National Industrial Conference Board, New York, 1965.

Mali, P. *Improving Total Productivity: MBO Strategies for Business, Government, and Not-for-Profit Organizations*. New York: John Wiley and Sons, 1978.

Patton, J. A. *Patton's Complete Guide to Productivity Improvement*. New York: AMACOM, 1982.

R & D Productivity, 2nd ed. Hughes Aircraft Co., Culver City, Calif., 1978.

Sumanth, D. J., and M. Z. Hassan. "Productivity Measurement in Manufacturing Companies by Using a Product-Oriented Total Productivity Model," Institute of Industrial Engineers Conference *Proceedings*, Norcross, Ga., 1980.

Sumanth, D. J. *Productivity Engineering and Management*. New York: McGraw-Hill, 1984.

Turner, W. C., and B. W. Tompkins. "Productivity Measures for Energy Management," Institute of Industrial Engineers Conference *Proceedings*, Norcross, Ga., 1982.

White, J. A. "Yale Management Guide to Productivity." Industrial Truck Division, Eaton Corporation, 1979.

QUESTIONS AND APPLICATIONS

1.(a) Working alone, compile a *comprehensive* list of measurement and evaluation techniques that can be applied at the individual, group, plant, function, or company level to assess performance. Then, by brainstorming and comparing techniques with the class, compile a master list.

(b) Refine the list of measurement techniques identified in part (a) by assessing how the techniques should be developed and implemented (list the steps, the do's and don'ts, the necessary preconditions, and so forth). Identify where and when the technique is most appropriate and applicable. Also list the costs, the risks, the benefits, and the training and development requirements.

(c) Which of the techniques listed measure and evaluate productivity? Which performance criteria do the techniques address either directly or indirectly?

2. Interview five managers. Have them list the techniques their organizations use to measure and evaluate performance at the following levels;
   (a) The individual level
   (b) The cost center level
   (c) The department level
   (d) The function level
   (e) The plant level
   (f) The company level
Ask them to discuss the problems with their measurement and evaluation systems. Also ask them if and how they measure productivity.

3. What is a "surrogate" productivity measurement system?

4. Design and develop a generic performance audit for an organization. Then design a

performance audit for one of the following organizations:

   (a) A florist shop
   (b) A clothing store
   (c) A manufacturing plant
   (d) A clerical office
   (e) An academic department
   (f) A hotel
   (g) An organization of your choosing

5. You have been provided with an example of a functional (materials handling) performance audit or measurement and evaluation technique in Table 7.2. Using this tech-

Table 7.3 Control System Evaluation Form

ORGANIZATIONAL SYSTEM (BRIEF TITLE OR DESCRIPTION) _____

BRIEF TITLE DESCRIPTION OF CONTROL SYSTEM COMPONENT	I. FOCUS: EFFECTIVENESS	EFFICIENCY	QUALITY	PRODUCTIVITY	INNOVATION	QUALITY OF WORK LIFE	PROFITABILITY	II. CONTROL SYSTEM EFFECTIVENESS (I.E. DOES IT FACILITATE ACCOMPLISHMENT OF DESIRED OUTCOMES)	III. CONTROL SYSTEM EFFICIENCY (I.E. DOES IT PRODUCE UNINTENDED & UNDESIRABLE CONSEQUENCES THAT OUTWEIGH EFFECTIVENESS?)
								NO NOT SURE YES How?	NO NOT SURE YES Explain:
								NO NOT SURE YES How?	NO NOT SURE YES Explain:
								NO NOT SURE YES How?	NO NOT SURE YES Explain:
								NO NOT SURE YES How?	NO NOT SURE YES Explain:
								NO NOT SURE YES How?	NO NOT SURE YES Explain:
								NO NOT SURE YES How?	NO NOT SURE YES Explain:
								NO NOT SURE YES How?	NO NOT SURE YES Explain:
								NO NOT SURE YES How?	NO NOT SURE YES Explain:

nique as a guide, develop a similar procedure for one or more of the following functions:
   (a) Quality control
   (b) Engineering
   (c) Marketing
   (d) Service engineering
   (e) Production control
   (f) Accounting
6. Work through these steps individually, then share findings in groups in the class.
   (a) Select a specific "organizational system" that you would like to concentrate on. It might be a department, an organization, a division, or a function.
   (b) Identify the most important "desired outcomes" for that organizational system. (Note: Do not list more than ten. Just identify the most important or critical. These should be neither too broad and/or vague nor too narrow.)
   (c) Think about the behaviors, individual and group, that cause or allow these desired outcomes to occur or be satisfied.
   (d) Think about the "roadblocks" that might hinder or even block accomplishments of these desired outcomes.
   (e) What resources are involved in accomplishing these desired outcomes?
   (f) Now, identify specific control systems managers of this system have at their disposal to measure, evaluate, and monitor performance.
   (g) Using the form on the facing page (Table 7.3), analyze first the effectiveness of each component of the measurement and evaluation system and then the overall measurement, evaluation, and control system itself.

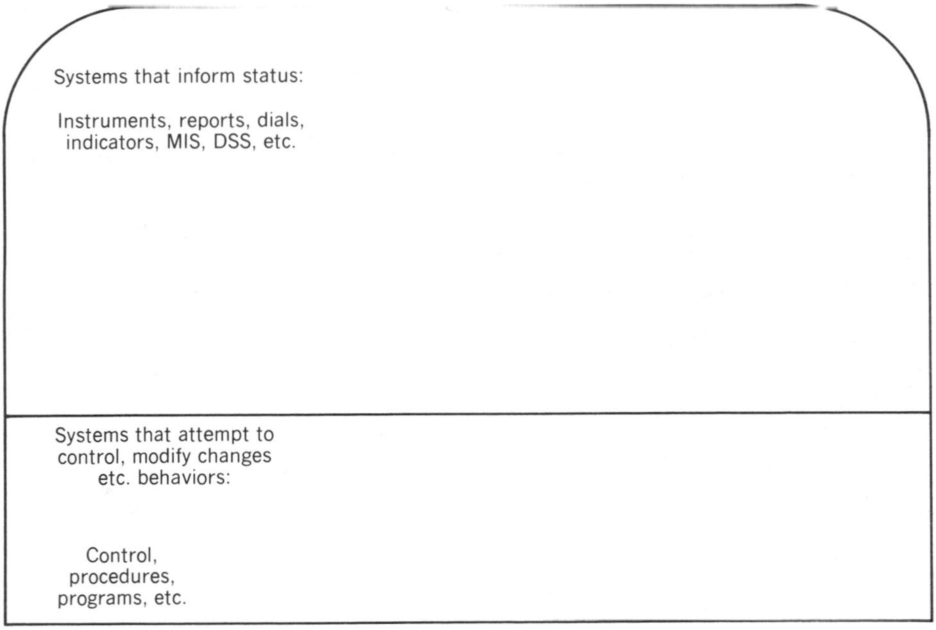

Systems that inform status:

Instruments, reports, dials,
   indicators, MIS, DSS, etc.

Systems that attempt to
control, modify changes
      etc. behaviors:

   Control,
   procedures,
programs, etc.

Figure 7.5   Control System/Control Panel Concept (Your System as it Presently Exists)

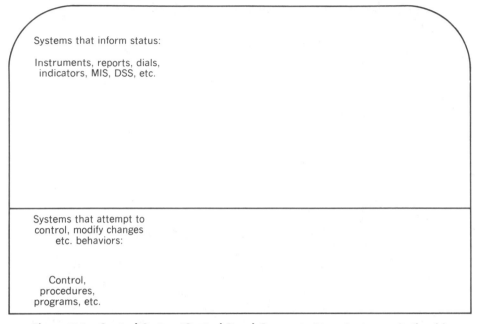

Systems that inform status:

Instruments, reports, dials,
  indicators, MIS, DSS, etc.

Systems that attempt to
control, modify changes
    etc. behaviors:

    Control,
  procedures,
programs, etc.

Figure 7.6   Control System/Control Panel Concepts (Your System as it Should Exist)

(h) Recall the aircraft instrument panel analogy discussed in Chapter 2, where the airspeed indicator, altimeter, and other instruments are the measurement and evaluation system, and the control stick, throttle, and so forth represent the control and improvement system. Using this analogy, draw a conceptual picture of what the control system actually looks like for the organizational system you are analyzing. Use Figure 7.5 to draw this. The size of the instrument and the location will reflect the relative importance given to the particular measurement technique or criterion.

(i) Draw a conceptual picture of what the measurement, evaluation, and control panel should look like in your opinion. Use Figure 7.6. How much deviation is there between the way the system looks and the way it should look?

7. Divide the class into teams. Have each team review and synthesize a different book from the references at the end of the chapter. Critique the books and compare and contrast the approach taken with that of this text.

8

SYNTHESIS AND CRITICAL REVIEW: IS PRODUCTIVITY A CRITICAL COMPONENT OF ORGANIZATIONAL SYSTEM PERFORMANCE?

HIGHLIGHTS

- Synthesis and Critical Review
- Performance Management Basics: "The Cube"
- Conclusion
- References

OBJECTIVES

- To present a capstone discussion for Part III of the book.
- To summarize, in a succinct fashion, the role of productivity in organizational system performance measurement, evaluation, and control.

Synthesis and Critical Review

Abstract

Productivity still appears to be a term and concept capturing the attention of managers in all types of organizations and at all levels within those organizations. However, the conclusion I have drawn after eight years of work in the field is that it is still a very confused and abused term.

PRODUCTIVITY ≠ PERFORMANCE

Conceptually, at least, I don't suspect there would be much argument with this statement. Yet my experience with over 500 managers and other professionals from a variety of organizations suggests that in practice, operationally, most managers evidently believe productivity = performance. The basic problem and dilemma with this is that the term performance in almost all applications is defined in a very broad, multiattribute fashion (the performance of an individual comprises a relationship between ability, effort, attitude, and so forth). In contrast, productivity is a rather well-defined and delimited term representing a relationship between the outputs produced from an organizational system over some period in time and the inputs required to produce those same outputs. Measures of performance for an organizational system include at least the following seven components or factors; (1) effectiveness, (2) efficiency, (3) quality, (4) productivity, (5) quality of work life, (6) profitability, and (7) innovation (product and process). So we see that productivity is actually simply a component of a performance measurement system. I am convinced that much of management's problems today arise from a lack of clarity and discipline in relation to what it is they actually want the system they are managing to do. This paper examines the relationship between these various components of organizational performance. Further, it discusses the role that productivity measurement can play in the design of an organization's performance control systems.

Basics

Productivity is an extremely abused and misused term and concept today. This occurs because there has been no disciplined attempt to stand up and say, "this is what it is and that's all it is." The "half-truth" rhetoric floating around about productivity is absolutely amazing. It has become such a significant buzz word that almost every discipline and profession imaginable have grabbed onto it and begun to use it in an attempt to further market and promote their own disciplinary and often myopic respective "solutions." The need for synthesis, clarification, disciplined definitions, and a generic conceptual framework is quite evident (Sink, 1980). In order to provide adequate conceptual development, the general management process, organizational systems performance measurement/control basics, and productivity basics will be briefly examined. The objective is to assist you in more clearly viewing the relationship between productivity, performance, and control systems design and development.

GENERAL MANAGEMENT PROCESS

Organizational systems performance measurement, evaluation, and control represents a critical component in the general management process. There are obviously many ways to classify control systems. They can be and are classified in terms of the resource they are intended to manage. Financial control systems (for example, accounting, comptroller, and budgets), production control systems, behavioral control systems (for example, performance appraisal, policies and procedures) are examples of this type. We could and do also classify control systems with respect to the type of organizational system performance they are attempting to control or manage. In general, there are at least seven distinct although not necessarily mutually exclusive measures of organizational systems performance. They are (1) effectiveness, (2) efficiency, (3) quality, (4) productivity, (5) quality of work life, (6) profitability, and (7) innovation. Every organization, every manager in one form or another monitors, evaluates, and controls one or more of these seven measures of organizational systems performance.

Note that productivity is only one measure of performance for a given system. It is <u>not</u>

clear it is necessarily the most important or even necessarily a critical measure of performance for all systems. The problem in designing a control system is a multiattribute or multicriteria one. No two organizational systems nor managers will likely equally weight these measures. Moreover, the characteristics of the actual development process of a control system will vary significantly from organization to organization and manager to manager. Some will be very explicit, rational, systematic, pragmatic while others may be very implicit, intuitive, subjective. Some will focus on only one measure of performance while others will truly be multiattribute in character. Some systems will be the result of an explicit, systematic, consciously thought-through design process while others will be more characterized by a "random walk" process comprising add ons to an inherited system.

In one respect, one important job of a manager is to determine (1) what the appropriate priorities or relative weights are for each performance measure; (2) how to measure, operationally, each performance measure; and, (3) how to link the measurement system to improvement (in other words, how to most effectively use the control system to cause appropriate changes or improvements). It is reasonably clear that the priorities or weightings for each of these performance criteria will vary according to several factors: (1) size of the system, (2) function or objectives of the system, (3) type of system (technology employed)—job shop, assembly line, service, and process, (4) maturity of the system in terms of management, employees, technology, organizational structure and processes, and (5) the environment (political, economic, and social) characteristics.

ORGANIZATIONAL SYSTEMS PERFORMANCE MEASUREMENT, EVALUATION, AND CONTROL BASICS

Managers at all levels and in all organizations are striving to achieve results through people. People are at the heart of successful accomplishment of objectives and of effective and efficient use of resources. Barnard (1939) states that "essential to the survival of organization is the willingness to cooperate, the ability to communicate, and the existence and acceptance of purpose." There is an appealing simplicity to his perceptions. In 1953, Peter Drucker identified seven key result areas (Figure 8.1). He pointed out that in order to be successful, in the long run, an organization would have to manage these seven key result areas.

In 1983, Peters and Waterman, in their book entitled *In Search of Excellence*, have identified attributes or criteria that "excellent" firms in the United States exhibit and manage carefully. Figure 8.2 summarizes those eight attributes.

One might conclude that the seven organizational system performance criteria presented at the outset of this section are at the heart of both Drucker's and Peters and Waterman's observations. Further, no matter what size, type, or kind the specific organizational system is, these seven performance criteria represent the basic areas that managers, supervisors, presidents, vice presidents, directors, and so forth should be focusing their management efforts on. Figure 8.3 reflect relations between these three conceptualizations of organizational system performance.

These seven performance measures or criteria are defined below.

1. *Effectiveness:* Accomplishment of purpose (Barnard, 1938). Accomplishing the "right" things:
    (a) On time (timeliness)
    (b) Right (quality)

- Customer Satisfaction
- Innovation
- Internal Productivity
- Operating Budget
- Employee Attitude & Development
- Management Development & Performance
- Social Responsibility

Figure 8.1 Generic Key Result Areas (Drucker, 1953)

Based on their study of how the best-run large corporations in America manage themselves, Peters and Waterman offer these eight lessons:

1. **Begin with a bias toward action.** The best companies encourage action over procrastination or extensive analysis. "Do it, fix it, try it," analysis doesn't impede action. Action research.
2. **Stay close to the customer.** The best companies cultivate their customers, are fanatics about quality control, and use customer suggestions for product improvement and innovation.
3. **Encourage autonomy and entrepreneurship.** At the most successful companies, all employees are encouraged to practice creativity and practical risk-taking during the execution of their jobs. Innovation champions, "make sure you generate a reasonable number of mistakes."
4. **Understand that people are responsible for productivity.** Rank and file are seen as root source of quality and productivity gain. They are treated as mature, adult people.
5. **Encourage "hands on," innovative values.** Winning companies have strong cultures. Values are maintained by personal enthusiastic attention from top management. Management stays close to the operations.
6. **Stick to the knitting.** The best companies know the ins-and-outs and singular qualities of their particular businesses and don't diversify into unfamiliar fields. Stay reasonably close to the businesses you know. Make your strengths decisive.
7. **Keep the form simple and the staffs lean.** Top staffs are kept small. The structures of the companies' organizations are kept simple and flexible. Elegant simplicity.
8. **Employ "simultaneous loose-tight properties."** The best companies maintain a paradoxical combination of centralized and decentralized properties in their organizational structures. They are tight about the things that are truly important and extremely loose about the rest. Autonomy pushed down to shop floor. Effective leadership and delegation. Tools don't substitute for thinking, intellect doesn't overpower wisdom.

Figure 8.2 Attributes of Excellent American Companies (Peters and Waterman, 1983)

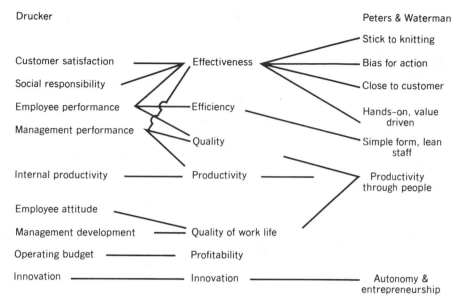

Figure 8.3   Organizational Systems Performance Criteria

(c) All the "right" things (quantity)
Where "things" are goals, objective, activities, and so forth.

2. *Efficiency:* Satisfying individual motives (Barnard, 1938), success at securing necessary personal contributions

$$\frac{\text{Resources expected to be consumed}}{\text{Resources actually consumed}}$$

3. *Quality:* Conformance to specifications (Crosby, fitness for use) where "specifications" can be identified as timeliness, various quality attributes, customer satisfaction, etc.

4. *Productivity:*

$$\frac{\text{Quantities of output from an organizational system for some period of time}}{\text{Quantities of input resources consumed by that organizational system for that same period of time}}$$

or

$$\frac{\text{Quality quantity}}{\text{Resources actually consumed}}$$

Hence, productivity is, by definition, a ratio and is a measure of effectiveness in the numerator divided by the denominator of the efficiency equation. Note that quite often productivity statistics for the nation are presented in terms of rates of change. These statistics are actually productivity indexes, where a productivity index is a particular productivity ratio for one period in time divided by that same productivity ratio for an earlier period in time. For example

Productivity Index =

$$\frac{\dfrac{\text{Quality quantity 1982}}{\text{Resources actually consumed 1982}}}{\dfrac{\text{Quality quantity 1981}}{\text{Resources actually consumed 1981}}}$$

or, rate of change of GNP to labor input, and so forth.

5. *Quality of work life:* Human beings affective response to working in and living in organizational systems. Those attributes of organizational systems that "cause" positive affective responses. Often, the focus is on ensuring the employees are "satisifed", safe, secure, and so forth.

6. *Innovation:* The creative process of adaptation of product, service, process, structure, etc. in response to internal as well as external pressures, demands, changes, needs, and so forth. The process of maintaining fitness for use from the customer's eyes.

7. *Profitability:* A measure or set of measures of the relationship between financial resources and uses for those financial resources.

$$\frac{\text{Revenues}}{\text{Costs,}}$$

Return on assets,
Return on investments, and so forth.

It is important to reinforce that whether we are speaking of a president of a firm or a first-line supervisor, this general management process just discussed still exists in a very fundamental way. Of course, the actual operational character of the management process differs drastically from top to bottom and within/across any organization itself. Allocation of specific time to the basic management functions is a primary difference. Time spent planning versus doing should be dramatically different. Many other actual behavioral changes are obvious in the character of the management process.

However, the point to be made is that the basic management functions of planning, organizing, leading, controlling, and adapting are existent at all levels of management and supervision. More specifically, every manager and supervisor has the task of controlling the systems performance they are responsible and accountable for. Further, they are responsible for and should be held accountable for performing this management function in a professional, effective, and efficient fashion. The control system they design, develop, implement, and maintain should not only monitor organizational system performance, but it should also indicate when, where, and perhaps why performance is or isn't what it should be.

Keep in mind that when we speak of control systems you likely will immediately respond by saying you have more than you need now. You can probably think of your quality-control system, your cost/budget control system, your inventory control system, and so forth. Notice that these control systems have been designed to control specific resources or results of production, and they have been designed so as to react to situations rather than manage them in a proactive sense. When we speak of control system design, do not allow yourself to develop a stereotype image of that system. Let's not fall into the same trap we did with quality control. Note that Japanese quality management control systems are designed to proactively build it right the first time. When I speak of control systems design, we infer a combination of monitoring subsystems integrated with proactive, in-process control systems that involve all levels of management, supervision, and employees in such a way as to increase the probability that behaviors will change in a positive direction. It is clear that these concepts are new and different and therefore will encounter resistance. The solution is not simple, these concepts can and will work if developed correctly.

At all levels in the organization, control system design, development, implementation, and maintenance need to start to take place. Most managers and supervisors inherit a "control system" and simply make it work or "band-aid" one together. A serious effort to rethink the purpose and logic of control systems needs to take place. Further, as mentioned, the control system is multidimensional, multiattribute, or multicriteria in character. The impact of these control systems on behaviors is a particular focus that surely needs examination in American organizations. We have a habit of "rewarding A while hoping for B". A simple analogy will help to clarify these concepts.

Imagine yourself as the designer for the control/instrument panel of a Boeing 747 jetliner. Think through some of the issues you would consider in terms of deciding:

1. What needs to be on the panel?

2. Where does each instrument, dial, knob, indicator need to be in relation to the pilot and copilot?

3. How big does each instrument need to be?

4. What will be and how can we ensure the relationship between system performance indicators on instruments and physical control by the pilot with knobs, dials, control stick, throttle, and so forth?

5. How can we design a control system for such a complex system and not overload the pilot (manager)?

6. What control aspects can we automate so as to relieve the pilot (manager) from the routine decisions?

Figure 8.4 illustrates a control panel layout for an aircraft. What do your control systems look like? Are they designed with as much care and interest? Is the performance of your organizational systems any less important than the performance of a single engine, general aviation plane? . . . than a commercial jet?

An airplane has categories of performance measures or criteria—navigational, communication, engine performance, aircraft attitude, and, of course, control response. Your organizational systems also have categories of performance criteria—effectiveness, efficiency, quality, productivity, quality of work life, innovation, profitability. Many of your organizations control systems look like Figure 8.5.

I don't presume to know what the optimum layout configuration is for each of your organizations or even each of your organizational systems. It is clear that no two managers will or necessarily should equally weight the seven performance criteria. However, I strongly suspect that in many of your organizations and in most of your organizational systems that the basic configuration (location and weighting) is not satisfactory. For example, conceptually at least, I feel a typical manufacturing control system should look something like Figure 8.6.

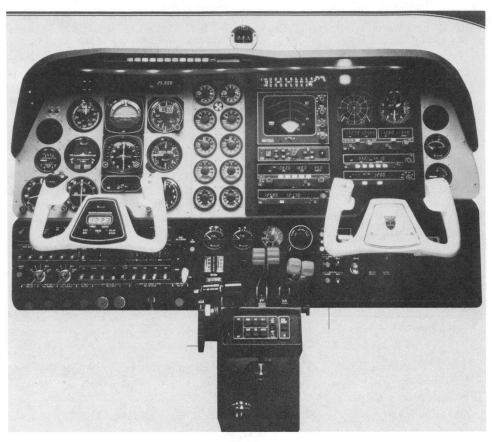

Figure 8.4 Aircraft Measurement, Evaluation and Control System

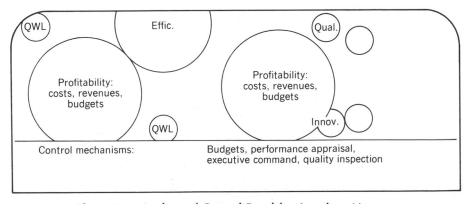

Figure 8.5 Analogy of Control Panel for American Manager

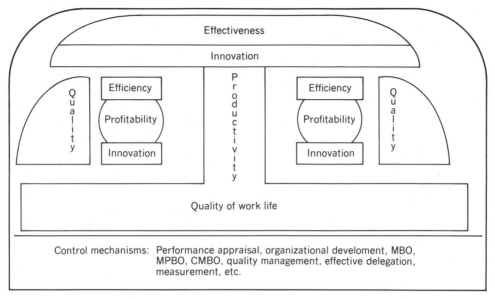

Figure 8.6   Analogy of Prescriptive Control Panel for American Manager

The differences between Figures 8.5 and 8.6 are somewhat obvious. The control panel in Figure 8.6 has more on it, makes better use of the entire panel space, should of course not overload the manager but should force him or her to be accountable for a broader range of performance criteria, and expands on the set of control mechanisms at the manager's disposal. In particular, you will notice that productivity takes a prominent location on the panel. I am convinced as a result of my interaction with over 1500 managers in the last three years that productivity, strictly defined, is absent from at least 95 percent of the control systems in American organizations.

To this point in the paper, the focus has been on clarifying basic concepts associated with the role of control in the general management process, specific components of organizational system performance, and basic productivity definitions. In conclusion, this paper will address the role that productivity measurement might play in the performance management process.

THE ROLE OF PRODUCTIVITY MEASUREMENT AND EVALUATION

Productivity measurement, as we have seen, is a reasonably simple process conceptually. Outputs from a given organizational system are placed in the numerator of an equation. Inputs from that same organizational system and for a matching period of time are placed in the denominator. The result is a ratio.

Prior to our discussion of specific productivity measurement techniques, there are several questions that need to be raised.

1. Why is the productivity criteria in organizational systems performance measurement and evaluation so important?

2. Is productivity measurement a necessary precondition for productivity improvement? Or, can we improve productivity without really measuring it?

3. If we are effective and efficient aren't we necessarily productive?

4. The bottom line is profits. How does productivity and profitability relate?

There is growing suspicion that productivity is in fact a neglected measure or criterion of performance. There is some evidence to suggest that American management has developed a concept of control systems that is too short term, profits oriented and too price driven. A hypothesis arising from these suspicions and evidence is that for many organizational systems, productivity measurement might significantly enhance management's ability to manage the relationship between outputs and resource inputs. So, what is being suggested in this presentation is that you consider the role that productivity might play in management of the performance of the organizational system you are responsible for. **Would the productivity relationships give you new insights into the performance of your organization? Could these productivity relationships help you to see areas where management intervention is most needed? Might these relationships improve the diagnostic capabilities of management?**

In Part III of this book we have reviewed three basic approaches for measuring productivity. Note that all three approaches actually allow us to measure and evaluate a number of components of the performance equation. For instance, the MFPMM provides data on price recovery, profit changes, and dollar impacts of productivity changes. Indirectly it provides insights into changes in efficiency and quality. The NPMM, depending on how the task statement in the NGT is worded, will provide either just productivity measures (ratios and/or indexes) or broader measures of performance. The MCP/PMT, like the NPMM, can focus broadly on the performance issue or more narrowly on just the productivity issue.

In a recent study completed for the Department of Defense (Industrial Productivity Office) and the Army Procurement Research Office, Sink, Tuttle, DeVries, and Swaim reviewed the existing theories and techniques of productivity measurement. The study revealed four primary approaches to measuring and evaluating productivity. Sumanth (1984) reveals a number of other approaches, although many are quite theoretical and not immediately applicable in industry. Our study revealed the following four basic approaches to measuring and evaluating productivity that are discussed in the literature and in some measure practiced in the field:

1. NPMM

2. MFPMM

3. MCP/PMT (Objective Matrix)

4. Surrogate Approaches

It is clear that there is a critical relationship between the seven performance criteria identified and discussed in Chapter 2 and elsewhere. Figure 8.7 presents

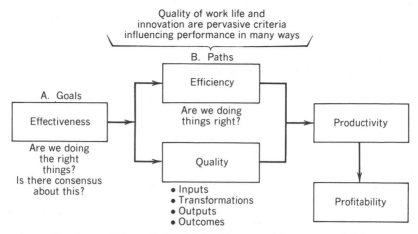

Figure 8.7   Interrelationships of Seven Performance Criteria

a conceptual picture of this interrelationship. Note that the preferred flow in terms of the sequence in which management addresses these criteria is from left to right.

If you think about the implications of this figure, I feel the concept will start to make sense. I submit that the Japanese have been driving the performance equation from left to right for the past 10 to 20 years. Further, I submit that they have had appropriate balance in terms of attention paid to the seven criteria. Think about it. When we think about the Japanese, we recognize their innovativeness, their attention paid to quality of work life, their effectiveness, their efficiency, certainly their quality, and, of course, their productivity. I assume many managers also think about their string of *successes*.

I believe many American managers and companies have been driving this performance equation from right to left. That is, we have allowed the "bean counters" to influence our decision making to the extent that we are sacrificing long-term survival for short-term efficiency and profits. Twenty years of affluence is about to kill us. Driving this equation from right to left, price-recovery, and cookbooks are easy and simple but do not represent the stuff from which those who are "in search of excellence" are made.

Performance Management Basics: "The Cube"

A recent development that borrows from the work of David Thompson (1967) is called the performance management cube. "The cube" although somewhat conceptual, has broad implications for the performance and productivity management process. The cube, of course, has three axes. Axis *A*, the *x* axis, is depicted in Figure 8.8. It represents clarity and consensus of goals, objectives, and activities for the work group, department, function, plant, firm, and so forth.

258   PART III MEASUREMENT AND EVALUATION

Uncertain,
lack of consensus

Certain,
consensus

x

Figure 8.8   Preferences Regarding Possible/Desired Outcomes

A simple scale divides the *x* axis into two parts. The left half of the axis represents unclear goals, objectives, activities, and/or lack of consensus. The right half of the axis represents clarity and consensus with respect to goals. This dimension is labeled *A* because it is the first component of the performance management process. It represents knowledge of what is right, what constitutes effectiveness, what is expected.

The second axis, axis *B* or the *y* axis, represents knowledge or belief about cause and effect. That is, it represents the extent to which we know or believe we know how to accomplish the "right" things. Figure 8.9 depicts this axis and, again, simply is divided into two components. The bottom half represents un-

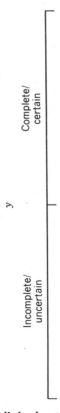

Figure 8.9   Beliefs about Cause/Effect

clear and uncertain knowledge or belief about cause and effect. The top half represents clear and certain knowledge or belief about this relationship. This dimension is labeled *B* because it represents the second component of the performance management process—knowledge or belief about how to do things right, how to be efficient, and how to maintain or achieve quality.

The third axis, axis *C* or the *z* axis, represents the extent or maturity of the development of performance measurement, evaluation, and control systems. It represents crystallization regarding the assessment criteria that will or are used to evaluate and control the performance of the system being managed. Figure 8.10 depicts this axis. The front portion of this axis represents a lack of crystallization regarding assessment criteria. That is, we do not know or have not agreed on how to measure and evaluate our performance. The back portion of the axis represents crystallized views on assessment criteria. This dimension is labeled *C* because it represents the third issue associated with the performance management process—a clarity, a consensus, and a crystallization of how we will assess and evaluate our performance. In expectance theory terms (to be discussed in Part IV), it represents knowledge of what constitutes performance. We can, therefore, infer that this dimension is critical because it also carries with it the inference of how we will be rewarded, sanctioned, or even punished.

The three main aspects of the performance management process are, then

1. *Goals: A*—Effectiveness: What should we be doing? What must we be doing?

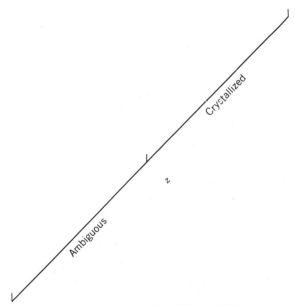

Figure 8.10   Standards of Desirability

2. *Path: B*—Efficiency, quality: How should we accomplish the "right" things? How can we accomplish the "right" things?

3. *Assessment criteria: C*—How should or can we measure, evaluate, and control our performance?

Figure 8.11 puts axes *A* and *B* together to form a four-quadrant matrix. Cell 1,1 represents a situation where there is no consensus about what to do. Even if there were, it is not clear we would know how to do it. Thompson suggests that axes *A* and *B* are the basic variables of decisions. Obviously, both dimensions or variables represent a continuum but have been dichotomized to keep the discussion simple. Thompson suggests each of the four quadrants calls for a different decision strategy. If you will, he is suggesting that performance-related decisions take on four different forms. These four types of decisions, depicted in the figure, are

Cell 2,2: Computational or programmed (DSSs, models, equations, control systems, procedures, policies)

Cell 2,1: Judgmental strategy (know what to do but not how to do it, experience, judgment, trial and error, experimentation)

Cell 1,2: Compromise strategy (know how to accomplish, have skills, abilities, and technology but don't know exactly what to do or don't have strong consensus. Suggests it is important to clarify goals, search for consensus, and so forth.)

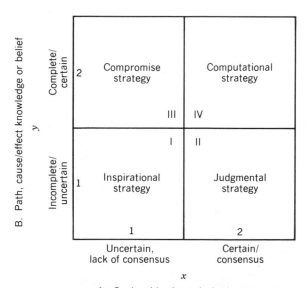

A. Goals, objectives, desired outcomes

Figure 8.11   Performance Management Decision Strategies

Cell 1,1: Inspirational strategy (don't know what to do or how to do it. Suggests effective leadership is a critical dimension.)

Figure 8.11 depicts the preferred movement from cell to cell. (Cell 1,1 → Cell 2,1 → Cell 2,2; Cell 1,2 → Cell 2,2) This prescription reinforces earlier statements regarding driving the performance equation (Figure 8.7) from left to right.

Figure 8.12 puts axes *B* and *C* together to form another four-quadrant matrix. Thompson suggests that this combination of dimensions presents four possible types of assessment situations, assuming we know what to do:

Cell 1,1: We don't know how to accomplish the "right" things or don't have the means by which to accomplish our goals and don't know or have not crystallized our thinking as to how to assess our performance.

Cell 1,2: We don't know how to accomplish our goals or don't have the means, but we have crystallized on how to assess our performance.

Cell 2,1: We know how to accomplish our goals and have the means, but we have not crystallized on how to assess our performance.

Cell 2,2: Path clear and assessment criteria crystallized.

The figure depicts selected assessment prescriptions for each of these situations.

Finally, Figure 8.13 depicts the performance management process cube. Each of the eight cells has a unique set of characteristics and hence a unique set of

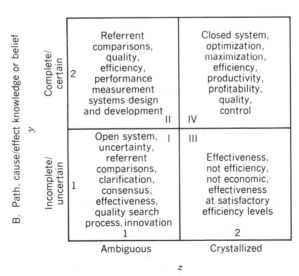

Figure 8.12   Assessment Situations

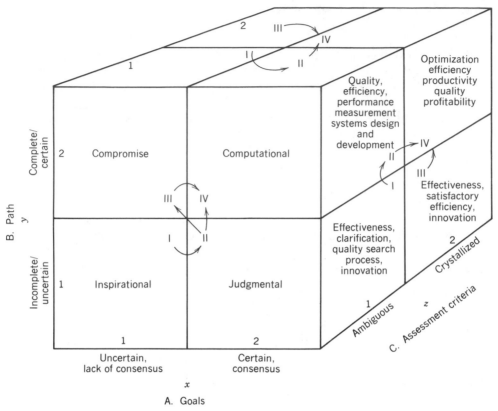

Figure 8.13   The Performance Management "Cube"

appropriate performance management strategies and behaviors relative to meas-urement, evaluation, control, and improvement. The implications for design and development of productivity measurement and improvement systems are critical. In some cells productivity measurement is a critical and appropriate component, while in other cells it is, on a relative basis, less critical. Note that an organizational system can exist in a certain cell by design, purpose, intent, default, or mistake. For example, a research and development function in an organization is very likely in Cell 2,1,2 by design and by purpose. The minute the R&D function develops a product or process (moves to Cell 2,2,2), a new project or assignment comes along. However, many R&D functions are in Cell 1,1,1 by mistake or as a result of poor management. We can and should do something about the latter but not necessarily the former.

Much more development needs to be accomplished with the cube. However, in its present state, it serves to clarify points made in Chapter 2 and to provide perspective on when productivity is an important organizational system per-formance criterion.

Conclusion

This part of the book and more specifically this chapter has examined the relationship between seven components of performance for "organizational systems." It has presented some management basics regarding (1) the general management process; (2) organizational systems performance measurement, evaluation, and control; (3) productivity in general. The question addressed in this chapter is whether productivity is a critical component of the organizational systems performance management/control process. As a partial answer to that question, these questions are raised:

1. Would productivity relationships give you new insights into the performance of your organization?

2. Could productivity relationships help you see areas where management intervention is most needed?

3. Might productivity relationships improve the diagnostic capabilities of management?

Chapters 4 through 7 have presented specific productivity measurement approaches that can be utilized to assist you in developing productivity measurement systems. The normative productivity measurement methodology (NPMM), a participative approach, the multifactor productivity/measurement model (MFPMM), the multicriteria performance/productvity measurement technique (MCP/PMT), and a few other approaches were reviewed.

Question 1: Can productivity be measured in all organizational systems?

Answer: Absolutely!

Question 2: Is it an important measure or criterion of performance in all organizational systems?

Answer: Not clear. In many organizational systems, it is more important to manage effectiveness, quality, innovation, and efficiency (for example, in service industries, in service functions, in education, and in R&D.) In others, like manufacturing, productivity is a much neglected measure of system performance.

Question 3: Is productivity a critical component of organizational systems performance?

Answer: Yes, because productivity is in reality a ratio of measures of effectiveness in the numerator and measures of efficiency in the denominator. However, how productivity is managed will vary from system to system. In some systems, as mentioned, it will be important but not emphasized as a criterion that is measured, monitored, and controlled directly. In other systems, productivity surely needs to be integrated into existing control systems. The challenge to management in the eighties and nineties will be to diagnose when to integrate productivity into control systems and to master how to do it effectively.

REFERENCES

Barnard, C. I. *The Functions of the Executive*, Harvard University Press, Cambridge, Massachusetts, 1938.

Drucker, P. F. *The Practice of Management*, Harper and Brothers, New York, 1954.

Peters, T. J. and Waterman, R. H. *In Search of Excellence*, Harper and Row Publishers, New York, 1982.

Sink, D. S. "Organizational System Performance: Is Productivity a Critical Component?", IIE Annual Conference *Proceedings*, Norcross, GA: Institute of Industrial Engineers, 1983.

Sink, D. S., Tuttle, T. T., DeVries, S. and Swaim, J., *Development of a Taxonomy of Productivity Measurement Theories and Techniques*, Oklahoma Productivity Center, Oklahoma State University, AFBRMC Contract No. F33615-83-C-5071, November, 1983.

Sumanth, D. J. *Productivity Engineering and Management*, McGraw-Hill, New York, 1984.

Szilagyi, A. D. *Management and Performance*, Scott, Foresman, and Company, Glenview, Illinois, 1981.

Thompson, J. D. *Organizations in Action*, McGraw-Hill, New York, 1967.

QUESTIONS AND APPLICATIONS

1. Refer to Figure 8.7. What are the implications of the concept of driving the performance equation left to right on the budget process? What are the implications for the strategic planning process?

2. Discuss and present arguments to defend the concept that managers will probably drive the performance equation right to left with respect to the budget process, but will also need to exercise a strategic planning and development process that drives the equation left to right.

3. Refer to Figure 8.13.

   (a) Describe the characteristics of an organizational system that would be represented by the following cells:

$$(1, 1, 1)—$$
$$(2, 1, 1)—$$
$$(2, 2, 1)—$$
$$(2, 2, 2)—$$
$$(1, 2, 1)—$$

   (b) As a manager, what would your focus of attention be in the following cells?

$$(1, 1, 1)—$$
$$(2, 1, 1)—$$
$$(2, 2, 1)—$$
$$(2, 2, 2)—$$

(c) What performance measurement and evaluation approaches and techniques would be most appropriate in the following cells?

(1, 1, 1)—
(2, 1, 1)—
(2, 2, 1)—
(2, 2, 2)—

(d) What performance control and improvement strategies, approaches, and techniques would be most appropriate in the following cells?

(1, 1, 1)—
(2, 1, 1)—
(2, 2, 1)—
(2, 2, 2)—

(e) What would be the character of strategic and action planning processes in the folowing cells?

(1, 1, 1)—
(2, 1, 1)—
(2, 2, 1)—
(2, 2, 2)—

4. Select an organizational system about which you have or can obtain some knowledge. Diagnose which cell most appropriately represents the situation this organizational system is currently in. Describe your diagnostic process and share it with others in the class.

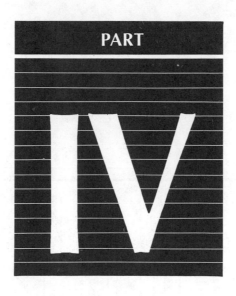

PART

IV

PRODUCTIVITY CONTROL AND IMPROVEMENT THEORIES, STRATEGIES, AND TECHNIQUES

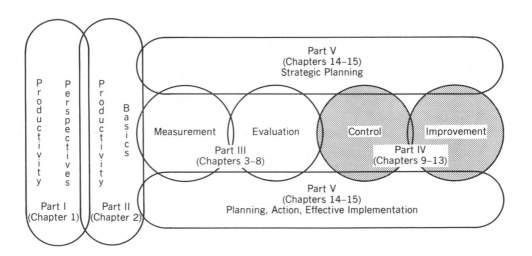

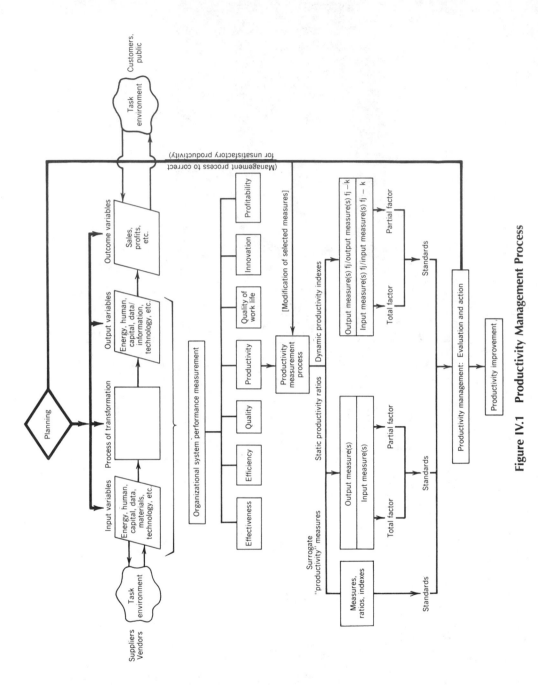

Figure IV.1   Productivity Management Process

The productivity management process involves (1) productivity measurement and evaluation; (2) productivity control and improvement planning and action; and (3) strategic productivity management planning. Figure IV.1 presents the productivity management process and highlights the areas or portions of the process that are predominant in productivity control and improvement activities. Conceptually, at least, it seems as if measurement, evaluation, control, and improvement are not necessarily mutually exclusive activities. Perhaps a Venn diagram (see Figure IV.2) can clarify the relationship among these activities. From this diagram you can see that control is partly an evaluation issue. You cannot have effective control without measurement and evaluation. Moreover, control is an improvement issue. You cannot have effective control without the ability either to maintain acceptable performance/productivity or to improve upon it.

Probably the major source of confusion regarding the term "productivity" is that it is used almost synonymously by many managers with the term "performance." Another source of confusion is the overly narrow view of productivity management as being oriented either totally toward measurement or, as is quite often the case, totally toward improvement. This book stresses repeatedly that productivity management is an *integrated* process. If one approaches productivity by emphasizing only one circle in the Venn diagram—either measurement or improvement—then in the long run the effort to manage productivity will be ineffective and inefficient.

Unfortunately, all too often, the improvement aspect of the productivity man-

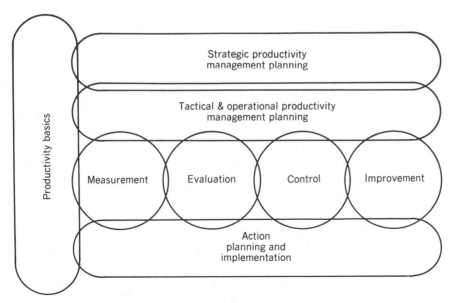

Figure IV.2 Productivity Management: Planning, Measurement, Evaluation, Control, and Improvement (Preferred Balance)

Strategic productivity management planning

Tactical and operational productivity
management planning

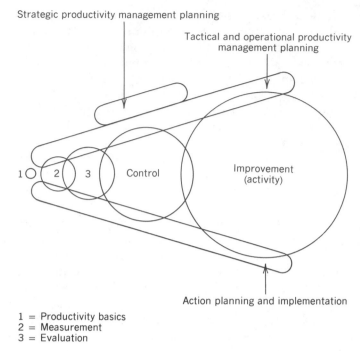

1 ◯ ⟨ 2 ⟩ ⟨ 3 ⟩ ⟨ Control ⟩   Improvement
                                   (activity)

Action planning and implementation

1 = Productivity basics
2 = Measurement
3 = Evaluation

Figure IV.3   Productivity Management: Typical Emphasis Placed by Most American Firms

agement process is overemphasized (see Figure IV.3). The reasons for this are
threefold:

1. Productivity improvement, like profits, is "bottom line."

2. Productivity improvement is action oriented, whereas measurement and evaluation are more passive; action is deferred until knowledge is obtained.

3. Productivity improvement literature outnumbers productivity measurement
literature by at least four to one.

Whereas productivity improvement is important, what is needed is better balance in a productivity management approach. There exists no conceptual, logical, comprehensive framework to enable managers to systematically think through ways to improve productivity in their particular organizational system. The literature that exists is fragmented and is based on localized instances of success. As a result, most productivity improvement efforts are piecemeal and poorly integrated, hence ineffective.

A firm premise developed throughout this book is that you cannot improve something you cannot measure and that you cannot measure something not defined. If one is to discuss and approach such a broad topic as productivity improvement, it will be necessary to develop definitions for the concept. With this in mind, we will now attempt to examine productivity improvement.

CHAPTER

9

PRODUCTIVITY CONTROL AND IMPROVEMENT: A TAXONOMY OR CONCEPTUAL FRAMEWORK

HIGHLIGHTS

- Introduction to Productivity Control and Improvement
- Productivity Control and Improvement Taxonomy
- Using the Taxonomy

OBJECTIVES

- To introduce the concept of productivity control and improvement.
- To present and discuss a productivity control and improvement taxonomy that may assist the reader in structuring thinking about the control and improvement process.
- To discuss how the taxonomy might be utilized.

Introduction to Productivity Control and Improvement

Productivity improvement in any organizational system can come about in a variety of ways. It is clear by now that productivity, as strictly defined, will improve and will be indicated through one of the five basic ratio changes depicted in Figure 9.1. The question facing managers is, of course, how to cause these various ratio changes to develop.

If one thinks carefully and systematically about these ratio types, it is possible to begin to identify strategies and techniques that apply to each ratio change. Let us first ask how output can increase. A cause-and-effect diagram with an increase in output as the effect is depicted in Figure 9.2. Note that any intervention that causes output to increase with either no or a smaller relative increase

271

$$\frac{O\uparrow}{I\downarrow}$$ Best of all worlds, output increases while input decreases.

$$\frac{O\blacktriangle}{I\uparrow}$$ Output increases at a faster rate than do inputs.

$$\frac{O\ \text{n.c.}}{I\downarrow}$$ Output remains constant while inputs decrease (for example, cost reduction programs).

$$\frac{O\uparrow}{I\ \text{n.c.}}$$ Output increases with no increase in inputs.

$$\frac{O\downarrow}{I\blacktriangledown}$$ Output decreases at a slower rate than do inputs.

Figure 9.1   Productivity Will Improve If One or More of These Relationships Is "Caused"

in inputs or costs will cause productivity to increase. As you can see from this simple cause-and-effect diagram, productivity improvement can come about in many ways.

Conversely, one can ask how inputs might be caused to drop. A cause-and-effect diagram with a decrease in inputs as the effect is depicted in Figure 9.3. Again, one can easily imagine and see that there are many ways or mechanisms for an organizational system to reduce its inputs while holding outputs constant, increasing outputs, or allowing outputs to decrease at a slower rate.

Because of the variety of options available, choosing actions that must be taken for one or more or all of these ratios to occur is typically frustrating. After countless discussions with managers from many types and sizes of organizations

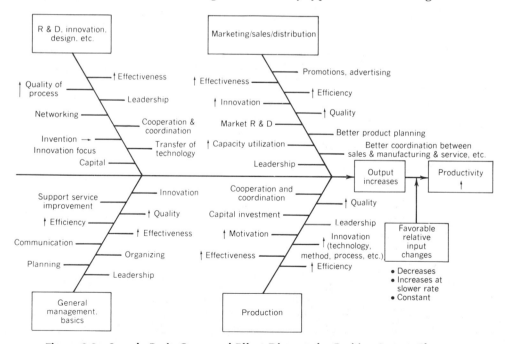

Figure 9.2   Sample Basic Cause-and-Effect Diagram for Positive Output Changes

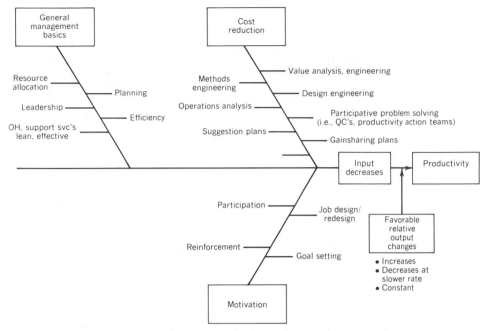

Figure 9.3   Sample Cause-and-Effect Diagram for Input Changes

and organizational systems, I have concluded that what is missing in our understanding of productivity improvement is a conceptual framework from which to develop a productivity improvement strategy. What is needed is some mechanism for

1. Identifying and categorizing available productivity improvement techniques
2. Relating these techniques to productivity improvement needs of a specific organizational system and unit of analysis
3. Relating this technique to implementation strategies and techniques

If management had a reasonably effective refined productivity measurement and evaluation system or even a general system, the ability to assess leverage points and critical areas for productivity improvement would exist. However, because the measurement system often has shortcomings, the process by which most management evaluates and determines how to control and improve productivity is simply not as systematic and logical a design process as it should be.

Productivity Control and Improvement Taxonomy

In an attempt to conceptually structure a logic for the techniques presented in this chapter and to assist managers with the process of more systematically searching for and selecting productivity improvement techniques to match their needs, the following taxonomy is presented. Taxonomy is the science of classification. It incorporates or utilizes laws and principles to guide the classifying of objects. In this case, a productivity improvement technique or approach taxonomy is developed to assist understanding of the process management must go through in shifting from measurement and evaluation (Part III) to evaluation, control, and improvement (Part IV).

Figure 9.4 presents the general taxonomy. There are many factors that could be utilized as axes for this taxonomy. And, of course, the taxonomy itself could be greater than three-dimensional, but we could not depict it graphically in that case. The three dimensions utilized for this taxonomy are

1. The unit of analysis, or the size and scope of the organizational system to be examined or, in this taxonomy, the focus of the improvement effort. Note that the scope of this book encompassed the group unit of analysis up to and including the firm level. This is highlighted in the figure.

2. The intervention type. This was the most difficult axis of the taxonomy to develop as there are numerous ways to categorize improvement and intervention types. The categories utilized reflect very broad, general types of intervention approaches. For instance, many more specific approaches, such as methods engineering, production control, and management procedures, would fall under the process category. It also becomes readily apparent that these categories are not mutually exclusive. An approach such as production control could easily fall under the process category as well as the technological category. The list of categories is reasonably comprehensive, however.

3. The controllable resources, or resource focus of the improvement effort. Note that this axis consists of the four basic categories broken out in the MFPMM in addition to a category for data or information.

This three-axis taxonomy now provides a conceptual framework within which to organize the specific types and locations of various productivity improvement techniques and approaches. One might argue that this taxonomy has little value for the practitioner. This taxonomy, however, is essential for systematic research and development, basic or applied, to take place. Hence, because it could be said that a job of all managers is applied action research and development on a continual basis, this argument has less merit than one would suspect initially.

This three-axis taxonomy has 210 cells, each of which has many productivity control and improvement approaches or techniques. In order to discuss them all in one book, the scope of the taxonomy for the purpose of this chapter will have to be significantly delimited. Figure 9.4 depicts this by way of the highlighted cells in the taxonomy matrix. Note that each cell on each axis is numbered. Hence, each of the 210 cells can be identified with a three-digit number reflecting its exact position in the matrix. For instance, 7, 3, 2 would represent the international unit of analysis, political intervention type, and capital resource focus.

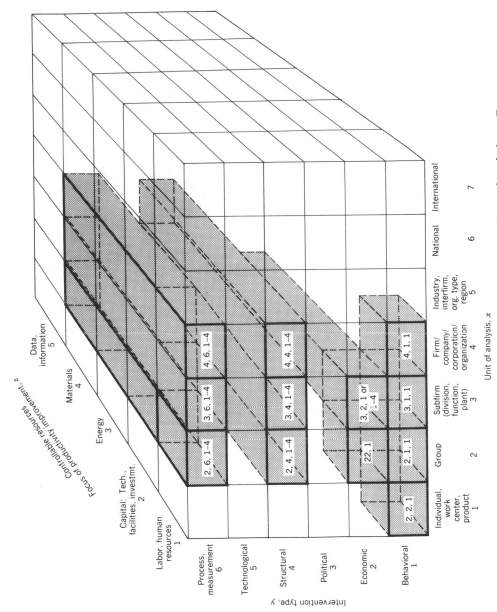

Figure 9.4 Productivity Control and Improvement Approaches and Techniques Taxonomy

Negotiation of long capital improvement loans between two governments might fall into this cell. The cells that either have been or will be addressed are the ones highlighted.

Cells 2, 6, 1–4:

- The NPMM (Chapter 4) is a group process that is measurement oriented and that focuses on potentially all resources.
- The performance/productivity action team process (Chapter 13) is a structured, participative, group process/program that is action and improvement oriented and that can and often focuses on all resources.

Cells 3, 6, 1–4:

- The NPMM developed in an aggregated and coupled fashion (along with the development of partial factor productivity measures—Chapter 4) can be a subfirm, measurement-oriented process that focuses potentially on all resources.
- The MFPMM can also be utilized at the subfirm level and is a measurement process that can focus on all resources. The implicit assumption being made by placing the measurement techniques in this productivity improvement taxonomy is that measurement itself often causes improvement.
- The performance/productivity action team process also falls potentially into these cells if it were implemented within an entire subfirm.

Cells 4, 6, 1–4:

- The MFPMM and TFPMM would fall directly into these cells. These techniques can be developed for the firm level, are measurement processes, and can focus on all resources. (TFPMM would be used if the measurement process focuses on all four resource cells; MFPMM would be used if the measurement process focuses on less than all the resources utilized by a given organizational system.) Again, the assumption is that the process of measurement in and of itself causes some improvement.
- If implemented firmwide, the performance action teams process would also fall into these cells.

Cells 2–4, 4, 1–4:

- Strategic planning for the design, development, and successful implementation of a productivity management program (Chapter 14) can be executed at the group, subfirm, or firm level; is a structural type of intervention; and, if comprehensive in character, would necessarily focus on all resources.

Cells 2, 2, 1 and 3, 2, 1 or 1–4:

- Group incentive and productivity gainsharing plans/programs, such as the Rucker, Scanlon, and Improshare plans (Chapter 12) are group or sometimes subfirm directed, economic and behavioral in type, and—depending on the plan—can focus on one or all of the resources.

Cell 1, 1, 1:

• Motivation basics (Chapter 11) are directed at the individual, are behavioral in focus, and are directed at or related primarily to the human resource.

Table 9.1 Common Productivity Improvement Approaches Classified into Taxonomy

PRODUCTIVITY CONTROL AND IMPROVEMENT TECHNIQUE	MOST APPROPRIATE TAXONOMY CELL
1. Quality circles	2, 1, 1
2. Robotics, automation, CAM	1–2, 5, 2
3. Value engineering	1, 6, 1–4
4. Energy audits	3–4, 6, 3
5. Production control	3–4, 5–6, 2 & 4
6. Quality control	1–4, 5–6, 1 & 2 & 4
7. Operations analysis, methods engineering	1, 6, 1 & 2
8. Operations research	3–4, 5, 1–4
9. Management by objectives	1, 1 & 6, 1
10. Job enrichment, design, redesign, etc.	1 & 2, 1 & 6, 1
11. Materials handling	3–4, 5, 1–4
12. Computer applications	1–4, 5, 1–4
13. Industrial engineering	1–7, 1–6, 1–4
14. Productivity measurement	2–5, 6, 1–4
15. Quality of work life programs	1–4, 1 & 6, 1
16. Orgnizational development	1–4, 1 & 3 & 4 & 6, 1
17. Traditional incentive plans/systems	1–2, 1–2 & 6, 1
18. Productivity gainsharing plans/programs	1–3, 1–2 & 6, 1–4
19. Management training	1, 1, 1
20. Employee training	1, 1, 1
21. Capital improvement incentives programs	4, 2, 2
22. Interfirm productivity measurement	5, 6, 1–4
23. Inventory control	3–4, 5–6, 4
24. Facilities planning and management	3–4, 5–6, 2
25. Cost accounting	2–4, 2 & 6, 1–4
26. Research and development	4, 5, 2–4
27. Certain regulation	4–5, 2–3 & 5, 1–4
28. Leadership	1–7, 1 & 4 & 6, 1
29. Design engineering	1, 5, 1 & 2 & 4
30. Service engineering	1, 5, 1 & 2 & 4
31. Strategic planning	2–4, 1–6, 1–4
32. Behavior modification	1 & 2, 1, 1
33. Performance appraisal and evaluation	1, 1, 1
34. Collaborative management by objectives	2, 1 & 6, 1
35. Work measurement and improvement	1, 6, 1 & 4
36. Decision support system design	1–4, 6, 5

Cells 2–4, 1, 1:

• Performance/productivity action teams and, generally, structured employee involvement, goal-setting, and problem-solving programs can be developed and implemented from the group level up to and including the firm level. Such programs are a behavioral process that focuses primarily on making the human resource more effective and efficient.

From Figure 9.4 you can see that this book has a reasonably comprehensive coverage of the productivity control and improvement taxonomy. Note, however, that within specific cells, this book typically covers only 1 or at most 2 or 3 of the 210 cells. There are numerous specific techniques that could be appropriately placed within each cell. To illustrate this, Table 9.1 lists common productivity improvement techniques and classifies them as to where they fit most appropriately in the taxonomy. A broad range of disciplines and functions are represented by this list of techniques, methods, approaches, and areas of study. For instance, design, development, and implementation of these techniques have involved and/or would involve industrial psychologists, mechanical engineers, production engineers, accountants, sociologists, staff analysts, business majors, quality-control personnel, statisticians, industrial engineers, production planners, electrical engineers, manufacturing engineers, technologists and technicians, hospital administrators, consultants, public administration majors, union officials, and labor relations specialists, among others. So you can see that productivity control and improvement is a truly interdisciplinary management activity.

Using the Taxonomy

The purpose of this taxonomy is to structure our perspective on productivity improvement approaches. It can help researchers identify particular cells that are underemphasized or for which technique effectiveness and efficiency are not what they should be. For managers, it may serve as an audit device relative to steps being taken in their particular organizational system(s) to control and improve productivity. For given organizational systems, certain cells are more critical, and the critical cells will change over time.

For example, from the boat manufacturing example and the MFPMM analysis (Chapter 5), it is clear that the priority cell is 3, 2, 4. That is, the primary leverage resource is material (fiberglass), the intervention type is resource cost reduction (economic), and the unit of analysis would be for the procurement function (subfirm). The manager of this boat manufacturing firm probably would know this intuitively and not require the taxonomy. However, in large organizations, the number and type of productivity improvement-related efforts is potentially very large. A device like the taxonomy could be constructed and completed for each major organizational system within the firm, and these completed taxonomies could then be checked for consistency, congruence, overlap, and so forth.

The taxonomies would help the manager to check to see that action planning and control and improvement strategies and tactics are consistent with the strategies developed out of measurement and evaluation. Figure 9.5 depicts this process.

The ability of the taxonomy to give perspective or structure to the broad and nebulous area of productivity improvement is perhaps its greatest strength. And that strength will now be used to develop some structure and support logic for the design of the remainder of this chapter.

A quick scanning of the literature specifically addressed at productivity improvement would reveal that most of it deals with the following subjects:

1. Automation, robotics, CAD/CAM, and capital investment in general

2. Human resource approaches: incentive systems, quality of work life, job design, quality circles, and other general management-related approaches

Perhaps this supports John Naisbitt's megatrend, called "Forced Technology—High Tech/High Touch" (*Megatrends*, 1983, Chapter 2). Naisbitt suggests that whenever new technology is introduced into society there must be a counterbalancing human response. For example

> The introduction of the high technology of word processors into our offices has led to revival of hand-written notes and letters.

and

> High-tech robots and high-touch quality circles are moving into our factories at the same time—and the more robots, the more circles.

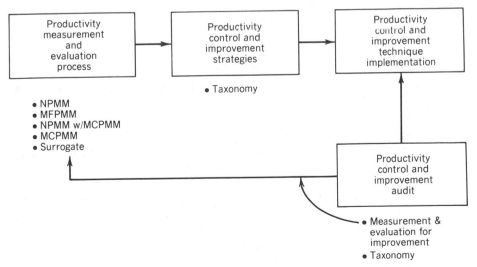

Figure 9.5   Productivity Control and Improvement Planning Process

So it seems that accompanying the increased attention being paid to automation is a counterbalancing effort on the human side. Interestingly, the automation, high-tech side is being driven by scientific and engineering advances for which there is so much momentum that they would be extremely difficult to slow down. The human resource focus, on the other hand, is being driven by the responses of humans in the affected systems. The technological advances, at this point, seem to be almost driving themselves whereas the human resource focus is perhaps being driven by reaction to the alienation that high technology can often induce.

The technological challenges facing industry in the coming decades will not be easy to surmount. Yet technology will still have an edge over the human resource challenges in that the scientific principles underlying many of the potential advances are well understood. For instance, we know that the future of computer-aided manufacturing lies in linking or integrating CAD with CAM and in tying together all the "islands of automation." The course is reasonably clear; it is basically just a matter of design, development, patience, and persistence. However, when it comes to the "high-touch," human resource-management questions, principles, techniques, and strategies for the eighties and nineties, competence is nowhere near the stage of development as it is in the high-tech areas. In short, in the coming years, our ability to manage the high-touch human resource issues will lag behind our ability to manage the high-tech areas. This lag will constrain continued progress into what Toffler calls "the Third Wave" and what Naisbitt calls "the Information Society." This is interesting because we have moved and will continue to move from labor-intensive industries to capital-, material-, and energy-intensive industries. Labor for many organizations will become a smaller percentage of the total costs. Yet labor has been and probably will continue to be a major portion of costs in service-related industries unless those services can be delivered via technology. In manufacturing, in particular, human resources will be a relatively small percentage of total costs (10–20 percent) and yet will be a constraining factor relative to technological progress in the eighties and nineties.

For example, in the aerospace industry, it is estimated that technology represents about 60 percent of the potential productivity gain; capital represents approximately 25 percent; and labor represents about 15 percent. However, without the abilities, cooperation, leadership, willingness, and so forth of *people*, the technological and capital-related productivity gains cannot be realized.

Figure 9.4 provides a structural framework for the task of productivity control and improvement, as well as a structural framework for Part IV. It is impossible for a text of this scope to be comprehensive in its treatment of productivity control and improvement approaches and techniques. The topics addressed in Part IV are critical relative to the successful, effective implementation of specific productivity improvement tactics such as capital investment, increased R&D, automation, computer-assisted decision making, and office automation.

The issue of effective delegation and decentralization of authority, responsibility, and accountability is central to change and innovation (Chapter 10). It is essential that American managers develop an improved understanding of motivation basics (Chapters 11 and 12). Effective and efficient participative management is absolutely essential for successful productivity management. Performance Action Teams (PAT) represent a well-designed, developed, and tested participative planning and problem-solving process that works well in American organizations (Chapter 13). Part IV, chapters 9–13, represents a small but important portion of the total range of topics that should be discussed relative to productivity control and improvement.

QUESTIONS AND APPLICATIONS

1. Develop your own taxonomy for classifying and understanding productivity control and improvement. Compare and contrast your taxonomy with the one presented in this chapter.

2. Identify some productivity control and improvement techniques you are aware of that are not listed in Table 9.1. Classify these techniques in the taxonomy presented in Figure 9.4.

3. Classify the following techniques using the productivity improvement taxonomy from Figure 9.6 as depicted below:
   (a) Organizational development
   (b) Methods engineering
   (c) Predetermined time systems
   (d) Standard data systems
   (e) Performance appraisal

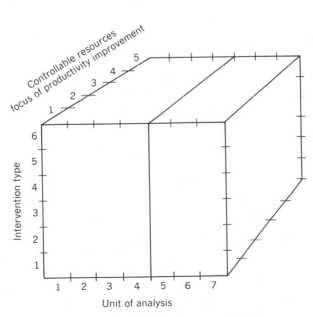

Figure 9.6   Productivity Improvement Taxonomy

(f) Selection and placement procedures
(g) Job analysis
(h) Automated storage and retrieval systems
(i) Linear programming
(j) Inventory control
(k) Productivity gainsharing plans (Improshare)
(l) Strategic planning
(m) Labor-management participation programs
(n) Collective bargaining
(o) Grievances
(p) Queuing theory
(q) Quality control
(r) Wage and salary administration
(s) Expectancy theory
(t) MBO
(u) CAD/CAM
(v) Robotics
(w) Personal computers
(x) Software/DSS
(z) Management/leadership

4. Critique the productivity improvement taxonomy. Are there variables you would have used for axes that are not taken into account? Develop a taxonomy for productivity measurement. Avoid preconceived notions of what a taxonomy looks like.

CHAPTER

10

Productivity Control and Improvement: Selected Techniques and Approaches

HIGHLIGHTS

- Introduction to Selected Techniques and Approaches
- Delegation and Decentralization
    Evolution of Management Thought
    Theory X and Theory Y
    Theory Z'
    Theories X_1, X_2, Y_1, Y_2
    Other management Style and Organizational Style Typologies Theory Z
- Conclusion

OBJECTIVES

- To introduce and outline the contents of Chapters 10, 11, 12, and 13.
- To expose the reader to the basic issue underlying the (r)evolution we are experiencing as American organizations become more participative—that of control and effective delegation.
- To confirm that different amounts and types of control are appropriate and effective in different situations, with different employees and groups, and for different purposes.
- To encourage the reader to consider Theories X, Y, and Z when designing and developing productivity management systems and programs.

Introduction to Selected Techniques and Approaches

In structuring a presentation on human resource-related approaches and techniques to productivity improvement, it seems logical to proceed in the following fashion:

1. *Chapter 10:* Delegation and decentralization (cells 1–2, 1 and 4, 1). This is a generic discussion on theory of and trends associated with participative management.

2. *Chapter 11:* Motivation basics—motivating employees in the eighties and nineties for increased performance (cell 1, 1, 1). This is a presentation on state-of-the-art theory, technique, and application in the area of motivation.

3. *Chapter 12:* Techniques for motivating improved performance. These include gainsharing, behavior modification/goal setting, job design/redesign, and participation and employee involvement.

4. *Chapter 13:* Performance/productivity action teams (cells 2, 1, 1–5). These teams are a uniquely American employee involvement process that is implemented top down in an organization and that is equally effective in manufacturing-related or service-type organizational systems.

This approach and presentation provides the reader with background and process for designing, developing, and implementing productivity improvement efforts that will be accepted and supported by the people in the focal organizational system. Peter Drucker's quote—"What we need to learn from the Japanese is not what to do, but to do it"—identifies the major roadblock to productivity improvement in this country. Americans in our type A (Ouchi, 1981) bureaucratic organizations (i.e., traditional American organizations) don't lack creativity or inventiveness; they lack the spirit, discipline, drive, and belief in innovation that are necessary for productivity improvement to occur. Innovation is applied invention—and, as Edison once said, it is 95 percent perspiration and 5 percent inspiration.

The material presented in this part on productivity control and improvement is obviously a small subset of the approaches and techniques that could be covered. However, it happens to be an important subset. The productivity control and improvement techniques presented herein represent an attempt to assure consensus about, commitment to, and motivation for effectively implementing other productivity improvement techniques. These approaches could be viewed as either stage setters or preconditions that should precede other attempts to improve productivity. Setting the stage with some or all of these techniques can make high-tech implementation more effective and efficient.

A simple equation for evaluating the effectiveness of implementing an idea for productivity improvement is (Maier, 1973)

$$Quality \times acceptance = effectiveness$$

This equation suggests that the quality of an idea, solution, design, program, or productivity improvement technique multiplied by the acceptance of that same item equals the effectiveness of implementing that item for productivity improvement. For example, if one has a high-quality idea for which there is low acceptance, the result is low implementation effectiveness. Similarly, if one has high acceptance for a low-quality idea for productivity improvement, the result again may be low implementation effectiveness. The key to productivity improvement in the future (much of which will involve technological changes) is to develop management systems and techniques that facilitate commitment on the part of the people in the organizational system to support the change.

Delegation and Decentralization

Productivity decline in the United States, productivity growth in Japan, and a search for answers and solutions has caused a rapid increase in the interest in and availability of books, articles, and even journals on Japan and Japanese management practices. William Ouchi's book, *Theory Z: How American Business Can Meet the Japanese Challenge,* has been by far the most prominent and successful to date. (The book was on the nation's best-seller list for over 15 weeks.) In 1960, Douglas McGregor presented his views on two bipolar perceptions of how people should be managed. Few organizational theories have won such widespread acceptance among managers and executives as "Theory X" and "Theory Y" presented in *The Human Side of Enterprise.* In 1967, Rensis Likert split both the X and the Y, in McGregor's original theory, into two additional subcategories. In 1973, Kurt Lewin, Ronald Lippitt, and Ralph White recognized anarchic leadership in which the dominant form of leadership was said to be "laissez-faire." Lundstedt, in comparing and contrasting these theories, used the term "Theory Z" to describe a "laissez-faire" management style. Confused?

When Douglas McGregor was president of Antioch College in Ohio, he asked Charles Kettering, then research director for General Motors, to make a contribution to the college. Kettering responded by funding a new library. When Kettering was asked to provide an inscription for outside the library, he suggested: "Enter here at your own risk." Libraries and, knowledge in general are a risky business. In reading this chapter, you may learn more than you wanted to know about Theories X, Y, and Z. So, to borrow from Charles Kettering, proceed at your own risk!

Evolution of Management Thought

There are a number of views on the evolution of thought regarding the management process. In 1961, Koontz presented his news on "the management theory jungle." Then, in 1980, Koontz "revisted" this jungle. Mintzberg presented classic views in *The Nature of Managerial Work* and "The Manager's Job: Folklore and Fact." Literally every general management text or book on organizational behavior in the past ten or more years has presented some framework depicting the evolution of thought on the management process. These conceptualizations are attempts to categorize certain consistent views on how managers do (descriptive) or should (perceptive) manage people, groups, resources, and organizations. One recent succint summary of this evolution that is worthy of attention is provided by Szilagyi (see Table 10.1). An interesting feature of this view is the environment column. Szilagyi notes, "The environment makes demands on the organization and the manager, and successful adaptation to the environment will facilitate attaining high performance levels" (Szilagyi, 1981, pp. 59–60).

The fundamental questions being addressed by many researchers during the past 20 years are: What causes what? Does the environment cause management to act in a certain way? Do academic theories cause management to act in a certain way? Do managers shape academic theories? Do managers shape or manipulate their environments?" So, as this section addresses Theories X, Y, and Z, it is important to consider the context within which these respective theories have developed.

For more background on the various "schools" of management thought, the reader is directed to any one of a number of general management texts: Szilagyi, 1981; Hellriegel and Slocum, 1978; Hicks and Gullet, 1981; Koontz, O'Donnell, and Weihrich, 1980. For specific "classic" readings from each of these schools, Matteson and Ivancevich have edited an excellent sourcebook of *Management Classics* (1981).

Theory X and Theory Y

Managers can directly or indirectly manipulate, modify, or control various aspects of behavior in the workplace, among the powers given to managers are control over

1. What tasks the subordinate does

2. How the subordinate performs tasks

3. When the subordinate performs tasks

4. When the tasks are expected to be completed

5. Resources the subordinate has at his or her disposal to accomplish the tasks

6. The work environment in which the subordinate performs the tasks

7. The instructions the subordinate receives as to the task

Table 10.1 Evolution of Thinking about Management Process

SCHOOL	SELECTED CONTRIBUTIONS	SELECTED CONTRIBUTORS	ENVIRONMENT
Classical School	Scientific management Control systems Time & motion studies Management functions Administrative theory	Gantt (1908) Taylor (1911) Gilbreth (1911) Church (1914) Fayol (1916) Mooney & Reiley (1931) Davis (1935) Urwick (1943)	Expanding size of organizations Market growth for goods & services World War I Depression Post-Industrial Revolution Decline of owner/manager Rise of professional manager
Behavioral School	Participation Motivation applications Professional managers Hawthorne studies MBO	Roethlisberger (1939) Mayo (1945) Barnard (1938) Drucker (1954) McGregor (1960)	World War II Unionization Need for trained managers Federal regulations Worker unrest
Management Science School	Operations research Simulation Game theory Decision theory Mathematical models	Churchman (1957) March & Simon (1958) Forrester (1961) Raiffa (1968)	Growth in corporation size Conglomerates Cold War Recession Military/Industrial complex
Contingency View	Dynamic environment Organic-mechanistic Technology Matrix designs Social responsibility Organizational change Information systems	Burns & Stalker (1961) Woodward (1965) Thompson (1967) Lawrence & Lorsch (1967)	Expanding economy Space race High technology products Vietnam War Civil rights International trade increase Social discontent Growth of skilled professions

SOURCE: Szilagyi, A. D., Jr., *Management and Performance*, Santa Monica, CA: Goodyear, 1981.

8. The subordinate's expectation or perception that he or she can complete the task on time and to the satisfaction of the supervisor

9. The subordinate's perception or expectation that valued rewards will follow successful performance on tasks

10. Which rewards the subordinate values

11. Which components of job-related problem solving the subordinate is allowed to become involved in

With these variables in mind, McGregor perceived that there were at least two distinct approaches a manager could take to manage his or her subordinates. These approaches are known as Theory X and Theory Y, two bipolar perceptions or beliefs about human behavior. Figure 10.1 depicts some basic assumptions implicit in each theory.

Theory X is an authoritarian view and one that "causes" direct intervention and tight control over the variables mentioned above. This theory assumes that most people require coercion, control, direction, and that they need to be threatened with punishment in order to be motivated to work hard.

Theory Y "causes" attitudes that encourage or support (1) delegating authority; (2) job enrichment; (3) improving communications; (4) participative problem solving; and (5) recognizing that people are motivated by a complex set of psychological needs and expectations.

The clarity and simplicity of McGregor's theory has caused both widespread acceptance as well as widespread criticism. Much as Herzberg's two-factor theory of motivation has been criticized as being overly simplistic, so has McGregor's

X

(1) Man inherently dislikes work and will avoid work.
(2) Because he dislikes work, man must be coerced, controlled, directed, threatened with punishment to get him to work toward achieving organizational goals.
(3) The average man prefers to be directed; he wishes to avoid responsibility, he has little ambition, he wants security.

Y

(1) Work is as natural as play.
(2) External control is not the only way for bringing about effort toward organizational goals. Man will exercise self-direction and self-control in the service of objectives to which he is committed, and commitment is a function of rewards associated with goal achievement.
(3) The average man seeks responsibility. His avoidance of it is generally a consequence of past frustration and caused by poor management from above. The average man has a high degree of imagination and ingenuity which is rarely utilized in modern industrial life, and this frustrates him; it leads him to rebel against his organization.

Figure 10.1   Implicit Theory X and Theory Y Assumptions Regarding Human Behavior in Organizations

bipolar theory of management style. The trend has been to increase the theoretical complexity of theories about types of management style.

Theory Z'

It is clear that there are more than two distinct categories of management philosophy and, therefore, more than two "management styles." It is also clear that one's management style can vary depending on the situation and other factors. Lewin, Lippitt, and White in 1939 perceived what they called "anarchic leadership," which they labeled Theory Z'. Compared to (Theory X, which is characterized by autocracy, and Theory Y, which is characterized by democracy, Theory Z' is characterized by an anarchic or "laissez-faire" style (Lundstedt, 1969). Managers adhering to this style would allow participation in problem solving to become an abdication of their management responsibility.

Tannebaum and Kahn describe three possible general distributions of control and influence reflected in the three forms of leadership and organizational style. In the autocratic form (Theory X), control and influence are concentrated usually in the upper levels of the organization. In the "laissez-faire" form (Theory Z'), they tend to be randomly distributed. And, finally, in the democratic form (Theory Y), they are somewhat uniformly distributed throughout the organization according to logical requirements set by a need for balance of effectiveness, efficiency, quality, productivity, and quality of work life.

Certain transient factors, such as day-to-day pressures, can affect the distribution of control and influences, although the actual long-term management philosophy may remain stable (Lundstedt, 1969).

Theories X_1, X_2, Y_1, Y_2

In a move to more accurately and validly depict the continuum of management attitudes and styles, Rensis Likert (1967) perceived four points as opposed to McGregor's two. Likert has basically split Theories X and Y into two additional subcategories. Likert developed an organization climate instrument that can be utilized to qualitatively and quantitatively identify the perceived "management or organizational style." Figure 10.2 depicts basic differences in the continuum presented by these various theories.

Other Management Style and Organizational Style Typologies

During the sixties, numerous theories, conceptual frameworks, and typologies emerged that relate to management style and organization style. Because all cannot be covered in detail here, a list of prominent theories, studies, conceptual frameworks, and so forth that relate or influence management attitudes and style will have to suffice. These include

1. *Environment types:* Thompson, 1967; Emery and Trist, 1965; Dill, 1958

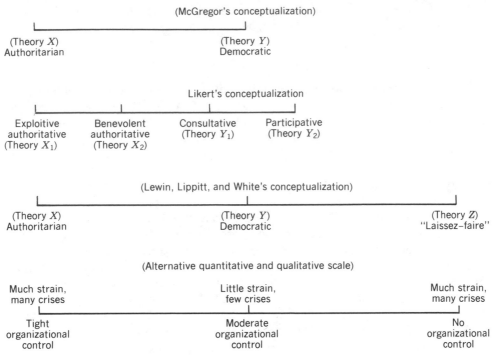

Figure 10.2   Organizational (Managerial) Style (S. Lundstedt, "Recognizing Organizational Diversity: A Problem of Finetuning", *Bulletin of Business Research*, Center for Business and Economic Research, Ohio State University, Vol. XLIV, No. 11, Nov. 1969)

2. *Organizational system types:* Burns and Stalker, 1961; Lawrence and Lorsch, 1967; Woodward, 1965; Blau and Scott, 1962

3. *Management style:* (leadership style, psychological types, problem-solving styles) Blake and Mouton, 1964; Myers, 1962, 1982; Myers and Briggs, 1962; Hellriegel and Slocum, 1975; Jung, 1923, 1971

Theory Z

Ouchi (1981) has proposed another approach to management style—Theory Z. Ouchi first identifies and discusses characteristics of the Japanese mode of management (type *J*) and then goes on to suggest a typology of American management modes, *A* and *Z*. Generally speaking, he indicates that type *A*, a typical American organization, is bureaucratic in structure, perhaps close to the left end of the scale depicted in Figure 10.2. He suggests type *A* organizations are supported by individualism and competition and often are accompanied by alienation and a lack of productivity. Type *Z* organizations, on the other hand, are the American version of the prototype Japanese organization. They are more

---

Table 10.2 Two Ideal Types of Organizational Control

TYPE A	TYPE Z
Short-term employment	Long-term employment
Individual decision making	Collective decision making
Individual responsibility	Individual responsibility
Frequent evaluation and promotion	Infrequent evaluation and promotion
Explicit, formalized evaluation	Implicit, informal evaluation
Specialized career paths	Nonspecialized career paths
Segmented concern for people	Wholistic concern for people

SOURCE: Ouchi and Johnson, 1978.

organic, adaptive, cooperative, and productive. Characteristics of both kinds of organizations are depicted in Table 10.2.

> Theory Z approach to management . . . quite simply suggests that involved workers are the key to increased productivity.

Theory Z suggests that better coordination of individual efforts performed under a philosophy of "trust, subtlety, and intimacy" results in higher productivity. Trust, subtlety, and intimacy are three important components of the culture of type Z, where subtlety is interpreted as acts that are sensitive to human sentiments rather than to rules, procedures, bureaucracy, single-minded focus on efficiency, and so forth.

Specifically, Ouchi's Theory Z maps reasonably well to our discussion of Theories X and Y except that his approach is more applications oriented and broader in concept and perspective. That is, Ouchi discusses the structure and culture of an organization as well as management attitude, philosophy, and style. In this respect, Ouchi's theory is more closely comparable to Likert's recent works (1967).

In his conclusion, Ouchi alludes to the notion that American organizations are in the midst of internal social structure change. The forces causing this change are basically the underlying properties of the individuals that make up these organizations. We all, for the most part, have been sensitized to the dynamic nature of needs, values, beliefs, and attitudes. The last two decades in the United States have heightened our awareness of the effects of change. A series of excellent books have explicated the factors at work (Drucker, 1980; Toffler, 1980; Naisbitt, 1983). Ouchi suggests the type Z organization, the industrial clan, approximates an organizational form that creates a change in internal social structure that simultaneously satisfies competitive needs for a new, more fully integrated form, as well as the needs of individual employees for the satisfaction of their individual self-interest.

Theory Z in the end, as described by Ouchi, is conceptually, at least, a much different phenomenon than McGregor addressed. McGregor was interested in explaining differences in management style. An organization filled with managers touting Theory X assumptions about human behavior would probably create a Theory X organization in structure and process. Similarly, an organization filled with Theory Y managers would likely be a Theory Y organization. And, of course, in reality we are all keenly aware that top management's views on control and influence on human behavior often are highly influential. Ouchi never really speaks about a manager with Theory Z assumptions about human behavior. Ouchi's conceptualization is at the organizational level while McGregor's is at the individual level. Conceptually, disregarding the focus of the theories, they both address how humans in organizations are controlled and influenced. In this respect Theory Z is a more refined and mature Theory Y.

Conclusion

Theory X and Theory Y are bipolar sets of assumptions managers hold about human behavior and, hence, about how human behavior can or should be controlled, influenced, or, in general, managed. Theory Z is a management mode in an American organization that is prototypical of a Japanese organization. Theory Z is a loosely structured description of an organizational type.

Figure 10.3 summarizes Theories X, Y, and Z. The horizontal axis of this matrix represents management control. Keep in mind that with respect to management style and its influence on organizational system performance, we are more concerned about the employees' perception of style as it relates to control than with the managers' perception. The focus of any analysis of leadership or management style must be on explicit behaviors and perceptions of those behaviors.

The vertical axis of this matrix represents factors that can be or are controlled by management. The question the matrix addresses is how are they controlled. Does management delegate very little and stay on top of things every minute? Does management share decisions and allow substantive participation on a broad range of issues regarding the performance of the organizational system? Or does management let things happen without intervening? Most managers with much experience have probably worked for or with managers that could be placed across the spectrum of control behaviors.

In theory and probably in practice, effective leaders and managers are able to adapt their style to fit the (1) situation; (2) subordinate; and (3) superior's type. These moderating variables appear in the center of the matrix and represent factors or contingency variables that probably should affect what position a manager operates from on the horizontal axis. For instance, consider an emergency situation or even a decision made under time constraints. It may be and often is quite appropriate, effective, and, hopefully, successful for a leader or manager to tightly control (1) what tasks are done; (2) who performs the tasks; (3) how the tasks are performed (in an emergency response, an organization's extensive training substitutes for leadership for many of these factors); and (4) resources allocated to tasks. The point is, under certain circumstances and for certain situations, even a Theory Y-oriented manager may have to act and behave as a Theory X type.

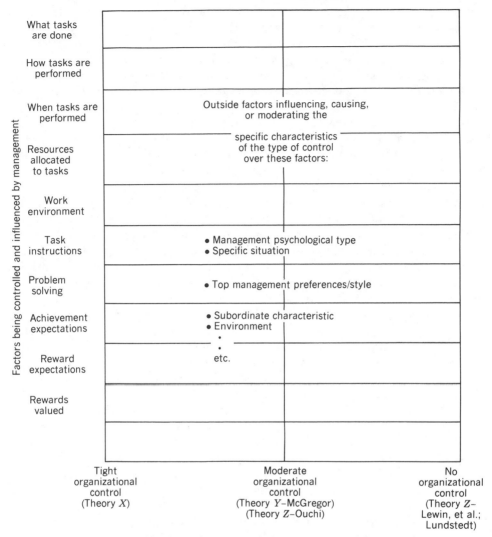

Figure 10.3   The ABC's of Theories X, Y, and Z

What we have seen in the United States in the past 20 years is a cultural leadership pattern. Early leadership research focused on identifying common traits of successful leaders. Similarly, American managers growing up in the forties and fifties developed an image of good leadership traits. Overly simplified, that image was one of a John Wayne style of leadership—decisive, self-confident, strong, independent, nonconsultative, and so forth. Ouchi's observation of the characteristics of type A organizations is actually a reflection of the perceptions of managers in those organizations as to what constitutes good leadership. Morris Massey, a (popular) speaker on the topic of "You are what you were when . . . ," has suggested that what we are is what we were when we went through our formative stages. We are a product of the environment that surrounded us as we developed our images of ourselves, others, our culture, our nation, our families, and, most relevantly, our organizations. Part of that environment consisted of observations of leaders in action. Television and movies began to play an increasingly important role in shaping America's images of life.

What "to-be" managers saw during that era were examples of Theory X-style leadership that worked. Note also that Theory X is much more in line with common stereotypical behavior for males. The leaders of the seventies and eighties grew up in an era where Theory X behaviors were in vogue, successful, and probably, more often than not, appropriate.

However, in the sixties and seventies, our nation changed drastically. So many things changed that it is hard to list them all. Geopolitical balance, values and attitudes, levels of affluence or standard of living, technology, information accessibility via the television and radio, economic expansion, and decline in the influence of the family are all significant changes that took place in about ten years in this country. A new image of the successful leader appeared, the "antihero." In fact, what happened is that there actually were either no images of leadership during the sixties and seventies or too many images—a state of affairs that paralyzed most managers. The "if it feels good, do it" attitude created a tendency for managers to adopt a Theory Z style of leadership. And, in general, the predominant leadership patterns during the sixties and seventies were X and Z. There were those managers who staunchly held their ground and perhaps became even more autocratic. There were also others who sensed a need for change but who overreacted and became Theory Z or "laissez-faire" managers.

The social science literature of the sixties on participative management, decision making, and so forth, facilitated a shift from X to Z in that it created movement and momentum in that direction (from left to right on the horizontal axis of Figure 10.3). Operationalizing Theory Y behaviors is not easy, and many managers misinterpreted or misapplied Theory Z thinking they were in fact exhibiting Theory Y behaviors.

What we are seeing in the eighties is exciting. We are beginning to experience increasing management awareness and maturity in terms of what has been termed contingency or situational management and leadership. This increasing awareness and maturity will enhance the potential for American organizations to adapt successfully to challenges they will face in the eighties and nineties.

The basic functions of management are planning, organizing, leading, controlling, and adapting. Recent emphasis in research and literature about organizational behavior has focused on "contingency" approaches to management—that is, having the flexibility and skill to adapt to changing situations in an appropriate way. McGregor sensitized us to the fact that underlying assumptions we have about human behavior may in fact strongly shape our management style. Massey has, in recent years, sensitized us to the fact that our style, our attitude, and our value system, are products of a formative period. (What you are now is what you were when.) And Jung, Meyers, and others have sensitized us to the fact that there may well be basic psychological types that predispose management attitudes and styles. The challenge facing managers is to know when to change style and to develop skills for adapting style. Someone once said we are much more likely to act our way into a new way of thinking than to think our way into a new way of acting. Let's hope we can all reverse this pattern!

REFERENCES

Blake, R. R., and J. S. Mouton. *Productivity: The Human Side*. New York: AMACOM, 1981.

Blau, P. M., and W. R. Scott. *Formal Organizations*. San Francisco: Chandler, 1962.

Burns, T., and G. Stalker. *The Management of Innovation*. London: Tavistock, 1961.

Dill, W. R. "Environment as an Influence on Managerial Autonomy." *Administrative Science Quarterly*, Vol. 2, 1958.

Drucker, P. F. *Managing in Turbulent Times*. New York: Harper and Row, 1980.

Emery F. E., and F. L. Trist. "The Causal Texture of Organizational Environments." *Human Relations*, Vol. 18, February 1965.

Hellriegel, D., and J. W. Slocum. *Management: Contingency Approaches*, 2nd ed. Reading, Mass.: Addison-Wesley, 1978.

Hellriegel, D., and J. W. Slocum. "Managerial Problem-Solving Styles." *Business Horizons*, December 1975.

Hicks, H. G., and C. R. Gullet. *Management*, 4th ed. New York: McGraw-Hill, 1981.

Jung, C. G. *Psychological Types*, New York: Harcourt Brace, 1923.

Jung, C. G. *Psychological Types*, Bollinger Series XX. The Collected Works of C.G. Jung, Vol. 6, Princeton, N.J.: Princeton University Press, 1971.

Koontz, H. "The Management Theory Jungle." *Academy of Management Journal*, Vol. 4, No. 3, 1961.

Koontz, H. "The Management Theory Jungle Revisited." *Academy of Management Review*, Vol. 5, No. 2, 1980.

Koontz, H., C. O'Donnell, and H. Weihrich. *Management*. New York: McGraw-Hill, 1980.

Lawrence P. R., and J. W. Lorsch. "Differentiation and Integration in Complex Organizations." *Administrative Science Quarterly*, Vol. 12, June 1967.

Lewin, K., R. Lippitt, and R. W. White. "Patterns of Agressive Behavior in Experimentally Created Social Climates." *Journal of Social Psychology*, Vol. 10, 1939.

Likert, R. *The Human Organization*, New York: McGraw-Hill, 1967.

Lunstedt, S. "Recognizing Organizational Diversity: A Problem of Finetuning." *Bulletin of Business Research*. Center for Business and Economic Research, The Ohio State University, Vol. XLIV, No. 11, 1969.

Maier, N. R. F. *Psychology in Industrial Organizations*, 4th ed. Boston: Houghton-Mifflin, 1973.

Matteson, M. R., and J. M. Invancevich, eds. *Management Classics*, 2nd ed. Santa Monica, Calif.: Goodyear, 1981.

McGregor, D. *The Human Side of Enterprise*. New York: McGraw-Hill, 1960.

Mintzberg, H. "The Manager's Job: Folklore and Fact." *Harvard Business Review*, Vol. 53, No. 4, 1975.

Myers, I. B. *Gifts Differing*. Palo Alto, Calif.: Consulting Psychologists Press, 1982.

Myers, I. B. *The Myers-Briggs Type Indicator.* Palo Alto, Calif.: Consulting Psychologists Press, 1962.

Myers, I. B., and K. C. Briggs. *Meyers-Briggs Type Indicator.* Princeton, N.J.: Educational Testing Service, 1962.

Naisbitt, J. *Megatrends.* New York: Warner Books, 1983.

Ouchi, W. *Theory Z.* Philippines: Addison-Wesley, 1981.

Szilagyi, A. D. *Management and Performance.* Santa Monica, Calif.: Goodyear, 1981.

Tannenbaum, A. S., and R. L. Kahn. "Organizational Control Structure: A General Descriptive Technique as Applied to Four Local Unions." *Human Relations,* Vol. 10, No. 2, 1957.

Thompson, J. D. *Organizations in Action.* New York: McGraw-Hill, 1967.

Toffler, A. *The Third Wave.* New York: Bantam, 1980.

Woodward, J. *Industrial Organization: Theory and Practice.* London: Oxford University Press, 1965.

QUESTIONS AND APPLICATIONS

1. Discuss how effective implementation can be strongly influenced by the *quality* of an idea, program, technique, or decision multiplied by the *acceptance* of same.

2. How does a manager manage change?

3. Read and discuss in class these two texts: *Implementation Strategies for Industrial Engineers,* by William T. Morris, and *The Changemasters,* by Rosabeth Moss Kanter. What do these two professionals have to say regarding the management of change and innovation in organization?

4. Summarize, in your own words, the ABC's of Theories X, Y, and Z.

5. Discuss the relationship between leadership and management style and productivity management.

6. What are the major roadblocks to productivity improvement in American organizations? How can these roadblocks be overcome? Perform a NGT session with a group using roadblocks to productivity as the task statement.

7. Discuss the impact of the moderating variables or factors on leader behaviors relative to control (see Figure 10.3). In other words, what factors, such as delegative skills, influence how management behaves in terms of control? What causes some managers to be good delegators and others to be poor delegators? What impact would this have on the likely success in an organization of techniques like quality circles or performance/productivity action teams? Are there some situations or even some organizational systems where group participative programs would not be appropriate? Discuss.

8. There are many instruments available to assist identification of psychology type, problem-solving style, managerial type, and so forth. Using a technique that is most readily accessible, divide the class into types. For example, the Meyers-Briggs Type Indicator would allow one to break the class into 16 potential groups. Or with the collecting information/processing information dimensions, the class could be broken into 4 groups— ST, SF, NF, NT. Once the instrument has been administered and the class is broken into "like types," give the following assignment:

   (a) Outline the design and execution of your group's perception of the ideal productivity management program.

(b) Outline the design and execution of your group's perception of the ideal productivity measurement and evaluation approach.

(c) Outline the design and execution of your group's perception of the ideal productivity control and improvement program.

Have each group present the highlights of its outline. Look for differences.

The instructor, armed with basic knowledge of the psychological type literature, can discuss and elaborate on hypotheses regarding the effect of psychological type/managerial style differences on management system design preferences. References on psychological types and the Myers-Briggs Type Indicator are:

"Myers-Briggs Type Indicator." Palo Alto, Calif.: Consulting Psychologists Press, 1977.
Myers, I. B. *Gifts Differing.* Palo Alto, Calif.: Consulting Psychologists Press, 1982.
Myers, I. B. *Introduction to Type.* Palo Alto, Calif.: Consulting Psychologists Press, 1981.
Hellriegel, D., and Slocum, J. W. "Managerial Problem-Solving Styles," Ohio State University, College of Administrative Science Reprint Series, Columbus, Ohio, 1976.

9. Acquaint yourself with the theory of "Situational Leadership" [see Hersey, P. and K. Blanchard. *Management of Organizational Behavior: Utilizing Human Resources* (4th ed.), Englewood Cliffs, N.J.: Prentice-Hall, 1982]. Discuss the implications of situational leadership theory on Theories X, Y, and Z (Figure 10.3) and on the appropriateness of quality circles in most American organizations. Remember that "maturity" of the follower is task specific. So, in the case of quality circles, we would be talking about the "maturity" (readiness and willingness) of American employees with respect to group problem solving. Also keep in mind that most quality circles programs in the United States are designed, developed, and implemented in less than one year. Recall the Musashi semiconductor program example. Discuss.

CHAPTER

11

Motivation Basics

OBJECTIVES

- To introduce the reader to state-of-the-art/science views on the relationship between motivation and performance.
- To differentiate between the concepts of acceptable performance levels and motivated performance levels.
- To provide a structured look at the primary factors that constitute performance and the interrelationships of those factors.
- To create an awareness of the importance of individual and group willingness, ability, and opportunity on group performance and, of course, on organizational performance.
- To expose the reader to current motivation theory. To encourage the reader (manager or prospective manager) to develop an understanding of motivation that transcends Maslow's hierarchy of needs.

Motivating Employees and Groups in the Eighties and Nineties for Increased Performance

Attention will now be shifted from management control-related behaviors to management behaviors and/or systems related to motivation. (This chapter would most appropriately be categorized in Cell 1, 1, 1 of the productivity improvement taxonomy.)

Motivating employees (blue-collar, white-collar, professional, clerical, and so forth) to assist with and be accountable for improving performance (effectiveness, efficiency, quality, innovation, productivity, quality of work life, and profitability) is and likely will be a significant challenge for most managers. Major difficulties managers have and will encounter are (1) constraints imposed on managers in terms of the type and availability of rewards or incentives that can be administered to employees; (2) continued inability on the part of managers to effectively and consistently apply and implement motivation techniques; (3) continued inability on the part of academicians to clearly and succinctly communicate and teach what they have learned about motivation, incentives, productivity, and so forth; and (4) the increasingly dynamic and complex nature of employee needs, demands, desires, and expectations of the organization and the job. Motivating employees may not be the only way to improve individual, group, and even organizational performance; however, it probably is a necessary condition for improving performance in most organizations in the long run. This chapter examines current thinking and techniques relative to motivating employees to improve performance.

Motivation and Performance

"In most manufacturing jobs, the best worker produces two to three times as much as the worst worker" (Lawler, 1972). A similar situation probably exists in most white-collar jobs. Figure 11.1 depicts this relationship.

Within an industry, the best organization typically outperforms the worst organization by at least two to three times. For instance, in "Direct Report Studies" (1946–54), the United States Bureau of Labor Statistics revealed that within two specific industries, mixed fertilizers and melting departments for gray iron foundries, the man-hours per unit of output varied significantly across plants. Figure 11.2 reveals that the ratio of average man-hours per unit of output for the highest group to that for the lowest group was 2.7 to 3.7. A more recent

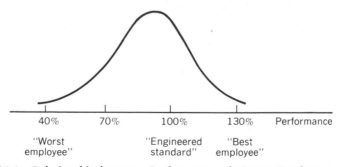

Figure 11.1   Relationship between Performance of "Worst Employee" and of "Best Employee"

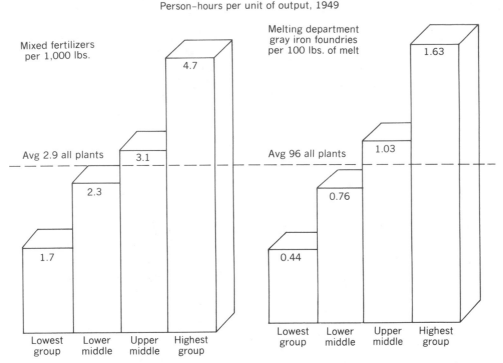

Figure 11.2 **Relationship between Performance of "Worst Organization" and of "Best Organization" within an Industry** (from Bureau of Labor Statistics, APC Productivity Perspective)

study of the baked goods industry made by the American Productivity Center (1983) reveals a similar relationship between the best performer and the worst performer in a given industry.

Managerial skill plays a large role in organizational performance. The following extract taken from *On the Art of Japanese Management* (Pascale and Athos, 1981) highlights the role of managerial skill:

In 1980, Japan's GNP was third highest in the world and, if we extrapolate current trends, it would be number one by the year 2000. A country the size of Montana, Japan has virtually no physical resources, yet it supports over 115 million people (half the population of the United States), exports $75 billion worth more goods than it imports, and has an investment rate as well as a GNP growth rate which is twice that of the United States. Japan has come to dominate in one selected industry after another—eclipsing the British in motorcycles, surpassing the Germans and the Americans in automobile production, wresting leadership from the Germans and Swiss in watches, cameras, and optical instruments, and overcoming the United States' historical dominance in business as diverse as steel, shipbuilding, pianos, zippers, and consumer electronics. Today, Japanese wages are slightly higher than those in the United States, and the cost of doing business in Japan— with imported raw materials, expensive real estate, and crowded highways—is decidedly higher. American executives complain of extra costs that stem from occupational safety regulations and pollution controls. While initially lagging, Japan's standards in these areas are now among the most stringent in the world. Some of us rationalize the disparity by emphasizing the problems stemming from the Arab oil crisis of 1974. While all other industrialized democracies have experienced inflation and a decline in productivity growth as a result of higher petroleum costs, Japan, which imports all of its oil, has maintained a very low rate of inflation, has increased productivity, and has by most accounts proven a more competitive trading partner in the past five years than ever before.

Despite the advantages of a homogeneous population, and those related to culture to be explored herein, there is no simple way to dismiss Japan's success. If anything, the extent of Japanese superiority over the United States in industrial competitiveness is underestimated. Japan is doing more than a little right. And our hypothesis is that a big part of that "something" has only a little to do with such techniques as its quality control circles and lifetime employment. In this book we will argue that a major reason for the superiority of the Japanese is their *managerial skill.*

One's first reaction to statistics about Japan's productivity and performance is often denial, and then a shrug of helplessness. The differences seem large and the task of "becoming more Japanese" seems impossible. But the striking finding of the research on which this book is based is that many of our most skilled American managers, and many of our most outstanding companies, do things that are surprisingly similar to what the Japanese do. Our problem today is that the tools are there but our "vision" is limited. A great many American managers are influenced by beliefs, assumptions, and perceptions about management that unduly constrain them

A pattern emerges—an interesting relationship between performance of the best and the worst, or between the best and the not-so-bads. As in sports, only the "acceptable" survive. Those individuals, companies, or nations that cannot perform at least at an "acceptable" level of performance will simply not be able to maintain growth or already high levels in standards of living.

From a management perspective, the critical questions are (1) How did the best become the best? (2) If we are not the best, how do we improve? (3) If we are close to or are the best, how do we maintain our levels of performance at the individual, group, organization, industry, and national levels? Different factors play various roles at each level. It is not possible to discuss all the factors and relationships at all levels here. The focus of this section is primarily at the group or individual level.

The underlying premise is that motivation is a major causal factor for individual performance. Further, motivating high performance at the individual level will have a direct and significant impact on group and even organizational productivity. Exactly what significance is attributable to this single factor is unclear. What is clear is that motivation may be necessary for high performance, but it is not alone sufficient for such performance. Other factors, such as leadership, communication, technology, innovation, capital, inflation, world economics, and political issues, certainly play a major role in determining productivity levels.

Evolution and Importance of the "Work Ethic"

Motivating employees to obtain certain levels of performance is currently a very misunderstood management function in this country. Management practice is emerging from an era (1900–50) when a manager told an employee what to do, how to do it, and when it was expected. The manager then performed the quality assurance on these assignments. (This is Typical Theory X-style management.) Generally, the employees did what they were told. During that period, the incentives were right—if an employee didn't follow, he or she was out of a job. Attitudes about following were different then.

Starting in the middle fifties, the economy in the United States improved substantially, and the United States entered an era of affluence. This era, which has leveled off at present, caused dramatic changes in work-related attitudes. Management became less disciplined as a result of "the good times." Employee expectations as to what workers should receive from their organizations and from the government soared. And employee perceptions of what they should contribute to these same organizations for commensurate rewards plummetted. In short, the old inducements/contributions ratio that Chester Barnard spoke of in 1939 increased substantially.

Then in the late sixties, national economic performance (productivity measured by GNP/labor hour) began to decline (see Chapter 1). Managers felt an uneasiness about an increasing inability to compete internationally. By this time the attitudes of the work force, and for society as a whole, were not conducive to a 1930s or 1940s management style. Young managers coming into the work force had been taught all about participative management, contingency leadership, situational leadership, quality of work life, and so forth. Their ideas on how to motivate a work force came in direct conflict with upper management's views. After all, "what you are is what you were when."

Today, the evolution of management philosophies and style is undergoing a rapid transition. Many managers are very concerned and, to some extent, confused. Widespread publicity of an apparent remarkably successful and drastically different Japanese management approach has created an incentive and drive for American managers to find out what it is all about. Combined with a productive past 20 years in the behavioral science and management science research fields, the market for continuing management education and for consulting has been favorable.

Unfortunately, the attitude of many of these managers is: "Tell me how to do it, step by step. Don't waste my time telling me how or why it works." This is likely a spinoff attitude from decision making associated with technology. Most managers cannot take the time to understand how the latest computer works. However, they should and must take the time to learn how and why the latest management science technique works. Managers must want to learn to become chefs instead of cooks with respect to behavioral science management techniques.

The eighties will be a critical period for managers. There are many uncertainties and risks, and things are changing rapidly. Further, there are so many techniques being touted as the productivity solution. These techniques are being applied with so little knowledge of if, why, or how they work that there is significant cause for concern. The eighties will require managers to reexamine how they manage the resources necessary to accomplish their goals and objectives. The need for effective strategic planning is ever more critical. Yet the importance of being able to motivate the necessary actions to bring strategic plans into existence is paramount. It will be increasingly important and perhaps more challenging for managers in the eighties to motivate high levels of performance and productivity.

Motivation Basics

Management comprises two major activities—planning and action. By action, it is implied that at higher levels of management, the delegation of responsibility and accountability for action become critical. Further, the motivation of action is often the most critical element. The general management process is depicted in Figure 11.3. In this process diagram, it is possible to view the role that motivation plays.

Motivation has been defined as a force to perform. A motive is some inner drive, impulse, or intention that causes a person to do something or act in a certain way. In general, the various views regarding motivation (and there have been many) have led to the following conclusions:

1. The analysis of motivation should concentrate on factors that arouse or energize a person's activities. These factors include needs, motives, and drives.

2. Motivation is process oriented and concerns behavioral choice, direction, goals, and the rewards perceived for performing (Szilagyi, 1981).

There have been two major thrusts in motivation research, both of which center on theoretical development. The first, content theory, focuses on the question of what arouses, energizes, or initiates behavior. The two most popular content theories are the need hierarchy theories (Maslow) and the two-factor theory (Herzberg). The second, process theory, focuses on choice behavior that can lead to desired rewards. Process theories basically state that individuals evaluate various behaviors that they perceive will lead to valued work-related outcomes. The two most popular process theories are expectancy theory (Vroom) and reinforcement theory (Skinner). (For an excellent review of these theories, see Szilagyi, 1981, Chapter 21.)

In general, the simple model shown in Figure 11.4 identifies the major components of performance. In the past, personnel psychologists have assumed

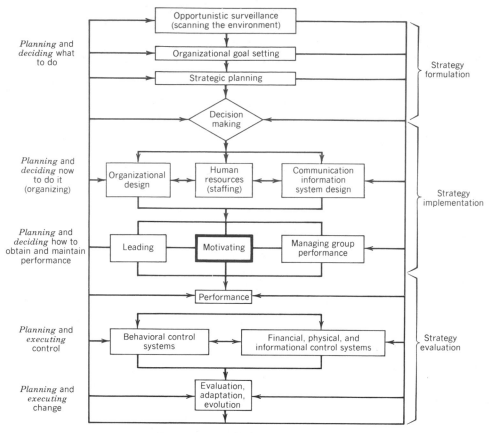

Figure 11.3   The General Management Process (Adapted from A. D. Szilagyi, *Management and Performance,* Goodyear, 1981.)

that performance is primarily a function of selection, placement, and training. Social and industrial psychologists have stressed the role of motivation in performance. A simple performance equation is:

$$\text{Performance} = f\,(\text{ability} \times \text{motivation})$$

More recently a broader model of human performance has incorporated the factors identified in Figure 11.4 (Blumberg and Pringle, 1982). Blumberg and Pringle suggest that capacity refers to the physiological and cognitive capabilities that enable an individual to perform a task effectively. In addition to ability, capacity represents knowledge, skills, intelligence, age, health, education, endurance, stamina, energy level, and motor and psychomotor skills. Willingness refers to the degree to which an individual is inclined to perform a task. This includes or represents the effects on behavior of motivation, job satisfaction,

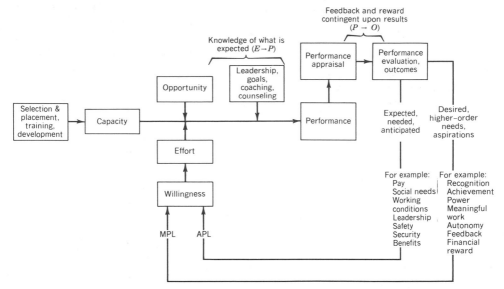

Figure 11.4   A Basic Performance Model

personality, attitudes, values, task characteristics, job involvement, need states, perceived expectation of ability to perform, and perceived expectation of rewards or outcome linkages as a result of performance. Finally, opportunity consists of those factors that moderate or influence a person's performance or task and that are beyond a person's direct control. Factors in this opportunity category would be tools, materials, supplies, working conditions, co-workers, leader behavior, policies, rules, information, time, and certain types of pay.

These three variables, capacity, opportunity, and willingness, come together in a sort of multiplicative fashion to shape or determine levels of actual performance. Of course, many other factors moderate or influence performance, such as leadership, role clarity on the part of the individual, and in particular the GOALS (goals, objectives, activities, learning, standards) or planning and action system. Note that a very capable employee (high capacity) in a very good and supportive system (high opportunity) with very low motivation (low willingness) will likely perform in a less than desirable or acceptable fashion (high, $1.0 \times$ high, $1.0 \times$ low, $.1 =$ low, .1).

Another scenario can be constructed that suggests that an individual with excellent capability, in a good and supportive system, and with high motivation but poorly set goals, objectives, and activities (due either to supervision or to the individual) would also perform at an undesirable level. Perhaps this individual does the wrong things right.

In a properly designed and executed performance management system, performance is always appraised. Performance appraisal is a process by which (1) what was agreed would be accomplished is reviewed against actual accomplishment (effectiveness); (2) the efficiency, resources actually consumed compared to resources expected to be consumed, of the individual is reviewed and discussed; (3) the quality of the accomplishment is reviewed; (4) individual innovation relative to a product or process, where important, is reviewed; (5) the individual's satisfaction with the job, tasks, assignments, organization, and so forth, is discussed (quality of work life); (6) the individual's contribution to group, department, or organization performance is discussed; and (7) goal setting for the future is accomplished. Note that no mention was made of evaluating and judging performance directly. Performance appraisal should be a performance review/discussion that leads the employee to do his or her own self-evaluation.

The next box in the model is performance evaluation. In systems with performance-based raises or merit evaluation, this is the position for "grading" performance. Note, however, that the outcomes provided to the individual as a result of and hopefully commensurate with some level of performance are not just financial. In fact, of the three basic kinds of rewards or outcomes that an organization can provide—financial, prestige (for example, office, parking space, title), and content (for example, job- or task-related assignments)—the financial category is often most constrained.

The standing disagreement between advocates of reinforcement theory (in industrial engineering, most notably Mitchell Fein) and advocates of job content theory (most notably Frederick Herzberg) centers primarily around the issue of whether money is a motivator. It appears that some forms of financial rewards are and some are not, depending on how the particular financial reward is administered. If a financial reward follows the principles of reinforcement theory, as it does in an Improshare Plan, then it can and will very likely motivate increased willingness.

In general, if the outcomes flowing to the individual as a result of the evaluation process are expected, needed, or anticipated, such as is the case with our paychecks, working conditions, leadership, security, and benefits, then we will likely see acceptable performance levels (APL) of willingness. Conceptually, APL is some level of effort close to an "engineered standard," if one existed. It is that level of effort that the organization (probably the supervisor or manager) at least subjectively views as being "acceptable."

If the outcomes flowing to the individual as a result of the evaluation process are desired, aspired to, or are what Maslow or Herzberg might have termed "higher order needs" as is often the case with recognition, achievement, the work itself, self-control or autonomy, feedback, and certain types of financial rewards (for example, gainsharing and incentives), then it is likely that motivated performance levels (MPL) of effort will be observed. Conceptually, MPL represents levels of effort above the 100 percent level for an engineered standard. Except for highly motivated individuals with a high growth need, the only way to consistently get an individual to exert MPL effort is to financially share productivity gains.

Note that APL and MPL levels of effort convert to APL and MPL levels of performance only if capacity, opportunity, and leadership (GOAL setting) are executed correctly.

Individual expectations for outcomes from the system vary significantly. There exists tremendous variance among and between individuals not only in terms of what they expect from a system in order to give APL effort, but also in terms of what they desire and what would really "turn them on" enough to provide MPL effort. One challenge facing managers in the area of motivation is to do a better job assessing and diagnosing specific expectations and desires of their employees.

This model, although very general, does assist in understanding the factors and relationships associated with employee performance. An important refinement to be made to this simple performance model is the distinction between the concepts of acceptable performance levels and motivated performance levels.

Obtaining Acceptable Performance Levels

Theories on needs, motivation, behavior, and so forth can very generally be summarized as follows:

From Figure 11.4 it can be observed that there are outcomes, outputs, and/ or conditions that most people *need* and expect from the systems/organizations/ groups within which they live and work. For instance, we all need and expect such things as security, safety, some degree of interpersonal relations, and some degree or level of working conditions. In short, we all have *aspiration levels* for these outcomes and/or conditions. When the system/organization/group provides these things at some degree or level near the aspiration level, there is a willingness to contribute to that system/organization/group at some level that is *perceived* to be *acceptable*. The degree to which the system/organization/group communicates *clear role* expectation will determine whether or not what is perceived to the individual is in fact acceptable to the system/organization/group.

Obtaining Motivated Performance Levels

People wish to obtain various higher-order desires or needs. The difficulty with satisfying these desires or needs is the amount of variance that exists from person to person. People exhibit broad differences in terms of the desire/need for achievement, power, recognition or self-esteem, meaningful work, identity with work, growth, responsibility, advancement, autonomy, feedback, role clarity and lack of ambiguity, and financial rewards. Figure 11.5 depicts the relationship between these desires and needs and/or outcomes and performance.

When a system, organization, group, or person provides outcomes that address these desires/needs, people typically respond by performing at some level in excess of what they perceive to be acceptable. It is important to point out that there are also some individuals who "reward themselves" with desired outcomes. These individuals are intrinsically motivated, or motivated from within.

Achieving a motivated performance level, then can be accomplished in two ways. It can be achieved by selecting people who are intrinsically motivated to

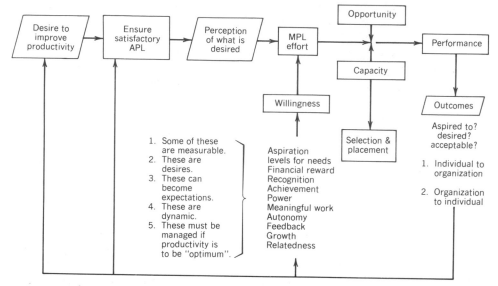

Figure 11.5   Achieving Motivated Productivity/Performance Levels

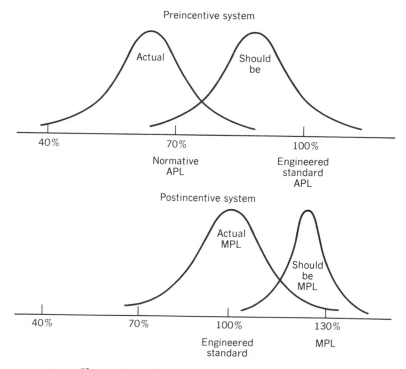

Figure 11.6   APL to MPL via Incentive Systems

perform above the level they perceive to be acceptable. Or it can be achieved by externally satisfying desires/needs. This is called extrinsic motivation.

The importance of the concepts of APL and MPL is understood when incentive system applications are examined. Incentive systems reward for improved performance. However, if the initial level of performance and any normative standards are below APL, then in essence management will be paying for something it should already be receiving. Figure 11.6 depicts this. The point is that management needs to learn how to obtain and maintain APL, which is what it is paying for, before it tries to increase the rewards in an attempt to achieve MPL.

State-of-the-Art Motivation Theory

It would appear that one of the biggest problems facing managers today with respect to motivating employees is that of outdated and overly simplistic concepts and models of motivation. In an audience of over 300 managers recently addressed on the topic of motivation, nearly all were familiar with Maslow's hierarchy of needs and Herzberg's two-factor theory. We would hope that this would be the case, for Maslow's theory is 30 years old and Herzberg's theory is at least 25. Yet less than 10 had heard of expectancy theory, or VIE (valence, instrumentality, expectancy) theory, although this theory is almost 20 years old.

Managers today can ill afford to have overly simplistic and often quite outdated thinking on the art of motivation. Motivating employees for improved performance is simply too important a management skill to let atrophy.

Expectancy Theory

In an attempt to provide additional foundation for the motivation basics just presented and in order to present at least one reasonably comprehensive, state-of-the-art theory on motivation, expectancy theory will be briefly described. As a process theory, expectancy theory attempts to focus on and provide some understanding as to why people choose particular behaviors. The theory presumes that in many work-related situations, individuals consciously evaluate alternative behaviors and then select the behavior that they believe will lead to valued work-related outcomes. This theory incorporates three main variables: (1) expectancy; (2) instrumentality; and (3) valence. Figure 11.7 depicts the basic model for this theory.

There are actually two similar expectancy models. Vroom's model, developed in the middle sixties, was the first. This model formulation is, as mentioned earlier, a behavioral choice model or a cognitive model. It assumes that cognition affects choice more than do other factors. The focus of this model is, therefore, on conscious choice.

Terminology utilized in the model includes the following:

Action: a conscious behavior controlled by the individual

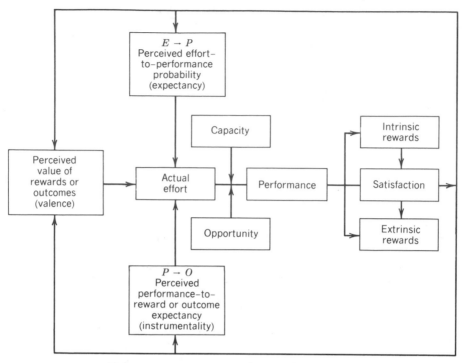

Figure 11.7   Expectancy Theory of Motivation

Outcomes: future events that may or may not be under the individual's control and for which the likelihood of occurrence is stochastic or probabilistic

Actions→outcomes expectancy: the perceived (cognitive) relationship between actions and outcomes. Expectancy can range, like probability, from 0 to 1.0. An individual can be very uncertain as to the possibility of completing a given task on time (0.0) or very certain that it can be completed (1.0). This expectancy is often labeled the $E \rightarrow P$, or effort $\rightarrow$ performance expectancy. Performance is often viewed as the primary outcome about which an employee may be developing subjective estimations.

Valence: the strength of an individual's preference for a given outcome. Each outcome an individual considers has some probable level of desirability. This level of desirability is a somewhat subjective satisfaction with the outcome(s). Valence can vary from -1.0 (highly undesirable) to $+1.0$ (highly desirable).

Outputs→outcomes expectancy, or instrumentality: the perceived relationship or correlation between two outcomes. The outcome that is perceived as perhaps causing other outcomes is termed a first-level outcome. The outcomes that might result from first-level outcomes are termed second-level outcomes. Instrumentality ranges from -1.0 (a negative relationship or correlation) to 0.0 (no correlation) to $+1.0$ (a strong positive relationship or correlation). This is often denoted as $P \rightarrow O$ or performance to outcome expectancy and, of course, instrumentality.

Motivation or force to perform is then formulated as:

$$F = \Sigma \, (E \times V_{1st})$$

or

the sum of expectancies times the first-level outcome valences
where

$$V_{1st} = \Sigma \, (I \, V_{2nd})$$

or

the valence of the first-level outcome expectancies is equal to the sum of the outcome$_{1st} \rightarrow$ outcome$_{2nd}$ instrumentalities times the respective second-level outcome valences.

The steps required to use this model in a diagnostic or analytical fashion are as follows:

1. Identify alternative actions
2. Identify perceived first-level outcomes
3. Identify perceived valences
4. Identify perceived expectancies
5. Predict an action with a higher force-to-perform number to dominate behavior

Note that needs have direct impact on valences. Hence, Maslow's and Alderfer's hierarchies of needs are reflected in this model. Applications of this model in real-world settings are as follows:

1. Diagnosis of low-motivation problems. Different levels of effort and performance are a first-level outcome. Second-level outcomes may include pay, raises, and relationship with co-workers.

2. Diagnosis of the willingness of an employee to accept a given task or assignment. This involves identification or anticipation of what must be done to ensure that an employee accepts a given task and works hard enough to perform it well.

Intrinsic motivation and an internal reward process have been built into the model also. One can treat intrinsic motivation as another second-level outcome or build it into the valence component of the equation:

$$V_{tot} = V_{int_{1st}} + V_{1st}$$

where:

$$int = intrinsic$$
$$V_{1st} = \Sigma (I \ V_{2nd})$$
$$V_{tot} = total \ valence$$

Force to perform is then:

$$F = \Sigma (E \times V_{tot})$$
$$= E \times [V_{int} + \Sigma (I \ V_{2nd})]$$

Consider the following example. An individual has been asked by his supervisor to work on a critical rush project that is very important to the company. The assignment, however, is difficult, and there is not much time to complete it. Expectancy on the part of this employee therefore will probably be low, say .20. In contrast, the valence for this outcome is probably high, say $+1.0$, because the supervisor has indicated that successful completion of the project may mean a promotion, something the employee has wanted for several years. However, the supervisor has not indicated that a promotion would be definite. Hence, $P \rightarrow O$ expectancy or instrumentality may be only 0.75. An individual's actual effort to accept and/or work hard on this project is predicted from the following equation:

$$(E \rightarrow P) \times (P \rightarrow O)(V) = \text{Predicted force to perform, or effort}$$

or

$$0.2 \times (0.75)(1.0) = 0.15$$

Note that the maximum possible value for this equation is 1.0. As a result, the supervisor might expect a motivation problem for this employee on this particular assignment.

The strength in this model is not in its exactness but in its structure. It provides a manager with a reasonably objective tool for evaluating motivation problems and for diagnosing and developing interventions. In this case, the supervisor with the right data from the employee can anticipate a potential problem and work to improve the employee's motivation for this project by increasing his expectancy.

There are more complex formulations for the expectancy theory model that incorporate multiple outcomes and intrinsic as well as extrinsic motivation considerations. An example will serve to explain the formulations. The basic motivation formula from the expectancy theory is:

$$M = [E \rightarrow P] \times [(P \rightarrow O)(V)]$$

However, in most real-life situations there typically are more than one perceived outcome for a given major work-related decision. Hence, the model has been altered to accommodate this. Figure 11.8 depicts this expanded formulation.

In this formulation, we have expectancies for both $E \rightarrow P$ and $P \rightarrow O$. $E \rightarrow P$ expectancy is the perceived probability of successful performance, given the effort put forth. $P \rightarrow O$ expectancy is the perceived probability of receiving an outcome, given successful performance. First-level outcomes are those occurrences that result directly from executing the task itself. If the individual gives these outcomes to himself or herself (for example, a sense of accomplishment, satisfaction), then they are intrinsic. If these outcomes come from or are mediated

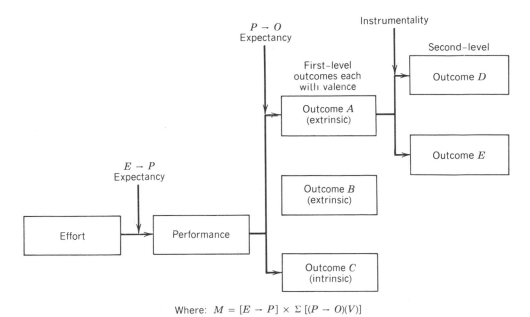

Where: $M = [E \rightarrow P] \times \Sigma [(P \rightarrow O)(V)]$

Figure 11.8   Expanded VIE Formulation: Second-level Outcomes, Instrumentality

by external factors, such as the supervisor, work group, or parents, then they are called extrinsic outcomes. Instrumentality is the concept associated with first-level outcomes being instrumental in the attainment of second-level outcomes. Certain outcomes are valued (or valent) because they have rather immediate and direct value. Others are valued because they are viewed as having the potential to lead to or cause the attainment of other valued second-level outcomes. Instrumentality can range from -1.0 (first-level outcomes will actually hurt attainment of second-level outcomes) to $+1.0$ (first-level outcome attainment correlates perfectly with attainment of second-level outcome).

An example will help bring this formulation and model to life. Let us utilize the same example as before, only this time we will expand it. An individual has been asked by his supervisor to work on a special project. Again, the project is critical but also a rush job. Table 11.1 sets up the example. From this table of data, we see that the following calculations need to be made in order to attain a measure of this individual's overall motivation or level of willingness to work toward successful completion of this project.

Table 11.1 Example of the Expectancy Theory Model

$E \rightarrow P$ EXPECTANCY	$P \rightarrow O$ EXPECTANCY	OUTCOMES W/VALENCE		$P \rightarrow O_{2nd}$ INSTRUMENTALITY $O_{1st} \rightarrow O_{2nd}$
		1ST LEVEL	2ND LEVEL	
		Extrinsic		
0.20	0.85 (1st level)	Get the job done in an acceptable fashion $V = +1.0$	Promotion $V = +1.0$	+0.75
			Raise $V = +0.85$	+0.90
			Maintain good relationship with co-workers $V = 0.95$	−0.50
			Job security $V = 0.20$	+0.75
			Professional growth $V = 0.90$	+0.30
	1.0	*Intrinsic* Satisfaction of meeting the challenge Vint. = +0.30		

1. The total valence for the first-level outcome (successful accomplishment of the project) needs to be calculated. The following formula is utilized:

$$\text{Valence (total)} = V_{int} + V_{1st}\text{-level outcome}$$

$$= V_{int} + \sum_{i=1}^{n} (I \times Vn_{2nd})$$

which from the table

$$= +0.30 + [(0.75)(1.0) + (0.90)(0.85) + (-0.50)(0.95) + (0.75)(0.20) + (0.3)(0.9)]$$

$$= +0.30 + 1.46$$

2. The level of predicted motivation or force to perform can now be directly calculated. Using the formula

$$M = [E \rightarrow P] \times \Sigma\,[(P \rightarrow O)(V)]$$

we find that

$$M = 0.20 \times [0.85(1.46) + 1.0(0.30)]$$
$$= 0.3082$$

Note that the maximum value of 6.0 was possible in this particular example as it was set up (if all values were 1.0).

There are a number of research and development questions that arise when one tries to develop a real-world example using these more complex formulations of the model.

1. Is the formulation for V_{total} valid?

2. Does it make sense to have a theoretically unlimited number of second-level outcomes and valences for those outcomes? Do they need to be mutually exclusive? Are they additive? Is the quantitative relationship between intrinsic and extrinsic outcomes correct?

3. Is the formulation for instrumentality valid?

The answers to these questions and many more are evidently not presently known. The continued development of this model represents an interesting and needed area for research.

Again, it must be stressed that the purpose of this model is not to develop precise numbers that represent levels of motivation for comparisons between employees. This is an intra-individual diagnostic model. Its power lies in its succinctness and explicitness about the basic factors and relationships that influence employee cognitive decisions about what to do, how well to perform, how much effort to exert, and so forth. A manager might simply develop a form, such as the one shown in Table 11.2, to use as a diagnostic device for specific major situations where the motivation of an employee may be a critical parameter in the success of a given endeavor. The form is self-explanatory. The potential in the form lies in its ability to force managers in certain critical situations to evaluate their perceptions of an employee's outlook on a given job, assignment, task, and so forth. The diagnosis part of the form would include an assessment on the part of the manager as to whether a low motivation problem is an $E \rightarrow P$ problem, a $P \rightarrow O$ problem, a valence problem, an instrumentality problem, or some combination of the factors.

Like most behavioral science models and theories, there exists a gap between the concepts and the actual application of the concepts. Managers and behavioral scientists have been busy researching and inventing new and more valid theories on motivation. What has been missing is innovation with respect to how to operationalize the new, more valid, and better theories, concepts, and models. This lack of innovation takes place because academic departments and journals in the field of business administration don't typically reward for innovation in practice only for invention in theory and basic research. What this has caused is an overabundance of journals, articles, and other literature that few practicing managers can or want to read. This alienates managers from the theory and concepts and simply enhances the already strong probability that new (less than 20-year-old) ideas and concepts on motivation will not be read. The young business school graduates have been exposed to the theory but lack experience to innovate and often meet resistance to change. As a result, we have and will continue to see a significant lag between new theories, concepts, and models in the management, behavioral, and social sciences and the application of these ideas in the real world.

The kind of innovation needed is that which builds decision-support systems for managers to utilize based on these theories and models. These systems need not be complex or automated. They may be as simple as a checklist, a form, or a graph that depicts the factors and relationships that a manager should consider while attempting to motivate employees. Most of these models of motivation are so simple that eventually they will become intuitive and a natural part of a manager's behavior. The form presented in Table 11.2 is an example of such a decision-support system.

We will now move from our discussion of expectancy theory as an example of state-of-the-art motivation theories to brief presentations of developed tech-

Table 11.2 Worksheet for Assessing Employee Motivation to Perform

EMPLOYEE _____

SITUATION _____

My view on the following is:

1. Individual's perceptions of potential outcomes or rewards associated with this project, assignment, job, etc.: (Extrinsic): _____

                         (Intrinsic): _____

2. Other persons who might value this activity: _____

_____

3. The approximate valence associated with each outcome by the individual:

              Outcome   *Valence* (-1.0 to $+1.0$)

            _____   _____

            _____   _____

            _____   _____

            _____   _____

            _____   _____

4. The approximate Performance to Outcome ($P \rightarrow O$) or instrumentality expectancy for each outcome (i.e., the employee's perception about the probability of rewards or outcomes being linked to performance):

              Outcome $P \rightarrow O$ (0.0 to $+1.0$)

            _____   _____

            _____   _____

            _____   _____

            _____   _____

            _____   _____

5. The approximate Effort to Performance ($E \rightarrow P$) expectancy (i.e., the employee's perception about the probability of being able to "successfully" perform on time):

              $E \rightarrow P =$ _____ (0.0 to $+1.0$)

Table 11.2 (*Continued*)

6. Predicted Force to Perform or level of motivation.

$$(E \rightarrow P) \ [\Sigma[(P \rightarrow O)(V)]] \ = \ \underline{\hspace{2cm}}$$

$$M_{max} \ = \ \underline{\hspace{2cm}} \qquad M_{actual} \ = \ \underline{\hspace{2cm}}$$

7. Diagnosis (mangerial behavioral interventions to be attempted): _____

_____

_____

_____

8. Follow-up evaluation: (Did the evaluation and improvement strategy work? Why? Why not?)

_____

_____

_____

_____

_____

niques for improving motivation and performance. Keep in mind that the reason for presenting a spectrum of behavior-oriented productivity improvement techniques is that human behaviors are frequently the constraining factor in the effective implementation of other types of productivity improvement efforts. The major productivity improvement technique to be discussed in chapter 13 is a process called performance/productivity action teams. However, this process, although reasonably simple to implement, is a complex design incorporating motivational elements of participation, goal-setting, reinforcement theory, and group process management. As such, it will be useful to present a brief discussion on four basic approaches and related techniques for motivating improved performance first.

REFERENCES

Blumberg, M., and C. Pringle, "The Missing Opportunity in Organization Research: Some Implications for a Theory of Work Performance. *Academy of Management Review*, Vol. 7, No. 4, 1982.

Direct Report Studies 1946–1954, Washington, D.C.: BLS.

Lawler, E. E. *Motivation In Work Organizations*. Monterey, Calif.: Brooks/Cole, 1973.

Pascale, R. T., and A. G. Athos. *The Art of Japanese Management*. New York: Simon and Schuster, 1981.

QUESTIONS AND APPLICATIONS

1. Discuss the rise and fall of civilizations. Incorporate the impact of such factors as affluence and rising standards of living on the work ethic. You may wish to utilize expectancy theory as you analyze these changes.

2. Investigate and discuss equity theory. How does it fit into the model of motivation presented in this chapter?

3. Investigate path-goal theory and discuss its insights concerning the model of motivation presented in this chapter.

4. Discuss the relationship between situational leadership theory and motivation.

5. Investigate the foundations of reinforcement theory. Discuss the moderating impact of reinforcement schedule on the effectiveness of a reinforcer.

6. Discuss the importance of management having a firm understanding of motivation basics in order to successfully manage productivity.

TECHNIQUES FOR IMPROVING MOTIVATION AND PERFORMANCE

HIGHLIGHTS

A. Reinforcement Theory
- Incentive Systems and Productivity Gainsharing
    Piece-Rate Plans
    Scanlon Plan
    Rucker Plan
    Improshare Plan
B. Behavior Modification
- Goal Setting
- Management by Objectives
- Managing Productivity by Objectives
C. Job Design, Redesign, Enrichment
- Job Characteristics Theory
- Job Diagnostic Survey
- Job Characteristics Inventory
D. Participation: Employee Involvement
- Introduction
- When is Participation Useful?
- Dilemmas of Managing Participation

OBJECTIVES

- To expose the reader to four generic categories of motivation, performance, and productivity improvement techniques.
- To emphasize the critical nature of the human resource (communication, co-operation, coordination, participation, commitment) component in any productivity and performance improvement endeavor.
- To focus on basic, fundamental human resource issues.

- To encourage the reader to take these concepts, theories, approaches, and techniques and incorporate them into a well-thought-through productivity improvement plan.

Table 12.1 Summary of Studies Comparing Four Motivation Techniques

TECHNIQUE	MEDIAN IMPROVEMENT IN PERFORMANCE	SHOWING 10% IMPROVEMENT	RANGE
Money (10)[a]	+30%	90%	+3% to +49%
Goal (17)[a]	+16	94	+2% to +57.5%
Job Enrichment (10)[a]	+8.75	50	-1% to +61%
Participation (16)[a]	+0.5	25	-24% to +47%

[a]Number of studies for each technique

A recent study by Locke et al. (1980) provides insights into the relative effectiveness of four broad and reasonably comprehensive categories or methods of motivation. Table 12.1 presents a summary of the results of this study. These results, although certainly not as conclusive as they may at first appear, will be useful in organizing the presentation of techniques to follow. The general category of reinforcement theory (money) will be presented along with representative techniques. These techniques include goal setting techniques, job design/redesign/enrichment techniques, and, finally, participatory techniques. Figure 12.1 depicts categorization of these theories and techniques of motivation.

A. Reinforcement Theory

Reinforcement theory is based on the principle that we can modify behavior by reinforcing desired behaviors and ignoring undesired behaviors. B. F. Skinner was a pioneer in promoting the reinforcement theory. Skinner's (1953) research on the effects of positive reinforcement received widespread publicity in psychological literature but was relatively ignored in application until the early seventies. At that time, a technique termed behavior modification, using money as a reinforcer, was developed and implemented in numerous organizations. This technique incorporates the theories associated with operant conditioning. It should be mentioned, however, that the use of money as a motivator probably can be traced to the origins of money itself (Locke et al., 1980, p. 363). The most systematic use of money as a reinforcer was by Frederick Taylor (1911), the father of scientific management. Taylor was a pioneer in the design, development, and implementation of scientific incentive plans. Incentive plans and such systems as the piece-rate or one-for-one plan, Merrick, multiple-price rate, point system, and Emerson plan were very popular through the late forties. During the fifties,

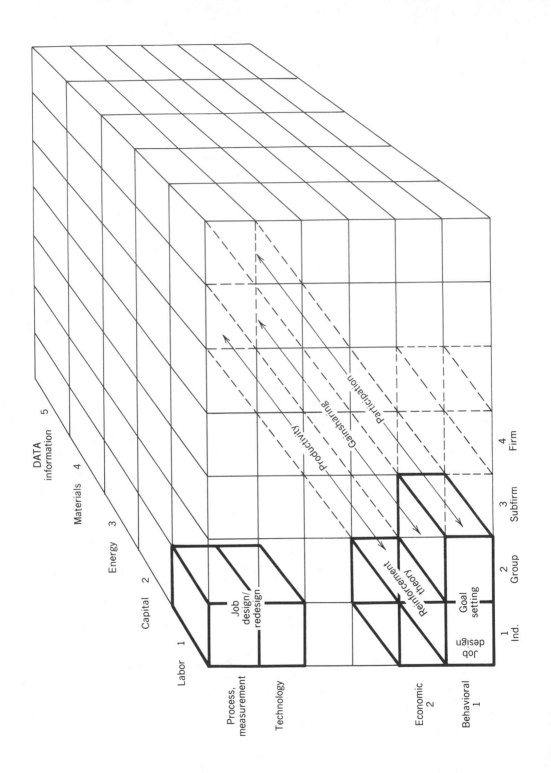

Figure 12.1 Techniques for Improving Motivation and Performance

however, incentive plans declined in popularity and use and were replaced by group plans, such as the Scanlon and Rucker plans.

Since the early fifties and even during the Hawthorne studies of the late thirties and early forties, the importance of money has been deemphasized by most social scientists. During the late sixties, however, when productivity began to be a national "buzz word" and concern, a revival of sorts occurred regarding incentive plans and what began to be called productivity gainsharing plans. Behavior modification programs and applications began appearing in the late sixties and early seventies, and a second generation group-type incentive or gainsharing plan titled Improshare appeared on the scene in the middle to late seventies.

For more background on the concepts and research underlying reinforcement theory, the reader is directed to the following references: Steers and Porter, 1983, and Lawler, 1973. A brief review of the Scanlon, Rucker, and Improshare plans will be provided in this chapter. Readers interested in more detail on the historical development of incentive plans or on the various reinforcement theory-type approaches should consult the reference section at the end of the chapter.

Incentive Systems and Productivity Gainsharing

The Scanlon, Rucker, and Improshare plans are the incentive systems and productivity gainsharing plans currently receiving the most attention and being implemented most frequently. The mechanics, formulas, and computations, of these plans, although important, represent a very small component of the overall implementation and maintenance concerns and efforts associated with these plans. Their implementation requires certain preconditions and especially certain

---

Theories and techniques (in order of relative effectiveness) (Locke et al., 1980)

1. Reinforcement theory
   - Incentive systems
   - Productivity gainsharing
   - Performance/merit appraisal & evaluation
   - Behavioral modification

2. Goal setting
   - MBO, CMBO, MPBO

3. Job design/redesign
   - Job characteristics theory
   - Job enrichment

4. Participation
   - Quality circles
   - Performance Action Teams

Figure 12.1 (Continued)

management philosophies and attitudes. Implementing these plans is both a science and an art. Relatively little detailed literature has been written about the Scanlon and Rucker plans because they have largely been in the hands of consultants who have carefully guarded the actual mechanics of the plans. Although also in the hands of a management consultant, Mitchell Fein, the Improshare plan has been much better documented as a result of Fein's attitude about sharing knowledge regarding his plan. Nevertheless, in general, implementation of a plan of this type is extremely difficult to execute successfully without skilled assistance.

Incentive systems and gainsharing systems are built upon a foundation of motivational principles and theories. Although these systems seem easy to implement in theory, in practice, developing and implementing reinforcement systems is extremely difficult and requires effective planning, sufficient background knowledge and experience, a supportive climate, a sound system, and a long list of necessary preconditions. In fact, implementation of such systems as the Scanlon, Rucker, and Improshare plans is practically impossible without skilled assistance.

Perhaps most critical in the development of a reinforcement system is the determination of what performance for the "system" under study actually is. An excellent piece written by Kerr entitled "On the Folly of Rewarding A While Hoping for B" addresses the problems that can result from not clearly defining "system" performance. If one wants effectiveness and it is defined and rewarded, one can expect performance improvement in this category (of course, assuming a number of other conditions are met). Much has been said recently about the lack of long-range planning on the part of American managers. You might want to contemplate the kinds of behaviors and outcomes that are rewarded in American businesses.

Basic performance models at the individual level depict performance as being influenced or caused by a multiplicative relationship involving ability, attitude, and effort. The transformation resulting from the combination of ability, attitude, and effort on a given task or job with performance (however it may be defined) is moderated strongly by the goals objectives activities learning standards system (GOALS) and the general management process planning, organizing, leading, controlling, and adapting (POLCA).

Performance in most jobs in most organizations is measured, appraised, and evaluated in one form or another. This process of performance appraisal and evaluation varies significantly from organization to organization and between levels within an organization. However, the basic objectives are similar—to measure, appraise (feed back), and evaluate performance of the individual. This process has been well researched and discussed in the literature (Latham and Wexley, 1981).

Performance appraisal is an underutilized and often ineffectively applied management process and strategy. Appropriate and effective performance measurement, appraisal, and evaluation is perhaps the most powerful source of motivation in an organization. The performance measurement process should force management to communicate knowledge of what is expected and to explicate and facilitate the GOALS system. If situational leadership principles are applied in this process, the effectiveness of the process for motivating acceptable performance levels will be increased significantly (Blanchard and Hersey, 1982). The performance appraisal or counseling process should force management to provide knowledge of results and to coach and counsel employee development. This builds capacity and capability for the future. And, finally, the performance evaluation process links outcomes, rewards, and reinforcers with performance. This process, most critical to employee motivation, establishes the logic to what the organization expects, desires, and is willing to reward. It should be consistent with the POLCA, GOALS, and performance measurement and appraisal processes. That is, the organization should consistently reward those things it has communicated it expects and desires.

Rewards fall into three basic categories: (1) economic; (2) prestige; and (3) work content. They can also be categorized as extrinsic—rewards coming from the environment—and intrinsic—rewards coming from within the individual. Table 12.2 identifies a variety of reinforcement techniques commonly in use. If the reinforcers are expected and/or needed, such as job security, safe working conditions, leadership training, and competitive and fair base wages or salaries, then satisfaction of these needs and expectations with appropriate outcomes should, if all other factors are managed correctly, lead to acceptable performance levels (APL). That is, levels of performance that are acceptable to the employer and employee will be achieved.

If an organization approximates APL and desires to strive for motivated per-

Table 12.2 Performance Reinforcement Techniques

	INDIVIDUAL	GROUP
Economic (Direct)	Piecework	Profit sharing
	Time payment	Scanlon
	Suggestion awards	Rucker
	Well pay (as opposed to sick pay)	Improshare
	Tuition refund	
Economic (Nondirect)	Cafeteria benefits	Fringe benefits
	Seniority benefits	Group awards
	Awards	
Nonfinancial	Earned time off	QWL programs
	Flex-time	Productivity action teams
	Job enrichment	Quality circles
	Performance evaluation	Zero defects
	Job posting	Safety
	Training	Survey feedback

SOURCE: Adapted from C. S. O'Dell, *Gainsharing: Involvement, Incentives, and Productivity,* New York: AMACOM, 1981.

formance levels (MPL), other reinforcers will need to be brought into the system. By and large, MPL means performance significantly above APL. If we had engineered standards for a job, MPL would necessarily mean consistent, sustained performance above 100 percent, perhaps 120 to 130 percent.

Incentive systems or productivity gainsharing systems are designed to reduce unit costs by encouraging higher levels of performance without increasing fixed costs associated with increased labor input. We are trying to do more with the same level of labor input. Productivity necessarily goes up (Q_i^o/Q_i^l increases), and unit costs go down because we obtain more performance (volume) while holding certain fixed costs constant.

Incentive systems and gainsharing systems attempt to motivate harder and smarter working (working more effectively and efficiently). Hence, good incentive systems and gainsharing systems do not, by definition, sacrifice quality for quantity. For an incentive system or a gainsharing system to be totally effective, it must be designed, developed, and executed as an integrated system. Motivating for MPLs must be done in combination with all other motivational systems and processes in the organization.

The remainder of this section will present specific incentive systems and productivity gainsharing systems that are prominent today. In general, the movement is away from individual, direct-financial incentive plans, unless the situation clearly dictates their appropriateness, and toward group productivity gainsharing plans. Traditional piece-rate or standard hour plans will be presented along with the Improshare, Scanlon, and Rucker plans. Remember that incentive plans and gainsharing plans are not appropriate in all situations. They are *not* necessary to obtain APL. Most organizations today are nowhere near APL. Attention to the basics is necessary in order to improve performance in that direction. Implementation of an incentive plan or gainsharing plan should be attempted only after the basics are mastered, APL is being attained, and considerable time and effort are devoted to the planning process. For a succinct, systematic approach to the development of a gainsharing plan, see O'Dell, 1981, and Graham-Moore and Ross, 1983. Also for more detail on motivation basics, see the following: Lawler, 1981, 1973; Luthans and Kreitner, 1975; Aldag and Brief, 1979; Steers and Porter, 1983; Latham and Wexley, 1981.

Once APL has been attained, a choice must be made between incentive and gainsharing plans. This choice must be based on the particular situation. There are numerous plans that can be used. The following discussion presents a review of prominent plans.

PIECE-RATE PLANS

Traditional piece-rate or piecework is an individual incentive plan, whereas the others are group plans. Group plans have increased in popularity because under such systems employees are motivated to achieve management goals through improved productivity and cooperation as well as to make more money. In individual incentive plans, employees are often encouraged only to make more money. Group plans also tend to include or incorporate additional employee involvement schemes, such as quality circles and productivity action teams.

Although many types of piece-rate plans exist, the most popular is the standard hour plan. The major advantage of the standard hour plan is that standards are expressed in time rather than money. Therefore, as rates of pay change, the standards will not need to be adjusted accordingly. It is applied to each employee on an individual basis. Calculation of the incentive bonus is based on the standard time required to produce 100 units of the product. This standard time is determined through traditional work measurement techniques. An employee's performance may be measured by calculating the standard hours earned. This is found as the product of the number of units produced by the employee divided by 100 and multiplied by the standard time established to produce 100 units. Expressed as a formula, this is (Niebel, 1976)

$$He = \frac{Pa}{100} Sn$$

where

He = Standard hours earned
Pa = Actual production

SCANLON PLAN

Relatively little has been written about the Scanlon Plan because it has been largely in the hands of consultants who often view the programs as proprietory. We can, however, compute gainsharing under it, although this plan does not have one specified formula for doing so. Instead, there are four common formulas used (O'Dell, 1981). Two of these, the multicost ratio and the allowed labor formula, are most appropriate after the plan has been in effect for some time. The single-ratio and split-ratio are best for newly installed plans. In addition, an organization may wish to develop a unique formula for specific applications.

The simplest of the four plans, the single-ratio formula, will be presented here. It is a good starting point. As a plan matures, conversion to other formulas will likely result. Complete coverage of the other formulas can be found in Moore and Ross (1978).

The single-ratio formula is based on a ratio of total labor cost to the sales value of production. For most companies, particularly in manufacturing, the ratio is relatively constant over time. If it is not, then the single-ratio formula should not be used. The following ratio is used to calculate an allowed payroll cost during each period:

$$\text{Base ratio} = \frac{\text{total payroll (labor) cost}}{\text{sales value of production}}$$

Using the base ratio, we can find the allowed payroll cost from the sales value of production during a particular period. The allowed payroll is then compared to the actual payroll. If actual payroll is less than allowed, there is a bonus. This bonus is divided between the company and employees, usually at 25 percent and 75 percent, respectively. Of the employee share, a portion may be placed in a bonus reserve, which serves two major functions. It is a buffer against future deficit months, which will be discussed shortly. It also serves as an extra incentive for long-term performance improvements. At the end of a year, the amount remaining in the bonus reserve, if positive, is distributed in the same manner as a monthly bonus, except none is placed back in the reserve. The year's bonus activity is wiped clean. If the reserve is negative, the loss (not an actual dollar loss) is absorbed by the company.

It should be beneficial to examine a sample bonus calculation using the single-ratio formula. Table 12.3 presents such a calculation. Lines 1 through 4 are used to arrive at a value of production, given on line 5. Notice that the value of production may not reflect actual production levels at a given sales price since returns, allowances, and discounts are explicitly considered. Accounting for changes in inventory can be used to absorb fluctuations or cyclical changes in sales. In this example, there was a net decrease in inventory, meaning more was sold than was produced. Using a base ratio of 0.283, the allowed payroll cost is found in line 6. Subtracting the actual payroll yields the total bonus pool. If this were negative, no bonus would be paid and the deficit would be subtracted from the bonus reserve. In this way, employees are encouraged to generate a

Table 12.3 Scanlon Plan Sample Calculation

1. Sales	$2,300,000
2. Less sales returns, allowances, discounts	− 74,000
3. Net sales	$2,226,000
4. Change in inventory (at cost or selling price)	− 441,000
5. Value of production	$1,785,000
6. Allowed payroll costs = Base ratio × line 5 = 0.283 × line 5	$ 505,155
7. Less actual payroll costs	− 470,000
8. Bonus pool	$ 35,155
9. Company's share = 25%	− 8,788.75
10. Employees' share	$ 26,366.25
11. Bonus reserve = 20%	− 5,273.25
12. Immediate distribution bonus	$ 21,093.00
13. Total participating payroll	$ 350,000
14. Bonus percentage = line 12/line 13	6%

bonus each month. The company's share of the bonus is 25 percent and the employees receive 75 percent. Of the employees' share, 20 percent is placed in the bonus reserve (line 11), and the remaining portion is immediately disbursed (line 12). The bonus percentage, line 14, is the percentage amount of wages paid to each participating employee. Notice that the participating payroll is not the same as the actual payroll from line 7. Actual payroll costs include such additional items as fringe benefits. Also, a new employee normally must undergo a probationary period before becoming a gainsharing participant. This period lasts approximately 60 days on the average (Moore and Ross, 1978). Thus, these employees appear in actual payroll costs but not in participating payroll costs. This form of calculation may be troublesome during steady periods of economic decline, such as is being experienced now.

THE RUCKER PLAN

Like the Scanlon Plan, Control of The Rucker Plan rests in the hands of a small group. In fact, "the Rucker Plan" is a registered trademark, and the plan is administered and marketed by the Eddy-Rucker-Nickels Company of Cambridge, Massachusetts.

The first step in implementing a Rucker Plan is to perform an accounting analysis necessary to determine an index of the firm's production value. The production value, or value added, is the difference between the sales value of the goods produced and the cost of materials, supplies, and services used to produce the goods. The value added to a product is used as the measure of productivity. This figure is quite sensitive to many factors, such as material cost and cyclical and seasonal factors. For this reason, the production value used for the plan is the average of the production value over the previous three to seven years (O'Dell, 1981). This provides a smoothing effect and a consistent value-added estimate.

The next step is to find the "Rucker standard." The Rucker standard is the percentage of the production value that is paid out to the workers in wages. It is essentially the value added to a product for each dollar of wage cost. The standard used should be the average over a number of years. Its historical accuracy lends credence to its use. It has been found that this ratio is stable over long periods of time for most manufacturing companies (O'Dell, 1981).

The bonus calculation is similar to the Scanlon single-ratio calculation but is slightly more complex because of the value-added concept. Table 12.4 presents a sample calculation. The value of production is calculated in the same manner by adjusting sales for returns, allowances, and discounts and by adjusting for changes in inventory. Materials, supply, and nonlabor costs, as well as other outside purchases, are subtracted from the value of production to obtain the value added. The Rucker standard is applied to the value added to yield the allowed employee labor cost. From this point, the calculations are again identical to Scanlon calculations with the exception of the company's share. In general, the company's share for the Rucker plan is 50 percent.

Table 12.4 Rucker Plan Sample Calculation

1.	Sales	$1,800,000
2.	Less returns, allowances, and discounts	− 60,000
3.	Net sales	$1,740,000
4.	Change in inventory	+ 360,000
5.	Value of production	$2,100,000
6.	Less materials and supplies $950,000 other outside purchases nonlabor costs $400,000	−1,350,000
7.	Value added	$ 750,000
8.	Allowed employee labor cost (Rucker standard = 50.2%)	− 376,500
9.	Actual labor cost	− 340,000
10.	Bonus pool	$ 36,500
11.	Company's share	− 18,250
12.	Employees' share	$ 18,250
13.	Bonus reserve = 20%	− 3,650
14.	Immediate distribution	$ 14,600
15.	Participating payroll	$ 220,000
16.	Bonus percentage	6.6%

IMPROSHARE PLAN

The Improshare Plan was developed by a management consultant, Mitchell Fein. Fein has written several publications, all of which form the basis of the material in this section. These publications are all included in the reference list at the end of the chapter. "Improshare" is a registered trademark.

Improshare differs from the other gainsharing plans in several respects. Productivity gains are not measured in dollars but in employee hours. Employee-hour standards and base productivity factors are used, and ceiling and buy-back principles are used to control time standards. There are considerable elements involved in an Improshare Plan, not all of which can be addressed here. However, general concepts will be presented and some sample calculations provided.

Calculating the bonus in an Improshare Plan begins with the employee-hour standard. This is the standard number of employee hours required to produce one unit of product. It does not include nonproductive work (down time, material handling, and so forth). It may be calculated as

$$\text{Employee-hour standard} = \frac{\text{total production hours}}{\text{units produced}}$$

There is an employee-hour standard for each product.

The employee-hour standard is used to determine the total standard value hours. Total standard value hours represent the standard number of production hours that should be used, given a particular production level. This is the product of the employee-hour standard and the number of units produced. For multiple products, these are summed to obtain the total standard value hours.

The base productivity factor (BPF) is found by dividing the total employee hours worked, including nonproduction hours, by the total standard value hours. The BPF provides a base measure for the number of total employee hours required to produce one unit of product. It can be expressed as

$$\frac{\text{Total employee hours, production and nonproduction}}{\text{Total standard value hours}}$$

Table 12.5 presents a numerical example to derive the BPF for a two-product company.

Bonuses are calculated on the basis of the BPF. This is used to derive the Improshare hours—the number of total employee hours allowed for a particular production level, given the base information. An example of bonus calculation is given in Table 12.6.

An important aspect of the Improshare Plan is the imposition of a productivity ceiling of 30 percent above standard. Consistent performance above this yields adjustments in time standards. This is not a disincentive to employees, however, since the company "buys back" the standard in the form of a single lump-sum payment to employees.

In order to track productivity sharing, four-week cumulative totals are found each week for actual employee hours and Improshare hours based on the BPF for each product. From these totals, hours saved and the employee productivity

---

Table 12.5 Derivation of the BPF

Production employees:	20 (A-8, B-12)
Nonproduction employees:	10
Number of hours worked by each:	40
Units produced:	A-80, B-60

Employee-hour Standards
Product $A = 8 \times 40/80 = 4$ hours
Product $B = 12 \times 40/60 = 8$ hours

Total Standard Value Hours
Product $A = 4$ hours $\times$ 80 units $= 320$
Product $B = 8$ hours $\times$ 60 units $= \underline{480}$
Total SVH $\phantom{= 8 \text{ hours} \times 60 \text{ units} = } 800$

$$\text{BPF} = (20 + 10)40/800 = 1.5$$

Table 12.6 Calculation of Improshare Bonus

Units Produced

Product A   = 100 units
Product B   =   50 units
BPF   =   1.5 (from Table 12.5)
Total employee hours   = 950

Bonus

Product A = 4 hours × 100 units × 1.5 =   600
Product B = 8 hours × 50 units × 1.5 =   600
    Improshare hours   = 1200
    Actual hours   =   950
    Gained hours   =   250
    Employee bonus hours
      (0.5 × gained hours)   =   125

Bonus = 125/950 = 13.2%

share (percent bonus) can be found. If the productivity share exceeds 30 percent, the hours saved above 30 percent are "banked" for future periods, for use when productivity is below the 30 percent ceiling. When both employees and management agree that productivity is consistently above the ceiling and are convinced that it will remain so, the buy-back procedure is used. Both parties benefit from this.

Productivity sharing calculations, including banking hours above standard, are illustrated in Table 12.7. If the employee productivity share (column 8) is above 30 percent (as in period 5), it is reduced to 30 percent. Based on this reduction, the hours saved are adjusted accordingly. The difference between the nonadjusted and adjusted hours saved is the excess above the ceiling and is banked for the following period (in this case, period 6). A steady rise in the number of excess hours points to consistent productivity above the ceiling.

If a buy-back is deemed necessary, Improshare provides a set of steps to update standards and calculate the buy-back percentage and dollar figures. These are

Table 12.7 Productivity Sharing Calculations

1	2	3	4	5	6	7	8	9	10
			CUMULATIVE		EXCESS (BANKED)	(5) +	HOURS SAVED (7) − (4)		EMPLOYEE
PERIOD (WEEK)	ACTUAL HOURS	IMPROSHARE HOURS	ACTUAL	IMPROSHARE	HOURS	(6)	100%	50%	PRODUCTIVITY SHARE (9)/(4)
1	360	490							
2	380	700							
3	360	500							
4	350	410	1450	2100		2100	650	325	22.4%
5	380	760	1470	2370		2370	882	441	30%
6	400	480	1490	2150	18	2168	678	339	22.8%
7	370	550	1500	2200		2200	700	350	23.3%

established to allow employees to remain at the ceiling after the buy-back so productivity earnings are not lost. The banked hours are not affected by the calculations. The terms used in the calculations are as follows:

M = multiplier used to revise all time standards, expressed as a decimal

V = ceiling-level productivity, expressed as a decimal (this includes both employee and company shares, so a 30 percent ceiling for employees at a 50 percent share would yield V = 1.6)

A = average actual productivity at the time of the buy-back, expressed as a decimal (this also includes both the employee and company shares)

S = employees' portion of productivity sharing, expressed as a decimal

R = employee base hourly pay rate

b = buy-back percentage, expressed as a decimal (portion going to the employees)

B = money buy-back total per employee

With these terms, the following calculations are made:

1. New product standards

$$M = V/A$$

2. Buy-back percentage (as a decimal)

$$b = (A/V)S$$

3. Buy-back total per employee

$$B = bR(2000 \text{ hours/year})$$

As an example, consider the following situation. The established ceiling is 30 percent, yet employees are averaging a 50 percent productivity share on a 50–50 Improshare Plan. The base hourly wage is $8.00. A buy-back decision would result in the following:

$$S = 0.5$$
$$R = 8.00$$
$$V = 1.0 + (0.3/S) = 1 + (0.3/0.5) = 1.6$$
$$A = 1.0 + (0.5/S) = 1 + (0.5/0.5) = 2.0$$
$$M = V/A = 1.6/2.0 = 0.8$$
$$b(A - V)S = (2.0 - 1.6)0.5 = 0.2$$
$$B = bR(2000) = (0.2)(\$8.00)(2000) = \$3200.00$$

The standards are revised through multiplication by M. For instance, if a standard time were 3 minutes, then after the buy-back it would be 3×0.8, or 2.4 minutes.

The Improshare Plan provides an effective productivity gainsharing measure with the added features of the ceiling and buy-back provision. There are also ways in which it can be applied to nonmanufacturing companies. As long as unit work counts can be established, an Improshare Plan can be used.

DESIGN CRITERIA FOR GAINSHARING AND INCENTIVE SYSTEMS

Well-designed and developed direct financial incentive and gainsharing plans tend to adhere to the following basic design principles:

1. Labor management communication, cooperation, and agreement on general principles of system

2. A foundation in a sound job evaluation system

3. Well-designed, valid, and acceptable criteria for measurement and evaluation

4. "Reasonable" standards

5. Rewards/payoff incentive clearly linked to performance

6. Measurement and reward of *all* tasks and responsibilities

7. Simplicity

8. Quality emphasis

9. Timely linkage of rewards and performance

10. Creation of cooperation rather than competition

11. Reward for MPL, not APL

12. Effective and efficient involvement strategy as mechanism for identifying ideas for productivity improvement

13. Maintenance of standards

14. Considerations for mechanism for changing standards

15. Guarantee of hourly rates or salary level

16. Incentives for indirect employees

17. "Guarantee" of job security

18. Volume of work forecasts.

Concern over a decline in productivity, the effects of inflation, and general problems in maintaining profitability and increasing levels of international competition have caused many managers to seek and experiment with a variety of "productivity improvement" techniques. Interest in and applications of group productivity gainsharing plans as well as more traditional incentive plans have been growing since the early seventies. Coupled with efforts in the areas of automation, participative problem solving, and quality of work life, productivity gainsharing holds much promise for improving labor, material, energy, and capital productivity. Integrated effectively with overall long-range productivity management planning efforts, gainsharing can significantly impact productivity and profitability. It would appear that the Improshare Plan is the best designed in that it incorporates what we have learned over the years about motivation theory. Failures with even the best-designed equipment, programs, plans, and so forth are likely if careful planning and execution do not take place.

B. Behavior Modification

Behavior modification attempts to motivate through identification of functional or desired behaviors and through reinforcement of those behaviors. This technique is developed based on certain key premises:

1. All behavior has consequences, which may be positive (satisfying), negative, or neutral.

2. Behavioral is a function of its consequences. Positive consequences increase the probability of the behavior. Negative consequences decrease the probability of the behavior. And neutral consequences decrease the probability of the behavior slowly.

3. What follows a behavior is more important than what precedes it.

A typical behavior modification process includes the following steps:

1. Define the problem or desired change behaviorally. A typical statement might be, "A change in which behavior will cause what I or we want?"

2. Develop measures with which to evaluate whether behavior is changing. Chart all behaviors.

3. Intervene; locate a consequence (reward or motive) that is relevant and potent and introduce it for the desired change only. Link rewards to performance.

4. Assess and evaluate.

5. Be patient, persistent, and consistent.

Behavior modification is an external or learning approach to the understanding, prediction, and control of organizational behavior (Luthans and White, 1971; and Nord, 1969). In this approach, observable behavior in the organizational system of interest and its consequences are the primary focus. Further, the focus is on what Skinner called operant or learned behavior as opposed to respondent (unlearned) behavior. Research suggests that operant behavior is a function of its consequences. In other words, the behavior occurs because it is rewarded and encouraged. A succinct summary of the historical development of behavior modification is presented in Figure 12.2.

Fundamental principles associated with behavior modification are (Luthans and Kreitner, 1975)

1. The necessity of dealing exclusively with observable behavioral events
2. The use of frequency of behavioral events as basic data
3. The importance of viewing behavior within a contingency context

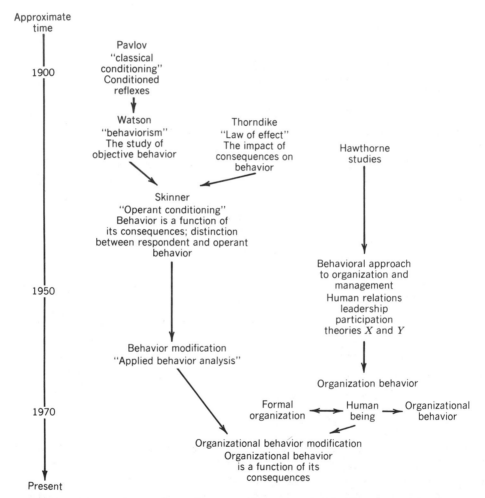

Figure 12.2 Historical Development of Organizational Behavior Modification (from Fred Luthans and Robert Kreitner, *Organizational Behavior Modification*. Copyright © 1975 by Scott, Foresman and Company. Reprinted with permission.)

For more details and instructions as to how to operationalize these principles, the reader is directed to the Luthans and Kreitner reference.

Intervention strategies that managers take are based on data collected about behaviors and are aimed at creating contingencies, expectancies, instrumentality, and valences for employees. These interventions take the form of both verbal and nonverbal behaviors. Specifically, positive and negative reinforcement, punishment, and extinction are behavioral control strategies that are systematically applied as consequences or outcomes to bring about changes in actions and action frequencies. Table 12.8 depicts the four basic behavior modification strategies and their combinations. All four strategies have three common characteristics: (1) They are used to change frequencies of objective behavioral events or responses; (2) the consequence or outcome must be perceived as being contingent on the specified response, and relative immediacy is important to ensure a contingent relationship; and (3) the type of effect a particular outcome has on an action's frequency of occurrence determines its strategic category.

Positive reinforcement is the contingent creation of positively valenced outcome(s). Positive reinforcement strengthens the probability that either a behavior or action will exist or that the frequency with which that behavior is exhibited will increase. Negative reinforcement is the contingent removal or elimination of some negatively valenced outcome(s). Negative reinforcement will also strengthen the probability of the desired behavioral response.

Punishment is the contingent and systematic application of a negatively valenced outcome or consequence. Note that punishment and negative reinforcement are not the same. This is a critical distinction that most people are not aware of. Punishment is intended to decrease the response frequency; however, it is a relatively ineffective strategy in most work situations as it often evokes many dysfunctional consequences, and its effect is often limited to the presence of the punishment itself. Extinction is the process by which learned responses, actions, or behaviors are not reinforced. That is, the manager attempts to communicate 0.0 valence, expectancy, and instrumentality. There are also, of course, combination strategies.

Research has shown that the schedule of the reinforcement plays a significant role in determining the effectiveness and efficiency of the behavior modification strategy. "When" and "how often" outcomes are attached to actions plays a critical role in the determination of performance. Table 12.9 depicts the various types of reinforcement schedules and their effects on responses.

More advanced topics associated with behavior modification that interested readers may investigate are shaping, modeling, and a review of various case study applications. As a productivity improvement technique, behavior modification, if practiced effectively and professionally, has significant implications for treating "work ethic" problems. Figure 12.3 depicts the potential relationship between behavior modification and organizational performance.

Operationalizing behavior modification theories and principles is a matter of developing a program that can be followed by managers in a step-by-step fash-

Table 12.8 Behavior is a Function of Its Consequences: Intervention/Reinforcement Strategies for Systematically Modifying Behavior

REINFORCEMENT STRATEGY	ANTECEDENT	BEHAVIORAL EVENTS	CONSEQUENCES	PREDICTED BEHAVIORAL OUTCOME
Positive reinforcement	Some given environmental conditions or states	A → B → C Certain Behavior(s)	Contingent presentation of an environmental condition that is perceived to have positive valence	Increase the frequency of response B
Negative reinforcement	"	"	Contingent termination or withdraw of a negatively valenced environmental condition (outcome)	Increase the frequency of response B
Punishment	"	"	Contingent presentation of a negatively valenced environmental condition (outcome)	Decrease the frequency of response B
Extinction	"	"	No outcomes administered	Decrease the frequency of response B
Combination strategies Extinction B/positive reinforcement C Punishment/positive reinforcement Punishment/negative reinforcement (others . . . exist)	"	→ B & C → incompatable responses/behaviors B & C	Combinations of consequences listed above	Decrease the frequency of response B; increase the frequency of response C

SOURCE: Adapted from Fred Luthans and Robert Kreitner. *Organizational Behavior Modification*. Glenview, Ill.: Scott, Foresman, 1975.

Table 12.9 Schedules of Reinforcement

SCHEDULE	DESCRIPTION	EFFECTS ON RESPONDING
Continuous (CRF)	Reinforcer follows every response.	(1) Steady high rate of performance as long as reinforcement continues to follow every response. (2) High frequency of reinforcement may lead to early satiation. (3) Behavior weakens rapidly (undergoes extinction) when reinforcers are withheld. (4) Appropriate for newly emitted, unstable, or low-frequency responses.
Intermittent	Reinforcer does not follow every response.	(1) Capable of producing high frequencies of responding. (2) Low frequency of reinforcement precludes early satiation. (3) Appropriate for stable or high-frequency responses.
Fixed ratio (FR)	A fixed number of responses must be emitted before reinforcement occurs.	(1) A fixed ratio of 1:1 (reinforcement occurs after every response) is the same as a continuous schedule. (2) Tends to produce a high rate of response which is vigorous and steady.
Variable ratio (VR)	A varying or random number of responses must be emitted before reinforcement occurs.	(1) Capable of producing a high rate of response which is vigorous, steady, and resistant to extinction.
Fixed interval (FI)	The first response after a specific period of time has elapsed in reinforced.	(1) Produces an uneven response pattern varying from a very slow, unenergetic response immediately following reinforcement to a very fast, vigorous response immediately preceding reinforcement.
Variable interval (VI)	The first response after varying or random periods of time have elapsed is reinforced.	(1) Tends to produce a high rate of response which is vigorous, steady, and resistant to extinction.

SOURCE: From Fred Luthans and Robert Kreitner, *Organizational Behavior Modification*. Copyright © 1975 by Scott, Foresman and Co. Reprinted with permission.

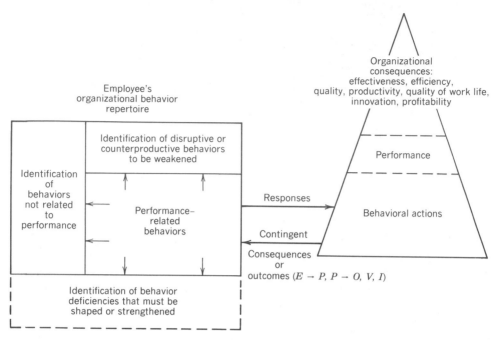

Figure 12.3   Organizational Behavior Modification Relationship to Organizational Performance

ion. A necessary precondition is, of course, training for the managers in terms of how and why the theory and model work. Luthans and Kreitner (1974) have developed a general organizational system behavior modification problem-solving model called behavioral contingency management (BCM). The model, described in the book of 1975, provides a methodology for identifying and contingently managing critical performance-related behaviors of employees in all types of organizational systems. The model is a five-step program: (1) identify; (2) measure; (3) analyze; (4) intervene; and (5) evaluate. Note the similarity between the behavior management process and the productivity management process, which consists of the following four steps: (1) measure; (2) evaluate; (3) control; and (4) improve. Evaluation and planning as superordinate functions are a part of all four stages. The BCM approach is presented in Figure 12.4.

Within their BCM approach, Luthans and Kreitner conceptualize a framework for identifying performance-related behavior problems. This is presented in Figure 12.5. After studying this figure, one can note a similarity between this framework and elements of the productivity improvement taxonomy. Note that the unit of analysis for this framework is the individual. Measurement, charting, visibility of behaviors to results, performance to action cause-and-effect relationship monitoring, graphs, and histograms are all critical components of the BCM process.

Behavior modification is an attempt to improve the clarity and visibility of the

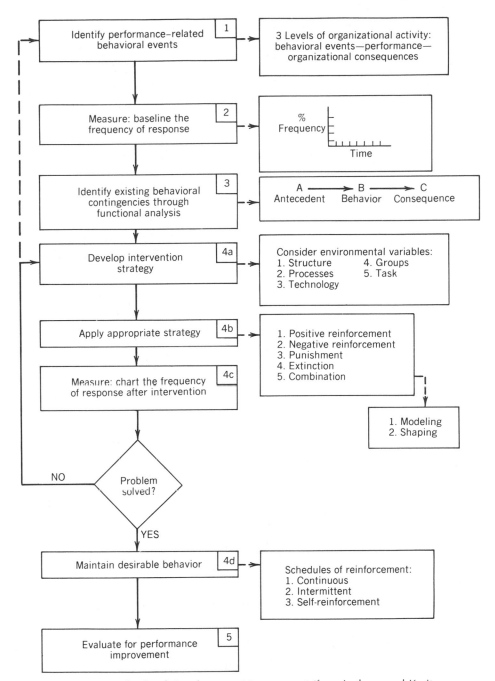

Figure 12.4 Behavioral Contingency Management (from Luthans and Kreitner, 1974, p. 13, Reprinted by permission of the publisher from *Personnel*, July-August 1974 © 1974 by AMACOM, a division of American Management Associations.)

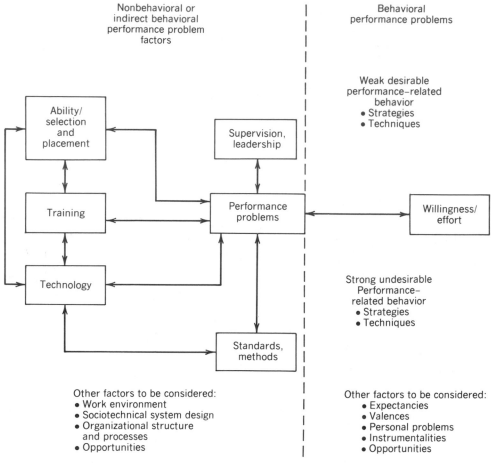

Figure 12.5   Framework for Identifying Performance-Related Problems (Adapted from F. Luthans and R. Kreitner, *Organizational Behavior Modification*, Scott, Foresman and Co., 1975.)

necessary goal and action congruity between the individual and the organization. Goal setting, reinforcement theory, expectancy theory, and change theory are the knowledge bases from which organizational behavior modification draws. Goal setting is the next topic of discussion.

Goal Setting

Goals can be conceptualized as desired future states for an individual or organizational system. They are statements regarding perceptions of desired outcomes (Etzioni, 1964). Goals also can be seen as often intentional constraints placed on present and future behavior based on an analysis of the past and projected needs, desires, aspirations, and so forth. (Cyert and March, 1963;

Simon, 1964). Therefore, goals are not only intended targets for future accomplishment, they are also behavioral and other types of resource commitments to make those desired states a reality.

> The concept of goals is a dynamic process by which individuals and organizations determine their future aspirations within certain known limitations. . . . Goal setting really becomes a process of allocating one's resources. . . . (Steers and Porter, 1983).

Goals guide and direct behavior, and herein lies the true motivational, performance, and productivity improvement power associated with goal-setting processes. Goals (1) focus attention and effort in specific directions; (2) can serve as standards against which effectiveness can be evaluated (recall that effectiveness equals successful accomplishment of goals, objectives, and activities); (3) can serve as a mechanism by which resource expenditures are justified; (4) can affect the structure and process of organizational systems; and (5) often reflect the underlying motives and character of both individuals and organizations ("First we make our habits [goals] and then our habits [goals] make us"). In addition, the process by which goals are established can, if developed properly, serve as a motivational device through development of an achievement orientation.

Extensive research has taken place during the past 15 to 20 years on the effect that goal setting has had on motivation and performance. The focus of much of the research has been on testing the following generalized hypotheses:

1. Difficult goals result in higher levels of performance than do easy goals.

2. Specific difficult goals result in a higher level of performance than do no goals or a generalized goal such as, "Do the best you can."

3. Goals mediate or moderate how performance is affected by monetary incentive, time limitations, knowledge of results, participation in decision making, competition, and so forth.

These hypotheses come from Locke's theory of goal setting presented in 1968. (For an excellent review of specific research that has been executed to test these and other hypotheses, see Steers and Porter, 1983.)

In general, research supports the following conclusions regarding goal setting as a mechanism for increasing motivation and performance. First, evidence suggests that specific, explicit, clear goals improve the probability that an employee will be more motivated to perform against those goals. There are, as one would expect, many other factors that would moderate this relationship, such as the goal-setting process, the specific nature of the goals, and an individual's ability or capacity. Management of objectives, an approach to planning and performance appraisal that attempts to clarify employee role requirements, relate employee performance to organizational system goals, improve manager-subordinate communication, facilitate objective evaluation of employee performance, and stimulate employee motivation is the major operational technique available to managers for accomplishing individual goal setting. Studies of management by objectives applications again tend to support the hypothesis regarding the positive effect of specific goal setting on motivation and performance (French, 1966; Ivancevich, 1972).

The second conclusion supported by research on goal setting is that difficult goals, as long as they are accepted, tend to increase the probability that motivation will be higher and performance greater. Figure 12.6 depicts this relationship. One way to operationalize goal difficulty would be to obtain the individual's subjective perceptions that he or she can perform against the goal (the $E \rightarrow P$ expectancies from expectancy theory). From Figure 12.6, we might ask the following question: What factors will affect the shape and position of the curve? The research on goal difficulty has suggested that as goal difficulty increases so does motivation, as long as the goal is accepted. Hence, the curve labeled "goal difficulty" is a rough approximation of the relationship. Note that when $E \rightarrow P$ is high, close to 1.0, the individual perceives there is a high probability that he or she can perform. At the same time, theory predicts that motivation is low. As $E \rightarrow P$ decreases, or as perceived difficulty in performing increases, motivation increases to a point where perhaps a "no chance" attitude causes goal acceptance and motivation to tail off drastically. So, research suggests that goal difficulty has a tendency to shift the curve to the left. Figure 12.6 summarizes the goal specificity relationship previously discussed. It suggests that as goals

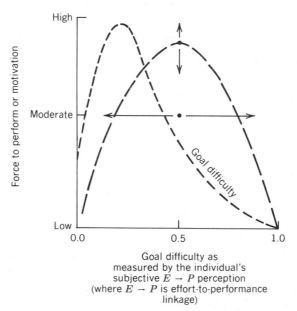

Goal difficulty as
measured by the individual's
subjective $E \rightarrow P$ perception
(where $E \rightarrow P$ is effort-to-performance
linkage)

Figure 12.6   Goal difficulty as measured by the individual's subjective $E \rightarrow P$ perception (where $E \rightarrow P$ is effort-to-performance linkage)

become more specific, motivation increases up to a point at which perhaps over-management begins to offset gains.

Research support for the mediating impact of goals between performance feedback or knowledge of results, time limits, and monetary incentives are less clear than previous findings. There is some evidence to suggest that frequent, relevant feedback is needed for a successful goal-setting program. However, the evidence is limited and more research is needed. Current leadership models, such as the "Situational Leadership Model" by Hersey and Blanchard would suggest that the maturity of the follower, subordinate, or individual moderates the need for performance feedback, at least from external sources. As psychological and physiological task-related maturity increases, the theory suggests that appropriate behaviors of leaders become less intervention oriented. Hence, specific leader-generated, task-related feedback would be less critical with more mature employees. Figure 12.7 depicts possible general relationships between performance feedback, ability and willingness of employee (maturity), and goal setting. The curve suggests that feedback in conjunction with goal setting will shift the hypothetical bell-shaped curve to the right; however, note that maturity of the follower will tend to shift the curve to the left. This suggests that a lesser degree of external performance feedback will go further or have the same impact

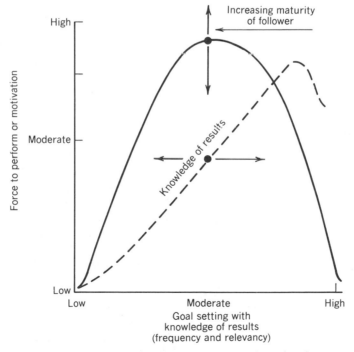

Figure 12.7 Moderate goal setting with knowledge of results (frequency and relevancy)

on a more mature employee as more performance feedback will have on a less mature employee.

Monetary incentives are more likely to increase goal acceptance and commitment than to induce a person to set a harder goal. It has been found that incentives can affect performance independently of goal level and goal commitment. Figure 12.8 depicts this possible relationship. It appears that the general relationship is one where incentives shift the hypothetical curve upward. In other words, at various levels of goal setting, the inclusion of monetary incentives will tend to increase the motivation to perform those goals. Incentives will increase the probability that an individual will *accept* a very difficult as well as very easy goal or very specific as well as a vague goal, and so on.

It appears that time limits affect performance only to the extent that they lead to goal setting. If time limits are established as part of goal accomplishment, then we find that as time limits increase (i.e., deadlines tighten), perceived

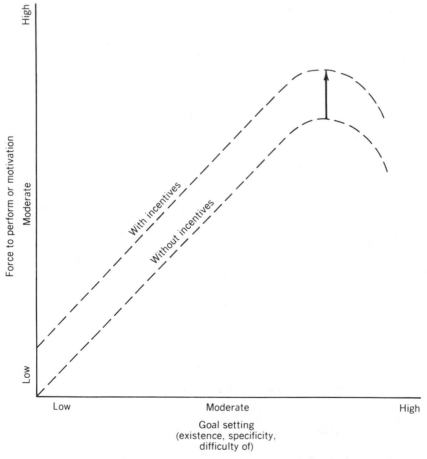

Figure 12.8   Goal setting (existence, specificity, difficulty of)

difficulty or decrease in $E \rightarrow P$ expectations cause corresponding shifts in motivation take place. When and if time demands are viewed as challenges or goals and are accepted, performance can increase. It would seem logical that if time demands and pressures were constant and not an occasional occurrence, stress would have a long-term impact on performance.

In summary, goal setting in one form or another can have a direct impact on employee motivation and performance. Goal setting involves a planning process, at the lowest level, between a supervisor and a subordinate. It helps to ensure that the meaning of the performance criteria is understood by the employee. Goal clarity, congruence, and commitment are essential to effectiveness, efficiency, and quality. Group goal setting can be equally, if not more, important than an individual goal-setting process. Traditionally, in fact, it is the group goal-setting and achievement process that is weakest in most organizations. Similar results, perhaps even more significant to organizational system performance, can be accrued through group goal-setting processes. (This will be discussed in more detail in a discussion in Chapter 13 on performance/productivity action teams, a form of collaborative management by objectives.)

Management by Objectives

As mentioned previously, management by objectives (MBO) is the major method or approach used by managers today to operationalize goal setting. As such, it is appropriate to present a brief discussion of MBO and a tutorial on how to execute an MBO process.

It has been said that there are two kinds of people in the world: those who like MBO and those who have tried it. Peter Drucker (1954) is typically credited with having at least recognized the value of establishing mutual goals for supervisor and subordinate. Drucker noted, "The job of management is to balance a variety of needs and goals." Douglas McGregor (1960) suggested establishing management by integration and self-control based on Theory Y assumptions of human behavior. The research and psychological foundations of MBO are numerous and varied:

1. *Need theory:* Maslow (1954, 1970) and Alderfer (1969, 1972), hierarchy of needs; Szilagyi (1981), changing patterns of human needs as related to time and circumstances

2. *Motives:* McClelland (1961), achievement motive, desire to be successful and avoid failure; White (1959), competency motive, desire for efficacy or self-power to produce desired results

3. *Dual-factor theory:* Herzberg (1966), hygiene factors, external to the job; motivation factors, internal to the job.

4. *Goal setting:* Vroom (1964), House et al. (1971), Locke (1968), Steers and Porter (1979), the effect of goal setting on performance, relationship of incentives to

behavior through effect on goals and intentions, effect of specific goals, effect of difficulty of goals to performance

5. *Feedback or knowledge of results:* Miller (1965), Leavitt and Mueller (1951), Kay, Meyer, and French (1965), Locke and Byran (1967), effect of feedback on performance, influence of quality and quantity of feedback, influence of timeliness and relevance of feedback, influence of positive or negative feedback contingent on comparison with some previously established goal or standard

6. *Participation:* Lawrence and Smith (1955), Likert (1961, 1967), Vroom (1964), Maier (1950, 1970), Hare (1953), conflicting results on participation and productivity, effect of type of participation, interacting effect with goal-setting activity, helpful in gaining subordinate's acceptance of decision as well as improving understanding between supervisor and subordinate

7. *MBO process itself:*
   (a) General Electric studies (Meyer, Kay, and French, 1965), effect of criticism on appraisees lower in self-esteem, effect of appraisee opportunity for influence in setting goals, goal setting more important relative to improved importance
   (b) University of Kentucky studies (Ivancevich, Donnelly, and Lyon, 1970), effect of top-down implementation on satisfaction of managers
   (c) Purex studies (Raia, 1965), production records, interviews, and questionnaires used to discover a significant contribution of MBO program in the area of performance appraisal
   (d) Mendleson study (Mendleson, 1967), seemed to indicate goal-setting techniques by themselves do not necessarily produce favorable outcomes
   (e) English case studies (Preston, 1968), initial resistance from managers to having responsibility for subordinate development, objectives at lower levels dependent on organizational objectives communicated along with company policy, reported cash savings in several areas that would not have occurred without the MBO process, assistance in problem identification area
   (f) Hospital study (Sloan and Schrieber, 1971), managers perceived improved planning and organization of work, clarification of mission, goals, and responsibilities; disadvantages and paper work were time consuming
   (g) *Management by Objectives* (Carroll and Tosi, 1973), intensive study of MBO summarized as follows:

These data show the importance of organizational commitment to MBO. Managers must feel that the MBO program is important, that the company is serious about it. It is necessary that organizational goals become clear, as goal setting at all lower levels is more difficult and perhaps even impossible without clear goals that can be fashioned into the departmental or individual goals. Managers must have time and resources so that they can utilize MBO. In addition, the time and energy requirements of the program should not be excessive.

An important finding of this research is that MBO affects different managers in different ways. The results obtained indicate that in many respects MBO should be tailored to the individual and his position rather than being presented as a single defined approach for all managers. Some of the findings indicate in what ways MBO should be tailored to fit certain types of managers.

As practiced and applied, MBO has gone through at least three distinct phases during the past 20 years. During the first phase, it was used primarily as a performance appraisal technique. Its emphasis was on jointly developing objective criteria and standards for individuals in a given job. This use for the technique tended to receive mild support from top management, be pushed by personnel, have limited involvement of line management, be a once-a-year appraisal, and be less effective than its potential would suggest.

The second phase of development for MBO focused on its use in planning and control. MBO, in this mode of application tied objectives to plans and in turn was a basis for budget control, utilized performance appraisal as an essential element, generated more top management interest and support, had more impetus from bottom-up than previously, and emphasized training and development of employees.

The third phase, which is basically the start of evolution to date, involves utilizing MBO as an integrative management process. The emphasis is on integrating key management processes and activities in a much more participative, delegative fashion. We find an increased need for and an emphasis on teamwork (hence, the development of collaborative management by objectives and performance/productivity action teams). The focus is more on action planning, action research, and dynamic performance review and evaluation. Emphasis is a more flexible system that focuses on individual and group growth and development.

The basic MBO process, excerpted from *Management by Objectives* by Raia, is depicted in Figure 12.9. Roughly 40 to 50 percent of all large firms in this country have used or are currently using some form of MBO. There are probably as many forms or MBO program designs as there are firms using the technique. Thus, although the basic process may be fairly consistent, the specific content and process probably varies somewhat. Typical elements of actual design that may vary are

1. Who is involved

2. Commitment to the approach and who makes the commitment

3. Degree of participation on the part of the superior as opposed to the subordinate

4. Frequency of the review

5. Type of goals or objectives that are discussed

6. Time frame involved in the goal-setting process

7. Linkage to reward system

8. One-on-one as opposed to group process

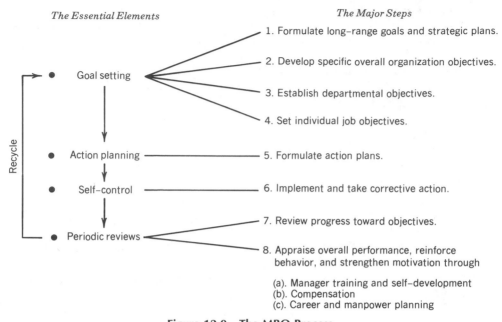

Figure 12.9   The MBO Process

9. Feedback mechanism used

10. Use of the results

11. Resolution of hierarchical and horizontal differences in goal targets

12. Resolution of general goal conflicts

13. Changing pattern of human needs.

Integrated into a larger scoped wage and salary administration, as well as a work planning and review program, MBO is typically sequenced as depicted in Figure 12.10. MBO and performance appraisal, if implemented correctly, have much potential as an individual performance improvement process. However, French and Hollmann (1975) suggest that a collaborative management by objectives (CMBO), participative, team-centered approach will minimize some of the deficiencies in more traditional one-on-one versions of MBO. CMBO will be addressed in more detail in the section on performance/productivity action teams.

Managing Productivity by Objectives

Finally, Mali (1978) presents an MBO hybrid termed managing productivity by objectives. He points out that developing objectives or, in this case, productivity objectives is a deceptively difficult process requiring discipline, precision of thinking, planning, and clear communication. He confirms a suspicion that most

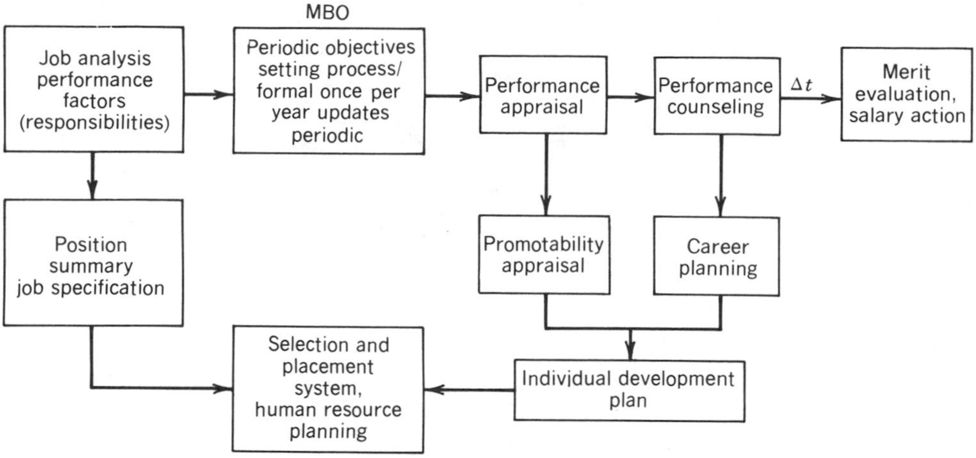

MBO

```
┌─────────────────┐   ┌─────────────────┐   ┌──────────────┐   ┌──────────────┐      ┌──────────────┐
│  Job analysis   │   │Periodic objectives│  │              │   │              │ Δt   │    Merit     │
│  performance    │──▶│setting process/ │──▶│ Performance  │──▶│ Performance  │─────▶│ evaluation,  │
│    factors      │   │formal once per  │   │  appraisal   │   │ counseling   │      │salary action │
│(responsibilities)│  │  year updates   │   │              │   │              │      │              │
│                 │   │    periodic     │   │              │   │              │      │              │
└─────────────────┘   └─────────────────┘   └──────────────┘   └──────────────┘      └──────────────┘
        │                                            │                  │
        ▼                                            ▼                  ▼
┌─────────────────┐                         ┌──────────────┐   ┌──────────────┐
│    Position     │                         │ Promotability│   │   Career     │
│    summary      │                         │  appraisal   │   │  planning    │
│ job specification│                        │              │   │              │
└─────────────────┘                         └──────────────┘   └──────────────┘
        │              ┌─────────────────┐                            │
        │              │  Selection and  │          ┌─────────────────────────┐
        │              │    placement    │          │ Individual development  │
        └─────────────▶│     system,     │◀─────────│         plan            │
                       │ human resource  │          └─────────────────────────┘
                       │    planning     │
                       └─────────────────┘
```

Δt: Appraisal and counseling should be temporally separated from the actual
      evaluation process (Meyer, Key, French, 1965)

Figure 12.10   MBO and the Performance Management Process

managers are not accustomed to such practices. He goes on to provide ten guidelines for the formulation of productivity objectives:

1. They must be measurable.
2. They must achieve single-ended results. (Keep them as simple as possible.)
3. They must incorporate deadlines.
4. They must be challenging, yet attainable.
5. They should focus on opportunities for improvement.
6. They must have motivating potential for those who must achieve them.
7. They must be supported by the organization.
8. They must be controllable.
9. They must have assigned accountability.
10. They must be evaluative; the desired outcome or results sought from the objective must be clear.

I would add that productivity objective setting for individuals or groups must invoke Pareto's Principle. Some mechanism for prioritizing and/or selecting the important goals and objectives to which to devote resources is essential.

> The purpose of a management system is to coordinate all the available resources and effort of people toward agreed-upon goals and objectives (Mali, 1978).

The goals, objectives, and purposes determine how relationships, activities, and processes need to be established. Managing productivity by objectives and

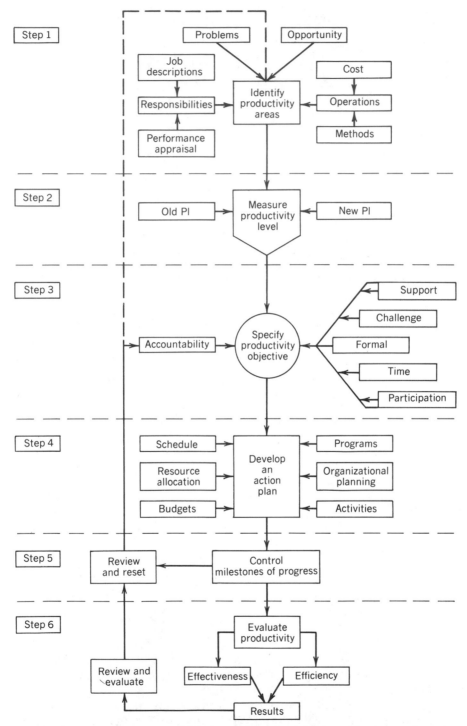

Figure 12.11   The Strategy of Managing Productivity by Objectives (Adapted from Mali)

a productivity measurement system like the MFPMM or NPMM create a management system to improve productivity. The NPMM can actually be utilized to generate the prioritized list of productivity objectives. The strategy for managing productivity by objectives is presented in Figure 12.11. Note that at key stages in this process, potential applications for the NPMM, MFPMM, and MCP/PMT are identified by Mali. Keep in mind also that this process is essentially an MBO process with a focus on productivity. The process can be developed and implemented at the individual or group level and can be integrated nicely into any ongoing or existing performance measurement and evaluation process at either of those levels.

Goal setting represents a potentially significant source of productivity improvement in American organizations. When goal-setting principles are developed and integrated with reinforcement theory, job enrichment or job content motivation theories, and participative methodologies in the form of management techniques, programs, and practices, and extremely powerful and potentially effective performance and productivity improvement approach results. The next kind of technique to be discussed is job design, job redesign, enrichment, and more generally the content theories and techniques of motivation.

C. Job Design, Redesign, Enrichment

The Locke et al. (1980) report of the study comparing the relative effectiveness of the four methods for motivating improved performance suggests that job enrichment would rank third behind money (reinforcement theory applications) and goal setting.

> Job or task design is the formal and informal specification of an employee's task-related activities, including both structural and interpersonal aspects of the job, with considerations for the needs and requirements of both the organization and the individual (Griffin, 1982).

Adam Smith, in his classic book *An Inquiry into the Nature and Causes of the Wealth of Nations* (1776), coined the phrase "division of labor" and described its advantages. A depiction of the general historical development of task design is presented in Figure 12.12. The Industrial Revolution and the ensuing development of scientific management created momentum for continued specialization of work. The fields and studies of wage and salary administration, job analysis, work simplification, methods engineering, motion economy, and certain incentive systems all were attempts to study, analyze, and design more efficient jobs and tasks. Making production methods more efficient was the overriding objective of management from the turn of the century through the early forties.

During the late thirties, research focusing on the relationship between the human being in the work system and specialized work began to attract consid-

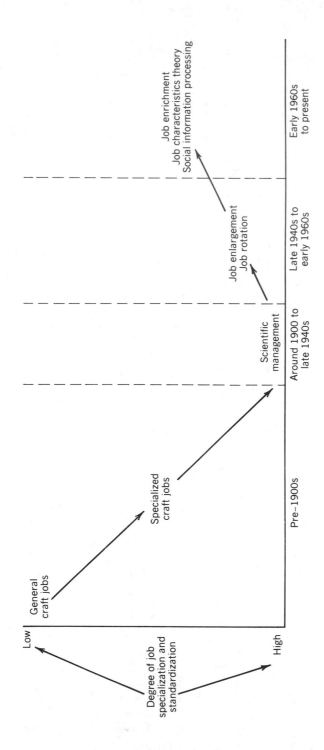

Figure 12.12 Historical Development of Task Design (From Alan C. Filley, Robert J. House, and Steven Kerr. *Managerial Process and Organizational Behavior*, 2nd ed. Copyright © 1976, 1969 by Scott, Foresman and Company. Reprinted by permission)

erable attention (Roethlisberger and Dickson, 1939). The human relations movement in management theory has overlapped with scientific management theory and practices for some 40 years now. Some evidence of both perspectives is still in existence today while hybrid, integrated management theories are also evident and will likely begin to dominate management thought in the eighties.

Problems associated with the effect of worker dissatisfaction on performance began to become more prevalent and pronounced in the fifties, sixties, and seventies. The fast-paced, highly regimented and monotonous routine of the asembly line has in particular been a primary area of problems and studies. Walker and Guest (1952); Herzberg, Mausner, and Snyderman (1959); Herzberg (1966); Davis (1957); and others developed early conceptualizations for both the problems and potential solutions.

Whenever the task of a job changes because of such things as new technology, a restructuring of the organizations, an employee's initiative, a supervisor's or manager's instructions, or an MBO process, it could be said that job/work redesign has taken place. The designation of such redesign may be formal or informal. Formal designation often takes the form of job specification, job descriptions, and supervisory communications. Informal designation of what a job or task entails, often a more influential form of communication and source of role perception, can take the form of supervisor-subordinate discussion, casual assignments, peer/co-worker communication, or employee initiative. Both formal and informal designation help shape the way a person performs a given task or job.

The goal of job/task/work redesign is to alter a specific job (or interdependent system of jobs) so that employees' perceptions of quality of work life improve along with their motivation to perform and their actual performance (effectiveness, efficiency, quality, innovation). The term "work redesign" necessarily includes such operational approaches and techniques as job rotation, job enlargement, human factors engineering and design, job enrichment, sociotechnical systems designs methods engineering and design, MBO, team-oriented MBO, quality circles, performance/productivity action teams, work simplification, and "work complification." Although much research has been done in this particular area, however, there is little conclusive evidence as to predictable effects of work redesign efforts. Much of the difficulty associated with studying this topic stems from:

1. The wide range and number of actual intervention techniques available to operationalize the approach.

2. The lack of specificity of technique and, hence, the lack of consistent application of technique.

3. The artistic, design orientation to work redesign. Diagnostic techniques and consistent evaluative procedure development and application have lagged behind the application of the techniques themselves.

Thus, there are no generally accepted criteria or performance measures with which to evaluate the quality of the design of a job. As mentioned, there exists no single strategy for improving the design of a job. Most jobs are never explicitly and systematically designed initially. Most jobs undergo a constant informal redesign process.

Systematic redesign efforts have tended either to utilize techniques for isolated job-related problems on a case by case basis or to utilize work redesign diagnosis, evaluation, and improvement techniques as a component of a larger program aimed at improving productivity and quality of work life. Hackman and Oldham (1975) suggest that work redesign differs from other behavioral science approaches in four ways that make it an excellent technique for initiating organizational change. First, work redesign alters the basic relationship between a person and his or her job. One could argue that this interface is perhaps the most critical one between an employee and the organization. Second, work redesign directly changes behavior by changing what people are expected to do on the job. Assuming assignments are accepted, it can be assumed that the behaviors of employees will be directed toward the new job activities. Third, work redesign often presents a variety of opportunities for initiating other organizational changes. The diagnostics and actual procedures associated with many job redesign approaches and techniques often provide additional insights and motivation for making further changes that can positively influence individual and group performance. For instance, methods, technology, critical interface, and interdependencies are opportunities for improvement that will often surface through a work redesign program. Finally, work redesign efforts, in the long term, can result in organizations that successfully balance social and human needs with the organization's technological needs. As Naisbitt (1982) puts it, "high touch" can be kept in balance with "high tech."

Job enlargement is horizontal task loading of a job, whereas job enrichment is vertical task loading. If one thinks of task demands in terms of a hierarchy, a secretary's general tasks tend to be at one level while a manager's tend to be at another. Job enlargement is giving a person more of the same type of work while job enrichment is delegation of "more complex" or "higher-order" work.

Work simplification represents sort of the inverse of job enlargement. It involves breaking work down into basic components and motions so that very specialized tasks can be identified, trained, and then performed efficiently. For instance, on an automobile assembly line it is not uncommon for the task of one worker to span only 30 to 40 seconds. The task has to be simple and well learned to allow the worker to keep up that pace for six to eight hours a day. It is important to keep in mind that work simplification isn't necessarily bad and job enrichment necessarily good. The question always is, "What is appropriate for the technology, task, and people in the work system?"

Scientific management, methods engineering, work measurement and improvement, motion economy, operations analysis, and human factors engineering and design were management's early systematic attempts to redesign work. During the early stages of the industrial revolution in the United States, there were several successful attempts to improve performance. Because management abused human rights and was unable to implement these techniques properly, however, these techniques developed an undeserved bad reputation.

In the late thirties, further misapplications of these techniques caused a human relations movement to develop. Unions grew in power and approaches directed at humanizing work began.

Job enlargement was management's first purposeful attempt to redesign tasks away from the specialization and standardization thrust of the turn of the century. The technique was first used in the late forties and has been attempted at IBM, Detroit Edison, American Telephone and Telegraph, Maytag, and other companies (Walker, 1950; Filley, House, and Kerr, 1976). In general, job enlargement, although a positive step in the field of job redesign, apparently has limited application. It can affect performance in cases of overstaffing and under-utilization of personnel and where demand for services and/or goals is either not met or increasing.

Job rotation was an early attempt to counter boredom and dissatisfaction caused by high levels of job specialization. During job rotation, assignments are changed, thus possibly affecting vertical and horizontal task loading. The short-term outcomes (efficiency) from rotation plans are disappointing (Miller, Dhaliwal, and Magas, 1973). As a longer-term effort, however, rotation may have promise in building organization communication, coordination and capacity. A key attribute of Japanese-managed organizations (type J and type Z; Ouchi, 1981) seems to be horizontal job rotation from function to function at all levels in the organization. This creates a slower promotion process, a deeper understanding of the integrated and interdependent characteristics of the organization, and apparently improved cooperation and coordination. Although job rotation as a job redesign technique appears to be in some disfavor among American managers, implemented effectively it would appear to have potential for improved long-term performance.

Job enrichment represents the most popular and probably most successful job redesign approach and technique. The foundation for this technique rests on Herzberg's two-factor theory (Herzberg, et al., 1959). The theory, although instrumental in this field's development, has been much criticized for its simplicity and methodological-related problems.

Job enrichment predicts that meaningful jobs can be characterized by at least the following six factors:

1. *Accountability:* the employee is held responsible for performance

2. *Achievement:* the employee senses the job is accomplishing something worthwhile

3. *Control over resources:* the extent of the employee's control over the task

4. *Feedback:* the extent to which a worker receives direct information regarding performance

5. *Personal growth and development:* the extent to which a worker has the opportunity to develop new skills

6. *Work place:* the control the worker has over the characteristics of the work place

Most research to date has focused on operationalizing and testing job enrichment theory (Dunham, 1977; Hackman and Lawler, 1971; Hackman and Oldham, 1975, 1976, 1980; Oldham, 1976; Oldham, Hackman, and Pierce, 1976; Pierce, Dunham, and Blackburn, 1979; Sims and Szilagyi, 1976; Sims, Szilagyi, and Keller, 1976; Wanous, 1974; Ferratt, Dunham, and Pierce, 1981). The reserch has focused on (1) testing the constructs represented by the six factors listed above; (2) determining the moderating effect of certain individual differences on job enrichment theory application; (3) developing and testing instrumentation for measuring job characteristics and job satisfaction; (4) continued development of theoretical models for the theory itself; and (5) applications and testing of models in real world settings.

Job Characteristics Theory

A significant theory that developed from much of the earlier work on job redesign is called the Job Characteristics Theory. The basic Job Characteristics Theory model is presented in Figure 12.13. It suggests that the presence and sense of five core job dimensions sets up the probability that positive critical psychological states will exist within an employee. The three job dimensions contributing to a job's perceived meaningfulness are skill variety, task identity, and task significance where

1. Skill variety represents the degree to which a job requires and allows a variety of different activities that involve a number of different skills and talents.

2. Task identity represents the degree to which the job requires and allows completion of a whole and identifiable piece of work, or doing a job from beginning to end with a visible outcome.

3. Task significance represents the degree to which a job is perceived by the employee as having substantial impact on the organizational system, the organization as a whole, and perhaps even society.

The job characteristic predicted to create a felt sense of responsibility for the work outcomes is autonomy where autonomy is the degree to which the job provides substantial freedom, independence, and discretion on the part of the individual in scheduling for, determining procedures for, and in actually carrying out work. The job characteristic that develops knowledge of results is feedback where feedback represents the degree to which executing work-related activities results in the individual obtaining clear and direct information about his or her performance.

These psychological states are theoretized as being primary determinants of employee motivation and satisfaction. The degree to which the critical psychological states exist is theoretized as establishing the probability that positive personal and work-related outcomes will exist or occur. The model suggests that a job with relatively higher degrees or amounts of the characteristics represented by the core job dimensions would be more highly motivating and satisfying than a job with lower amounts of the core job dimensions. The theory also

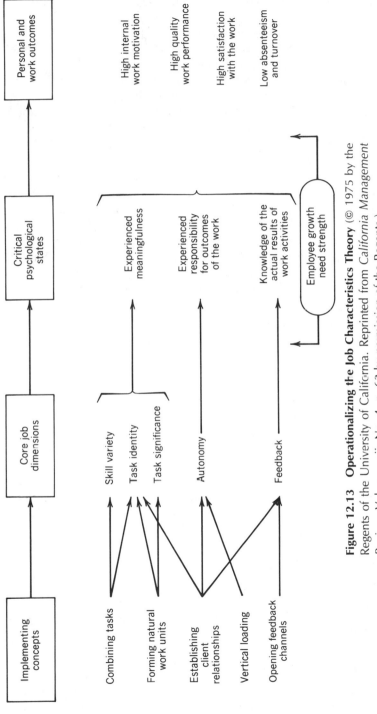

Figure 12.13 Operationalizing the Job Characteristics Theory (© 1975 by the Regents of the University of California. Reprinted from *California Management Review*, Volume xvii, No. 4, p. 62 by permission of the Regents.)

proposes an important moderating variable between the core job dimension (independent variable) and personal and work-related outcomes (dependent variable)—that of employee growth-need strength.

Operationalizing the job characteristics model has primarily involved obtaining worker reactions (self-reports) the job characteristics via perceptions gathered using such instruments as the Job Diagnostic Survey (JDS) (Hackman and Oldham, 1975) and the Job Characteristics Inventory (JCI) (Sims, Szilagyi, and Keller, 1976). The most common dependent variable has been job satisfaction (a weakness in this area of research), which has been evaluated through such job-related attitude surveys as the Job Descriptive Index (JDI) (Smith, Kendall, and Hulin, 1969), the Index of Organizational Reactions (IOR) (Smith, 1976), the Minnesota Satisfaction Questionnaire (MSQ) (Weiss, Dawis, England, and Lofquist, 1967), and the Stogdill Satisfaction with Work Scale (SWS) (Stogdill, 1965). Recent research has addressed the question of whether it is possible to discriminate between common measures of perceived job design (the JDS and JCI) and those of job satisfaction (the JOI, IOR, MSQ, and SWS).

Scores from the JDS, for example, relative to the five core dimensions are combined to obtain a motivating potential score (MPS). The formula utilized to combine scores from the JDS and to obtain a single MPS index is

$$(MPS) = \frac{\frac{skill}{variety} + \frac{task}{identity} + \frac{task}{significance}}{3} \times autonomy \times feedback$$

Jobs with low MPS are potential candidates for task redesign. Figure 12.14 depicts an example of diagnostics possible with JDS. Once target jobs for redesign efforts are determined, job characteristics theory is further operationalized by redesigning tasks according to five basic principles (Hackman, Oldham, Janson, and Purdy, 1975). The principles are

1. Form natural work units.
2. Combine tasks.
3. Establish client relationships.
4. Establish vertical loading.
5. Open feedback channels.

Figure 12.13 depicts the relationship between the core job dimensions and these change principles.

Job Diagnostic Survey

The Job Diagnostic Survey was first presented in the literature by Hackman and Oldham (1975). The JDS was intended "to be of use both in the diagnosis of jobs prior to their redesign, and in research and evaluation activities aimed at

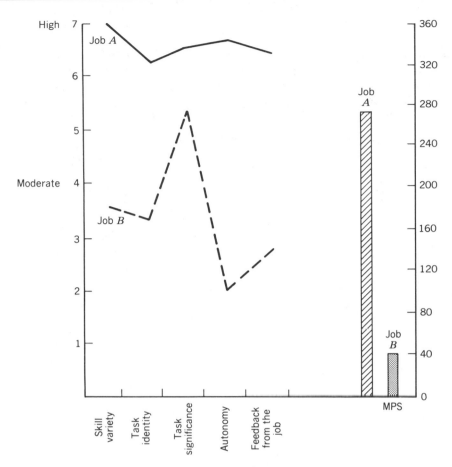

Figure 12.14   JDS Profile of a "Good" and "Bad" Job (From J. R. Hackman "Work Design" in J. R. Hackman and J. L. Suttle eds., *Improving Life at Work: Behavioral Science Approaches to Organizational Change*, Santa Monica, Calif.: Goodyear, 1977. Reprinted with permission.)

assessing the effects of redesigned jobs on the people who do them" (Hackman and Oldham, 1975). The total JDS includes 83 items. The five core job dimension scores are obtained by averaging the scores from the three items and then utilizing the MPS formula to obtain an index. Hackman and Oldham (1980) present average data obtained from the JDS for nine different job families. Hence, an advantage of the JDS is that the beginnings of a pool of normative data with which to make comparisons exist.

Job Characteristics Inventory

The Job Characteristics Inventory consists of 30 items. The JCI does not tap the significance characteristic or dimension. However, preliminary research on the JCI indicates it may have certain advantages over the JDS.

These diagnostic instruments provide one with necessary data with which to evaluate and assess the degree to which job design or the content of the job itself may either be enhancing performance or detracting from it. They do not represent the only sources of data on the character of work. Job analysis and evaluation, as components of a wage and salary administration program, also represent a focus on the analysis of jobs. Interested readers are directed to

Table 12.10 Steps for Implementing Task Redesign

MEASUREMENT AND EVALUATION	PROBLEM ANALYSIS	**Step 1:** Recognition of need for change (a) JDS, JCI (b) Low performance (c) QWL problems (d) Intuitive
		Step 2: Evaluation, diagnosis, selection of task redesign as the appropriate performance/productivity improvement technique or approach
		Step 3: Diagnosis of the work system and context (a) Diagnosis of existing jobs (b) Diagnosis of existing work force (c) Diagnosis of technology (d) Diagnosis of organization design (e) Diagnosis of leader behaviors and style (f) Diagnosis of group and social processes
PLANNING FOR CONTROL AND IMPROVEMENT	DECISION ANALYSIS	**Step 4:** Cost/benefit analysis of proposed changes
		Step 5: Go/no-go decision
	DESIGN ANALYSIS	**Step 6:** Establishing the strategy for redesign (a) Design (b) Test (c) Evaluate (d) Modify
ACTION PLANNING AND EFFECTIVE IMPLEMENTATION	IMPLEMENTATION ANALYSIS	**Step 7:** Implementation of task changes
		Step 8: Implementation of any supplemental changes in other peripherally related systems
		Step 9: Evaluation of task redesign effort(s)

Henderson, 1979 for more detail on specific concepts and techniques associated with job analysis.

Griffin (1982) provides a picture of what an integrated system of task measurement might look like. Coupled with a reasonably comprehensive view of compensation management provided by Henderson (1979), Griffin's work can help the reader view how job characteristics measurement and evaluation fit into the bigger picture.

Of course, the major concern or focus of most managers is on improvement— How can one improve the motivation and performance of the employee through job or task measurement and evaluation? In other words, how does one effectively link measurement and evaluation to control and improvement? Of course, the answer is that based on the measurement process, the manager must design and develop intervention strategies and techniques aimed at alleviating or overcoming deficiencies found in the job designs. Specific detail on task design and/or redesign approaches is beyond the scope of this text. (For more detail, see Griffin, 1982; Aldag and Brief, 1979; and Henderson, 1979.

A general and reasonably integrated framework for implementing task redesign is provided by Griffin. The steps are outlined in Table 12.10. Note that this approach is a systematic process of planning, measurement, evaluation, control, and improvement cycled and sequenced according to this specific job redesign focus.

Although the research on the success of job enrichment and other task redesign approaches is inconclusive, it would appear that management's attention to the motivational potential of what we do on the job is warranted. Selection and placement, reinforcement theory applications, effective leadership, and training and development are all factors that will have strong influence on performance levels. Efforts to systematically evaluate and consider the nature of the work itself will play an important role in productivity improvement in the future as well.

Participation or employee involvement has also been a factor or variable examined for its role in affecting employee performance and motivation as well as productivity in general. This topic will be examined in the next section.

D. Participation: Employee Involvement

Introduction

In Locke's (1980) study, referred to and summarized in Table 12.1, the findings suggested that participation was the least effective of the four methods for motivating improved performance. Median percentage improvement in performance for the 16 studies examined was 0.5 percent, with only 25 percent of the studies showing greater than a 10 percent improvement in performance. The range of results was -24 percent to $+47$ percent. Based on these findings, one might question the wisdom of even considering participation as an effective motivation strategy or technique.

Participation, however, perhaps even more so than the other three motivational strategies just discussed, comes in a tremendous variety of shapes, sizes, and forms. Locke's study was limited to studying participation in decision making or in joint decision-making interventions. Thus, many other forms of participation were specifically excluded such as delegation, job enrichment, the Hawthorne studies, and System 4 studies (Likert). These studies were confounded with many other factors, such as incentives and technology.

Locke et al. also excluded quality circles for at least two reasons. First, few if any systematic studies of quality circles exist today, much less in 1980 when the researchers published their study. Second, the impact of quality circles on motivation and performance is also confounded because quality circles are a mixed motivational strategy. They are perhaps first recognized as a participative problem-solving technique, although they also have distinct elements of job enrichment, delegation, goal setting, training and development, and sometimes even reinforcement theory.

Participation as a strategy is probably going to be a necessary but not sufficient element of any productivity improvement program in the eighties and nineties In the United States and in several developing countries, the maturity and expectations of employees are increasing rapidly. Many employees today, particularly professionals and those in service sectors, expect if not demand to participate to a certain extent in job-related problem solving. It seems, despite Locke's findings, that the question of whether participation should be used in the work place, is being answered for managers. Participation, in structured, systematic designs, is increasingly coming to be viewed as necessary process for facilitating and enhancing the probability that other productivity improvement techniques will be implemented effectively and efficiently. The consensus and commitment to accept, support, and even cause positive productivity improvement-related changes that can be generated with properly designed participatory programs is seen as being essential to the survival of most organizations in the coming decades.

It seems therefore that the appropriate questions relative to participation are

1. Why use participative strategies and techniques?

2. When should a participatory approach or technique be used?

3. What form of participative technique should be used?

4. Who should be involved?

5. What focus should be used for the participation (for example, to solve a problem, design a solution, clarify an issue, and make a decision)?

6. How should the participatory approach or technique be integrated into a broader-scoped productivity improvement program?

7. How should the participatory approach, technique, or program be lead or facilitated in an appropriate fashion so as to ensure its success?

Some of these critical questions regarding participation as a performance and productivity improvement strategy will be addressed here; others will be addressed in the discussion of performance/productivity action teams.

It would appear that employee involvement strategies and techniques are mechanisms for involving select individuals (employees, citizens, students, faculty, managers, executives, supervisors, clients, customers) either by necessity or choice in one or more of the following categories of activities:

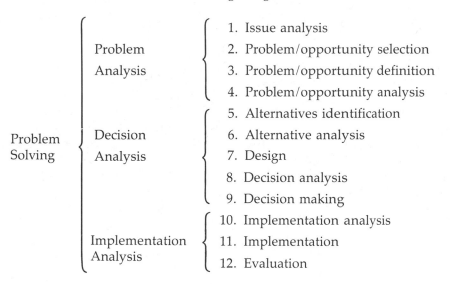

Further, employee involvement most frequently incorporates small-group activity relative to one or more of the 12 categories of activities or focuses.

Employee involvement or participation should

1. Involve participants in an effective and efficient fashion

2. Involve the "right" participants in the "right" tasks

3. Create an awareness of the need for change

4. Facilitate effective and efficient decision making in terms of how, what, where, and where to change

5. Facilitate self-awareness and self-direction on the part of the participants and groups in terms of change

6. Facilitate beneficial change

7. Improve goal congruency between individual, group, and organization

8. Tap the human resource

9. Make implementation more effective, more timely, and in general easier

10. Develop participants' skills

11. Improve communication, cooperation, and coordination

12. Create commitment

13. Improve work-related attitudes

14. Improve the quality of effort

15. Create improved group harmony

16. Lead to potential improved productivity

17. Lead to potential improved overall performance.

Employee involvement participation programs and techniques, when designed properly and integrated effectively into managerial and organizational processes, can accomplish many of the outcomes listed above. Any technique or approach, regardless of its quality and appropriateness, if not implemented correctly, can obviously fail and perhaps even produce dysfunctional consequences. This seems to be particularly true of behavioral and social science techniques and approaches, since so much of their success depends on the skills of the person or persons executing them. In fact, in the United States today there are countless organizations attempting to implement quality circles with untrained and unskilled personnel. The result is a high failure rate as well as the significant negative impact on the morale of people in those organizations.

Involvement and participation is increasingly coming to mean:

1. A voice in problem solving;

2. Consultation, consensus seeking;

3. Consent in final decision;

4. Disciplined, structured, systematic attempts to capture and utilize individual and group wisdom;

5. Shared decision making;

6. Effective delegation;

7. Shared conceptions of problems and appropriate action steps;

8. An opportunity to create winning situations and attitudes;

9. A mechanism for improving labor-management cooperation.

Involvement and participation are increasingly coming to be viewed as necessary but not sufficient ingredients of a productivity improvement effort or program.

Managers in the United States and in other developed and developing countries are using involvement strategies because:

1. The maturity (capacity and willingness) of employees is increasing;

2. It is becoming more popular and appropriate to delegate downward decentralize the logic and responsibility for decisions in organizations;

3. Involvement strategies enhance the effectiveness of decision implementation; (The Japanese are relatively slower to arrive at a decision but significantly superior at implementing the decision once made.);

4. They increase the understanding of the decision and the implementation issues;

5. They tend to increase the information and skills that the group and individuals may require for future organizational assignments;

6. They build the capacity of groups, organizational systems, and organizations to solve problems proactively and effectively;

7. They often provide a mechanism for employee and supervisor development as well as improving communication at that vital interface;

8. They can and often improve interdepartment, interfunction communication and cooperation. Vital internal interface communication is often a critical performance and productivity problem that can be attacked with appropriately designed employee involvement program;

9. They pave the way for needed innovations in organizational systems.

When the "right" task or problem is presented to the "right" group of individuals, and the "right" group process is utilized, the results are exciting and represent a tremendous source of productivity improvement.

An example of the vast potential for effective participation in most organizations is illustrated by a recent article describing a small-group activity program within the Musashi Semiconductor Works in Japan (Davidson, 1982). Musashi is Hitachi Corporation's oldest and largest semiconductor factory and is an independent profit center accounting for 9.8 percent of Hitachi's total sales of $13.4 billion.

Despite the level of computer-aided manufacturing systems utilized in the production process, the plant is still highly labor intensive. Labor costs and productivity, measured in terms of effective yield, are critical determinants of success in the industry. All of Hitachi's 27 Japanese factories, including the Musashi factory, employ formal small-group activity programs. The Japanese concept of small-group activity encompasses a much broader range of activities than normally associated with quality circles.

Musashi began its small-group activity program in 1971. The initial stage of implementation occurred between 1971 and 1975 and was called the enlightenment period. The focus during this period was almost entirely on orientation of management in the principles, philosophy, strategies, structure, techniques, and functions of small-group activities. Figure 12.15 depicts the project plan for the Musashi program.

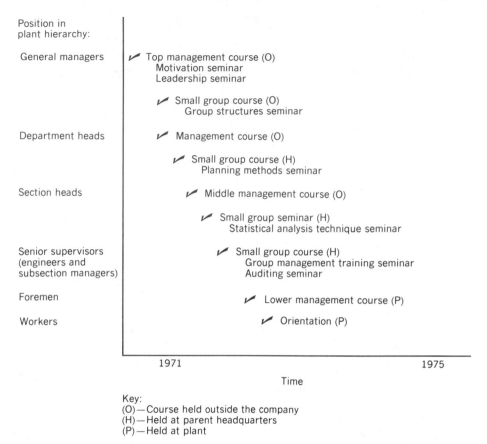

Figure 12.15   Education Support Program for Small Group Systems Implementation

The 2700 workers at Musashi are organized into 360 groups of 8 to 10 persons per group. Group formation is largely by work stations, and each group elects a leader. The administration and coordination of the program is discussed in some detail in Davidson's article, and interested readers are directed to that reference for more program management details.

The results of the program are impressive. The first formal improvement proposals from groups were filed in 1977. Table 12.11 depicts the results. Between July and December 1980, the average group completed about 45 improvements per month. The completed improvements can be broken down into the following categories:

Of the 98,347 improvements completed in the second half of 1980:

• 26 percent resulted in a reduction in standard times at individual work stations
• 27 percent resulted in the reduction of inventory

Table 12.11 Results of Musashi Program

PERIOD	IMPROVEMENT PROPOSALS SUBMITTED	IMPROVEMENT PROPOSALS IMPLEMENTED	PERCENTAGE IMPLEMENTED	NUMBER PROPOSALS/ SMALL GROUP
1978	26,543	Data not provided		73
1979 1st half	47,347	Data not provided		131/half yr.
1980 2nd half	112,022	98,347	87.8%	311/half yr.

- 6 percent resulted in safety improvements and overhead cost savings
- 24 percent resulted in efficiency improvements in office and clerical functions

The remaining proposals were devoted primarily to increasing yields at various stages of production.

The effectiveness of this small-group activity is incredible by most western standards. Yet the results achieved by Musashi as compared with other Japanese firms with similar programs are not that impressive. Managers in the United States need to take heed!

The dilemma facing most American managers is how to develop small-group activities, employee involvement, and participative programs in such a way that they can be equally effective, obtain the potential outcomes just presented, and create results even close to the Musashi experience. If we allow employees to become involved in appropriate elements in the management process, we capture their minds. If we capture their minds, we capture their hearts. If we capture their hearts, we ensure their commitment. If we have hired the right people and have created the right opportunities and supportive climate, their commitment will ensure performance.

When Is Participation Appropriate?

Participation is not always the most appropriate process for accomplishing tasks. There are circumstances under which autocratic, authoritative, unilateral problem solving or delegation to a single individual is appropriate. The use of small groups as a form of participation is most appropriate for the following purposes (Kanter, 1982):

1. To gain new sources of expertise and experience

2. To obtain collaboration that multiplies an individual's effort by providing assistance, support, and stimulation for better performance

3. To allow those who feel they have knowledge about a subject or problem to become involved

4. To develop consensus on controversial issues, problems, and ideas for performance and productivity improvement, goals, action programs, and so forth

5. To allow representatives of those organizational systems who are going to be affected by an issue, problem, decision, or implementation a chance to influence the approach and results and to build commitment

6. To identify and tackle a problem or an opportunity that no one person "owns" or feels responsible for by virtue of organizational assignment

7. To allow for a wider range of creative discussions/solutions than are typically available by normal means

8. To balance or confront vested interests and points of resistance to needed change

9. To address and provide a forum for conflicting and different perspective views on an issue, problem, decision, or implementation

10. To avoid precipitous or poorly thought-through action and explore a variety of possible effects

11. To create an opportunity and enough time and other resources to study a problem in depth

12. To develop and educate people through their participation in creating new skills, new information, new perspectives, new contacts, and so forth

However, there are also times when participation is not appropriate. Among them are (Kanter, 1982):

1. When one person clearly has greater expertise on the subject than others

2. When those to be affected by a decision acknowledge and accept that expertise

3. When there is a "hip pocket" solution—an easy, clear, acceptable solution the manager of the company already knows is the "right" answer

4. When the issue, problem, decision, or implementation is part of someone's regular job assignment, and it is not clear that the person will accept a group approach

5. When there is not much salience regarding the topic—it isn't viewed as being a critical issue, problem, decision, or implementation

6. When there is no time for participation

7. When the affected people work more happily and productively alone

8. When there is no potential benefit from involving others in the process

Vroom (1973), Vroom and Yetton (1973), and Vroom and Yago (1974) have developed a model (decision tree) that asks managers to address a sequence of questions regarding certain situational variables associated with the choice of decision-making processes. The intent is to force managers to think through alternative decision-making processes (autocratic, consultative, participative) and to select one that is most appropriate given the situation. In other words, such factors as quality requirements, time pressures, information dispersion, and need for acceptance may dictate choosing an alternative process.

A prevailing bias in the literature for the past 10 to 20 years is the preference for democratic processes like participative problem solving and decision making. The popularity of quality circles in the past ten years has certainly heightened this bias. The bias suggests that participation is somehow always better than an authoritarian or autocratic process. Of course, we examined this bias earlier and learned that the issue is one of control and how that control is exercised. We also stressed that tight control (authoritarian, autocratic) is not necessarily bad, moderate control (consultative) better, and loose control (participative, delegative) best.

Vroom et al. have developed a very promising and potentially powerful approach for looking at certain situational variables and their relationship to alternative problem solving and decision-making processes. Note that the focus is on group-type problems and decisions and the intent is to diagnose the appropriate leadership style. The model will sort of operationalize the lists Kanter provided on when and when not to use participation.

Five types of managerial problem-solving or decision-making styles are presented in Table 12.12. The model suggests three major categories of variables that should play a role in determining whether a particular problem-solving and/or decision-making style will result in desired outcomes:

1. The quality requirement for the decision

2. The extent to which subordinate acceptance and commitment to the problem or decision is important to effective implementation

3. The amount of time within which the decision has to be made

A fourth variable that has been written about (Hersey and Blanchard, 1982) and that is predominant in situational leadership theory and models is the follower's (in this case the group's) maturity (willingness and capacity).

In this model, outcomes from specific problem-solving and decision-making processes depend primarily on the first two categories of variables. The time variable primarily affects the selection of a leadership (problem-solving and/or decision-making) style when there are two or more appropriate styles for a given situation.

As you can see from Table 12.12, AI and AII are autocratic styles and neither really involves subordinates. CI and CII are termed "consultative styles" and extend the level of involvement or decrease the amount of control exerted by the leader on the problem or decision. GII is a group or participative leadership style and provides for a great deal of subordinate involvement. If you refer to Figure 10.3, you can see that conceptually AI is on the far left side of the horizontal control axis, CI and CII is near the center, and GII is near the right side of the scale.

Table 12.12 Models of Leader Behavior

FOR INDIVIDUAL PROBLEMS	FOR GROUP PROBLEMS
AI You solve the problem or make the decision yourself, using information available to you at that time.	AI You solve the problem or make the decision yourself, using information available to you at that time.
AII You obtain any necessary information from the subordinate, then decide on the solution to the problem yourself. You may or may not tell the subordinate what the problem is, in getting the information from him. The role played by your subordinate in making the decision is clearly one of providing specific information that you request, rather than generating or evaluating alternative solutions.	AII You obtain any necessary information from subordinates, then decide on the solution to the problem yourself. You may or may not tell subordinates what the problem is, in getting the information from them. The role played by your subordinates in making the decision is clearly one of providing specific information that you request, rather than generating or evaluating solutions.
CI You share the problem with the relevant subordinate, getting his or her ideas and suggestions. Then you make the decision. This decision may or may not reflect your subordinate's influence.	CI You share the problem with the relevant subordinates individually, getting their ideas and suggestions without bringing them together as a group. Then you make the decision. This decision may or may not reflect your subordinates' influence.
GI You share the problem with one of your subordinates and together you analyze the problem and arrive at a mutually satisfactory solution in an atmosphere of free open exchange of information and ideas. You both contribute to the resolution of the problem, with the relative contribution of each depending on knowledge rather than formal authority.	CII You share the problem with your subordinates in a group meeting. In this meeting you obtain their ideas and suggestions. Then you make the decision, which may or may not reflect your subordinates' influence.
DI You delegate the problem to one of your subordinates, providing him or her with relevant information but giving the subordinate responsibility for solving the problem above. Any solution that the person reaches will receive your support.	GII You share the problem with your subordinates as a group. Together you generate and evaluate alternatives and attempt to reach agreement (consensus) on a solution. Your role is much like that of chairperson, coordinating the discussion, keeping it focused on the problem, and making sure that the critical issues are discussed. You do not try to influence the group to adopt "your" solution and are willing to accept and implement any solution that has the support of the entire group.

SOURCE: V. H. Vroom, and A. G. Yago, "Decision Making as a Social Process: Normative and Descriptive Models of Leader Behavior," *Decision Sciences*, Vol. V, 743–69, 1974.

Based on theory and research in this area, Vroom and Yetton have developed a set of situational characteristics or "problem attributes" and associated diagnostic questions, which appear in Table 12.13. Associated with these attributes and questions are a set of rules used for selecting among alternative leadership (problem-solving and decision-making) styles. The rules are designed to ensure decision quality (rules 1, 2 and 3), and acceptance (rules 4, 5, 6, and 7).

1. *The information rule:* If the quality of the decision is important and if the leader does not process enough information or expertise to solve the problem alone, AI is eliminated from the feasible set. (Its use risks a low quality decision.)

2. *The goal congruence rule:* If the quality of the decision is important and if the subordinates do not share the organizational goals to be obtained in solving the problem, GII is eliminated from the feasible set. (Alternatives that eliminate the leader's final control over the decision may jeopardize the quality of the decision.)

3. *The unstructured problem rule:* In cases where the quality of the decision is important, if the leader lacks the necessary information or expertise to solve the problem alone, and if the problem is unstructured, the method used must provide not only for the leader to collect the information but also to do so in an efficient and effective manner. Methods that involve interaction among all sub-

Table 12.13 Problem Attributes Used in the Model

PROBLEM ATTRIBUTES	DIAGNOSTIC QUESTIONS
A. The importance of the quality of the decision	Is there a quality requirement such that one solution is likely to be more rational than another?
B. The extent to which the leader possesses sufficient information/expertise to make a high-quality decision by himself or herself	Do I have sufficient information to make a high-quality decision?
C. The extent to which the problem is structured.	Is the problem structured?
D. The extent to which acceptance or commitment on the part of subordinates is critical to the effective implementation of the decision.	Is acceptance of decision by subordinates critical to effective implementation?
E. The prior probability that the leader's autocratic decision will receive acceptance by subordinates.	If you were to make the decision by yourself, is it reasonably certain that it would be accepted by your subordinates?
F. The extent to which subordinates are motivated to attain the organizational goals as represented in the objectives explicit in the statement of the problem.	Do subordinates share the organizational goals to be obtained in solving this problem?
G. The extent to which subordinates are likely to be in conflict over preferred solutions.	Is conflict among subordinates likely in preferred solutions?

SOURCE: Victor H. Vroom, "A New Look at Managerial Decision Making," *Organizational Dynamics*, Vol. 1, No. 4, Spring 1973, p. 67. Reproduced with permission.

ordinates with full knowledge of the problem are likely to be both more efficient and more likely to generate a high-quality solution to the problem. Under these conditions, AI, AII, and CI are eliminated from the feasible set. (AI does not provide for the leader to collect the necessary information, and AII and CI represent more cumbersome, less effective, and less efficient means of bringing the necessary information to bear on the solution of the problem.

4. *The acceptance rule:* If the acceptance of the decision by subordinates is critical to effective implementation, and if it is not certain that an autocratic decision made by the leader would receive that acceptance, AI and AII are eliminated from the feasible set. (Neither provides an opportunity for subordinates to participate in the decision, and both risk the necessary acceptance.)

5. *The conflict rule:* If the acceptance of the decision is critical, and subordinates are likely to be in conflict or disagreement over the appropriate solution, AI, AII, and CI are eliminated from the feasible set. (The methods used in solving the problem should enable those in disagreement to resolve their differences with full knowledge of the problem. Accordingly, under these conditions, AI, AII, and CI which involve only "one-on-one" relationships and therefore provide no opportunity for those in conflict to resolve their differences, are eliminated. Their use runs the risk of leaving some of the subordinates with less than the necessary commitment to the final decision.)

6. *The fairness rule:* If the quality of the decision is unimportant and if acceptance is critical and not certain to result from an autocratic decision, AI, AII, CI, and CII are eliminated from the feasible set. (The method used should maximize the probability of acceptance, as this is the only relevant consideration in determining the effectiveness of the decision. Under these circumstances, AI, AII, CI, and CII are eliminated from the feasible set. To use them is to run the risk of getting less than the needed acceptance of the decision.)

7. *The acceptance priority rule:* If acceptance is critical, if it is not assured by an autocratic decision, and if subordinates can be trusted, AI, AII, CI, and CII are eliminated from the feasible set. (Methods that provide equal partnership in the decision-making process can provide greater acceptance without risking decision quality. Use of any method other than GII results in an unnecessary risk that the decision will not be fully accepted or receive the necessary commitment on the part of subordinates.)

In order to facilitate the use of the rules and problem attributes in diagnosing situations, a decision tree has been developed. Figure 12.16 presents the Problem-solving or decision-making model. The problem attributes are presented along the top of the figure. The rules are applied to determine which path to take. For example, if you answer "no" to question A, you proceed to question (mode) D. If you answer "no" to that question also, you reach an outcome, in this case AI. This suggests that based on your diagnosis of the situation, style AI is most appropriate.

There are 14 outcomes or terminal points, each of which represents a problem

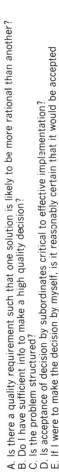
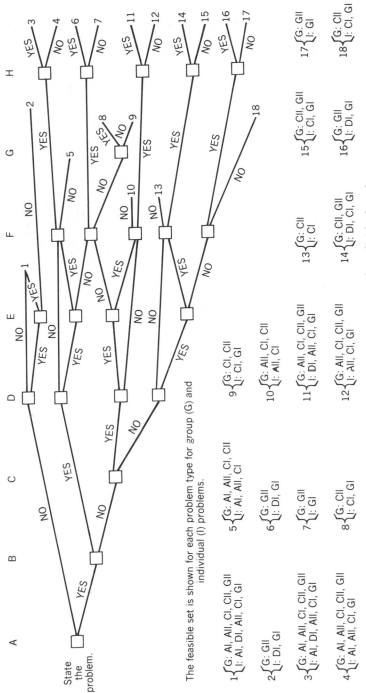

A. Is there a quality requirement such that one solution is likely to be more rational than another?
B. Do I have sufficient info to make a high quality decision?
C. Is the problem structured?
D. Is acceptance of decision by subordinates critical to effective implementation?
E. If I were to make the decision by myself, is it reasonably certain that it would be accepted by my subordinates?
F. Do subordinates share the organizational goals to be attained in solving this problem?
G. Is conflict among subordinates likely in preferred solutions? (This question is irrelevant to individual problems.)
H. Do subordinates have sufficient info to make a high quality decision?

The feasible set is shown for each problem type for group (G) and individual (I) problems.

1. G: AI, AII, CI, CII, GII / I: AI, DI, AII, CI, GI
2. G: GII / I: DI, GI
3. G: AI, AII, CI, CII, GII / I: AI, DI, AII, CI, GI
4. G: AI, AII, CI, CII, GII / I: AI, AII, CI, GI
5. G: AI, AII, CI, CII / I: AI, AII, CI
6. G: GII / I: DI, GI
7. G: GII / I: GI
8. G: CII / I: CI, GI
9. G: CI, CII / I: CI, GI
10. G: AII, CI, CII / I: AII, CI
11. G: AII, CI, CII, GII / I: DI, AII, CI, GI
12. G: AII, CI, CII, GII / I: AII, CI, GI
13. G: CII / I: CI
14. G: CII, GII / I: DI, CI, GI
15. G: CII, GII / I: CI, GI
16. G: GII / I: DI, GI
17. G: GII / I: GI
18. G: CII / I: CI, GI

Figure 12.16 Decision-Process Flow Chart for Both Individual and Group Problems

type. These are depicted in Table 12.14. For example, type 2 represents a situation in which quality is not important but acceptance is. In this situation, the subordinate will likely accept your decision or solution. It is probable that more than one style will be effective for many of the problem types. If this is the case and time is a factor, use the style that results in the quickest decision.

This model provides a structured approach to determining when to use participation. Vroom and Yetton (1973) found that about 30 percent of the variance in style used by the leader stemmed from the situation and only about 10 percent was attributed to individual preferences for being autocratic or participative. There is evidence to suggest that the more successful leaders and managers are effective at changing their style appropriately as the situation changes. This model provides a nice diagnostic tool for determining which style is more appropriate and suggests a more careful diagnosis to ensure that a participative strategy is in fact appropriate.

> Many American managers have fallen into the trap of assuming participation, often in the form of quality circles, is the "one best way for today." This is reinforced by the turbulent, competitive times we face and the Japanese success with small group activity. This is analogous to the pilot of a jet liner in a terrible storm with instrument panel out going out of the cockpit back to the passengers and asking them to circle-up and help him get out of the storm (anonymous).

Dilemmas of Managing Participation

In her excellent article entitled "The Dilemmas of Managing Participation," Kanter cites six sets of management considerations that she feels need to be ad-

Table 12.14 Problem Types and the Feasible Set of Decision Processes

PROBLEM TYPE	ACCEPTABLE METHODS	METHOD SELECTED USING THIRD CRITERIA
1	AI, AII, CI, CII, GII	AI
2	AI, AII, CI, CII, GII	AI
3	GII	GII
4	AI, AII, CI, CII, CII[a]	AI
5	AI, AII, CI, CII, GII[a]	AI
6	GII	GII
7	CII	CI
8	CI, CII	CI
9	AII, CI, CII, GII[a]	AII
10	AII, CI, CII, GII[a]	AII
11	CII, GII[a]	CII
12	GII	GII
13	CII	CII
14	CII, GII[a]	CII

[a]Within the feasible set only when the answer to question F is "yes."

SOURCE: Victor H. Vroom, "A New Look at Managerial Decision Making," *Organizational Dynamics*, Vol. 1, No. 4, Spring 1973, p. 67. Reproduced with permission.

dressed in order for participation to produce the best results for everyone. They are:

1. Initiation, or how to begin

2. Structure and management

3. Choice of topics, issues, problems, decisions, designs, implementations, and so forth

4. Problems or dilemmas of teamwork or small-group processes

5. Methods of linking groups or teams to their environment, or how to integrate group activities

6. Evaluation, continuation, and evolution

It is neither appropriate nor possible to discuss all of Kanter's insights regarding participatory processes. However, a few points will be highlighted.

First, the question of how to establish or initiate a participative process arises. Kanter warns against forcing participative strategies on an unwilling or unprepared group of employees, subordinates, or individuals. The question of participation appropriateness should not be limited to the decision tree analysis just presented. Participant readiness is a critical element often overlooked. Management in the United States often falls into the trap of invoking Theory Y programs in Theory X ways. This can lead to resistance on the part of employees and to what Kanter calls the paternalism trap or the "why aren't they grateful syndrome."

Second, the question of structure and management of participation needs to be addressed. "True freedom is not the absence of structure—letting the employees go off and do whatever they want—but rather a clear structure that enables people to work within established boundaries in an autonomous and creative way" (Kanter, 1982, 1983). Controlled dispersion of power to identify, analyze, act, decide, and implement is evident in the Musashi example. It seems that participative programs in the United States are either too structured or lack structure entirely. Quality circles will leave behind a legacy that the technique came along at exactly the right time. Managers had been hearing about participation for 10 or 15 years. Many managers bought the premises, philosophies, and theories but could not operationalize them. Most managers would really rather be cooks than chefs. When the quality circles came along with nice neat steps, they really filled a tremendous need. Quality circles, when implemented correctly, can represent fairly appropriate levels of structure.

Delegation does not equal abdication of responsibility, authority, and accountability. Again, participative strategies and techniques, like quality circles, are not Theory Z ("laissez-faire" style). Sometimes intentionally and sometimes ignorantly management will develop a quality-circle-type program in six months and expand rapidly within a year. The lessons are there to be learned if only American top management will listen. Participation cannot succeed in an unsupportive, unprepared cultural climate. Musashi spent three years laying the foundation for a small-group activity program. Yet most American managers will not even wait three months before implementing a quality circles program. On the one hand, management can report that quality circles have been implemented. On the other hand, programs are failing at a very high rate. Short-term views on planning, impatience, lack of building for the future, lack of well-thought-through strategic plans and programs, and not doing one's homework, all reflect characteristics I personally see in American management. We had better start thinking our way into a new way of acting.

Third, the process by which the topics for participation are chosen needs to be examined carefully. As was indicated in Part III and as will be expanded on in the next chapter, the nominal group technique represents an excellent structured group process and consensus-seeking device. Employees at all levels have unique views on strategic planning, productivity problems or opportunities, quality problems, objectives for selected organizational systems, issues, decisions, and performance problems. The range of topics that employees in the organization can contribute to is immense. We need only ask them to supply their input. The nice aspect about the nominal group technique is that it forces groups to prioritize the topics, issues, and problems. An agenda can then be developed from which to devote resources.

Fourth, teamwork or group process is a phenomenon that is developed and not a natural skill in most people. This is why group behavior topics play such an integral role in the training preceding Japanese small-group activity and quality circles. I find that the nominal group technique helps American employees to be effective in groups. Just by letting American managers and employees experience the process, many sense important effective group process characteristics. Couple the affective experience from an NGT session with a little group behavior material, and a potentially effective group process training module is developed.

Of particular practicality is the concept of using a group-process or productivity observer during a meeting. This observer is responsible for evaluating the group process independent of the content. To do so, the observer sits back at a meeting and evaluates the quality of the process from a distance. Forms can be provided to help the person take notes and organize data. In regularly scheduled meetings comprising the same group of people. The job of observer can rotate among the group members. This helps all members of the group to learn to evaluate group processes. Simple group-process observation instruments and diagnostic guidelines are also available in various references (Bradford, 1978; Shaw, 1976; Bales, 1950; Hirota, 1953; Guetzkow and Dill, 1957; Maier, 1950; Smith, 1972).

The group-process observer is looking for:

1. Appropriate leader behavior;

2. Existence of structure when appropriate and/or needed;

3. A situationally appropriate mix of necessary role behaviors (challenger, initiator, opinion seeker, information provider, elaborator, energizer, evaluator,

critic, recorder, coordinator, opinion giver, encourager, harmonizer, compromiser, gate-keeper or expediter, standard setter, organizer, and so forth);

4. Existence of conflict management;

5. Existence of clear purpose for the meeting;

6. Closure mechanisms for the meeting;

7. Effective, nonjudgmental search processes;

8. Appropriate size (most group meetings with an expressed goal should range in size from 3 to 12—much beyond 12, groups do not function very effectively and efficiently);

9. Effective group decision-making processes when called for, and in general effective decision making that ensures appropriate levels of quality and acceptance.

The ability of groups to manage their own performance depends on the ability of group members to collect data; process that data and provide constructive feedback to the group; be objective in problem solving; develop a team perception and an awareness of how personal behavior and style affect the group; provide a systematic, explicit mechanism for self-evaluation (group and individual); develop team skills in time management, process observation, creative problem solving, structured problem solving, communication, group behavior and performance, group goal setting, and group decision making and consensus seeking; try out new behaviors; reinforce new behaviors; and experiment with new techniques for improving group performance. The bottom line, I believe, is that an organization can be no stronger than the groups that comprise it. If an organization comprises work groups that function poorly, the organization is obviously going to function and perform poorly. Many Japanese organizations have been successful because of a focus of teamwork, group effort, and group problem solving. There are certainly other factors, although the focus on the group is one that should not be overlooked by American managers.

The group-process observer should provide process-related feedback to the group immediately or shortly after the meeting. There are 12 basic guidelines to be followed when providing feedback on the data collected:

1. Give the group only what it is ready to use.

2. Do not overload the group.

3. Stress what the group did right as well as what it did wrong.

4. Report rather than evaluate the process and behaviors that were observed.

5. Discuss or focus on role behavior as opposed to individual behavior.

6. Go lightly on personality clashes.

7. Keep in mind that the focus is on group performance, not individual performance.

8. Confront apparent conflict openly; be objective and open.

9. Deal with things that can be changed.

10. Be specific, not general.

11. Always examine intentions and motives. Remember, we tend to judge others by their behaviors and ourselves by our intentions. Focus on reasons that behaviors did not match intentions.

12. Give feedback as soon as possible; the end of the meeting is ideal.

Keep in mind that effective, efficient, productive, innovative, satisfying group behavior and functioning doesn't just happen automatically when you put a leader in a room with a group of people. Those of you who have failed with quality circles know this quite well. Good group behavior is developed. It takes commitment, knowledge, skill, patience, and persistence.

The fifth dilemma of participation that Kanter discusses is that of linking teams to their environment. Within this particular dilemma, she discusses six issues: turnover, fixed-decision problem, turf protection problem, too much team spirit, not-invented-here syndrome, and the question of the life cycle of a group.

Turnover in groups is inevitable. And the reliability of groups goes through four stages in newly formed groups or in groups that accept new members:

1. Forming or entry of a new person

2. Storming (Who are you? Why are we here?)

3. Norming (What role will you play in the group? What role will I play?)

4. Performing

Newcomers, outsiders, late comers have not shared the group's experiences; they don't have a sense of the group's history or evolution. The group-process observer must watch carefully to see if someone integrates new members or orients a new group successfully.

The fixed-decision problem relates to the fact that group involvement and the scope of group problem solving may often be or need to be highly constrained. Participation need not always mean that a group start from scratch to "invent the wheel." Sometimes a group's purpose is highly delimited in terms of what it has done in the past, what other groups have done, or what needs to be done next.

Managing intergroup problem-solving processes, like quality circles, can become a challenge in terms of the problems associated with turf protection, overreaching or extending boundaries, or power and politics issues. The concerns that must be dealt with range from supervisors, managers, and unions who feel that participation threatens their base of power to overzealous groups who solve other groups' problems but are unable to implement the solutions. Too much team spirit and the not-invented-here syndrome are related to this potential problem. Large participative group programs will likely need to be integrated in some fashion. The task is essentially equivalent to integrating a decoupled productivity measurement system (see Chapters 3 and 4). Figure 12.17 illustrates several of the dilemmas associated with linking teams to their environment.

The final problem Kanter discusses concerning the dilemma of linking teams to their environment is that which she calls "a time to live and a time to die." A major commercial airline has developed a significant productivity improvement strategy. Within the strategy there exist several generic, long-standing programs focusing on pride in the company, productivity, cost reduction, and safety. These programs have a lot of visibility and are clearly communicated to employees on a consistent basis. In addition, further annual productivity or performance improvement operational programs are developed to facilitate accomplishment of the year's goals and objectives. These programs have extremely explicit performance targets that are, again, clearly communicated to worker groups. The programs tend to focus on group performance and organizational performance, not individual performance. A variety of feedback mechanisms and reinforcement devices, many financial in nature, are used to attain various levels of performance. Progress against goals and objectives is tightly monitored and evaluated. It is communicated clearly from the beginning that the program will last for a fixed period of time or until performance goals and objectives are met and maintained for a specified period of time. This forms closure on the program.

As Kanter points out, "While it is easy for teams to take on a life of their own, participation needs constant renewal, for the sake of both team members and the organization." It is quite common today for organizations to fail to think through how a program like quality circles should evolve. There needs to be a greater sense of continuity and longer-range planning in American organizations, particularly with respect to behavioral and social science program interventions. This is not to suggest that programs do not have life cycles. On the contrary, the programs themselves have life cycles; the philosophy, principles, goals, and desired outcomes for the programs do not. There needs to be a central, consistent program theme with shorter-term, logically sequenced, evolutionary sets of operational programs moving the groups and organization toward constantly higher levels of performance.

Kanter's sixth and final dilemma of participation is that of evaluation. Regardless of how well participation is implemented, it will not solve all organizational problems. It is essential that management develop realistic desired outcomes, primary and secondary, for participation programs. Experience and research suggest that participative programs will not automatically and instantaneously improve productivity. Further, participation will also not automatically counter the "alienation," low morale, absenteeism, turnover, poor quality, and lack of commitment found in many workers and managers in this country. Increased satisfaction is a more likely result from participatory programs than is increased performance, and studies show that satisfaction and performance are not very highly correlated.

Perhaps most importantly, management must manage expectations in a participatory program by being realistic about what can be expected in a given time

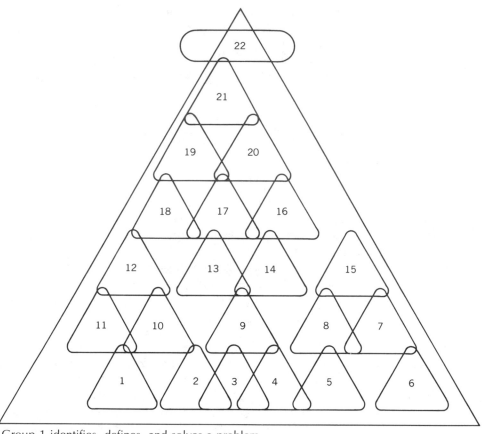

Group 1 identifies, defines, and solves a problem.
Group 2 has the same problem but doesn't recognize it.
Group 10 must approve and implement the solution.
Group 12 must obtain budget approval from Group 19.
Group 6 develops a drastically different and better solution to a problem similar to the one Group 1 worked on, but Group 10 doesn't know about it.
Groups 16–18 feel threatened by this participatory program because Groups 19–22 developed and sold the program to Groups 12–15, trained Groups 7–11, and implemented it in Groups 1–6, but neglected to involve Groups 16–18.
Groups 19–20 are hot on the trail of a new productivity improvement technique that they want to sell to Groups 21 and 22 as a replacement for this current program. Groups 1–6 haven't had a chance to work the problems out of the current program and are starting to become disenchanted because of resistence to change from Groups 12–18 and frustration with very slow implementation of resolved problems.

Figure 12.17

frame. "If people are given realistic information about exactly what they can expect in the beginning and are not promised everything, then they can calibrate their own personal goals accordingly (Kanter, 1982, 1983)." In short, evaluation of the effect of participatory programs is not easy but is the beginning of developing realistic statements about both short-term and longer-term desired outcomes.

Participation appears to work most effectively and efficiently when it is well managed. This entails

1. A clearly designed management structure and involvement throughout the organization

2. Assignment of meaningful and manageable tasks with clear boundaries and parameters

3. A time frame; a set of accountability and evaluation targets and standards that a group must meet

4. Training, development, and information available to facilitate the program

5. A mechanism for involving all appropriate persons in a given issue and for integrating their efforts and contributions

6. A mechanism for providing visibility, feedback, recognition, and rewards for a group or team's efforts and performance

7. A clearly understood mechanism for the continuing evolution of participatory programs

Guidelines for organizations wishing to experiment with participative programs are

1. *Begin the program by testing concepts, principles, and processes on top management.* Top management can be involved in a strategic planning process that embodies processes similar to those employees at lower levels will experience. After top management has experienced this process, they can be briefed on how similar processes would be utilized at lower levels in a participatory program. Any uncomfortableness with program design on the part of top management can be either resolved or accommodated in a new design and process. This top-down implementation process helps to ensure top management legitimization and support. The processes and techniques associated with the specific participatory program (NGT, Delphi technique, problem solving) should be exposed to successively lower levels of management over a period of six months to a year or longer. This exposure provides all levels with training and development as well as allowing strong points or areas of resistance to emerge. At each level of management, the processes and techniques are presented in a situationally relevant and purposeful fashion.

2. *At all levels, start with local, relevant, and salient issues.*

3. *Allow participants to identify and prioritize for themselves the issues, problems, decisions, designs, or implementations on which they want to focus.* This does not mean that the group might not be "primed" with prioritized lists from customers, management, peers, subordinates, and so forth. However, final decision as to what will be addressed by the group should rest with the group itself. No constraints, guidelines, or limitations should be placed on any topics although management reserves the right not to act on certain topics.

Participation is not an abdication of management's responsibility and accountability. It is simply an opportunity to expend involvement in the management process. Employees and managers at all levels should be encouraged to be open about all concerns, but also to realize that some things fall outside the realm of group decision making.

4. *Neither promise nor expect too much.* Manage expectations carefully.

5. *Involve parties whose power might be at stake and attempt to give them important, rewarded roles in the new system or integrate them right into the process.* Seek to find points of major resistance to the participatory program and confront the causes of resistance directly. Resistance resulting from a conflict between a manager's style and the participatory program often cannot be overcome. Such incongruity is a significant factor in program failures and should signal a problem to the program designer. Because of such potential conflict, it is so critical to introduce the program top-down in a truly participative fashion.

6. *Be prepared to provide education and training on both the skills associated with participative programs and the topics to be addressed.*

7. *Maintain strong leadership for the program.* Leadership must be effective at creating visions of future states, and also helping to reach those states.

8. *Make sure the teams or groups apply good motivational principles.* Set goals and deadlines for accomplishing those goals. Keep an action orientation. If the process gets stale or stagnant, modify it or give the process a rest. Sometimes groups need a period where they just perform reactive, day-to-day tasks and activities and are not proactively looking to improve. Recognize the need for this.

9. *Provide rewards and feedback, as well as tangible, explicit, measurement, evaluation, and reinforcement systems.* Be creative and perceptive in terms of potential reinforcers or incentives.

10. *Participative teams supplement rather than replace organizational system management processes and hierarchy.* They are a proactive process by which the members in an organizational system can be looking for ways to improve and act on their own ideas.

> Managing participation is a matter of balance—and patience. It takes longer to weld people into a team than to order them around; it takes longer to teach people a variety of jobs than to give them one simplified task. . . . So wherever there is a pressure for quick results, people will be unlikely to support participation (Kanter, 1982, 1983).

We have now completed a reasonably brief presentation of four methods of motivating improved performance. Reinforcement theory, goal setting, job design/redesign/enrichment, and participation have been presented as four somewhat generic and mutually exclusive approaches to motivation and performance improvement. Recall from Figure 11.4, however, that motivation and the effort that hopefully ensues are only one of many factors that shape or cause various levels of performance. Capacity of the individual, opportunity provided the individual, the goal setting and communication process, leadership, performance $\rightarrow$ outcome linkages, and technology are all factors that play a critical role in performance equation.

Attention is now directed to a specific group-oriented performance and productivity improvement technique. The performance/productivity action team process (PAT) is a technique that integrates elements of the four methods. Reinforcement theory applications are built into the process to enable goal-congruent behaviors of the individual to be reinforced. Feedback, charting, and various performance-contingent reward types are possible in the process. The process directly affects goal setting by prioritizing ideas for performance/productivity improvement. Consensus facilitates assurance that employees in the group are committed to successfully implementing the ideas. The PAT process is an excellent example of a technique that invokes or operationalizes job characteristics theory. Core job dimensions are directly and positively changed by the PAT process. And, finally, the PAT process is a participative program utilizing various forms of structured group processes to ensure its effectiveness and efficiency.

REFERENCES

Aldag, R. J., and A. P. Brief. *Task Design and Employee Motivation*. Glenview, Ill.: Scott, Foresman, 1979.

Alderfer, C. P. "An Empirical Test of a New Theory of Human Needs." *Organizational Behavior and Human Performance*, Vol. 4, 1969, pp. 142–175.

Alderfer, C. P. *Existence, Relatedness, and Growth: Human Needs in Organizational Settings*. New York: The Free Press, 1972.

Bales, R. F. *Interaction Process Analysis: A Method for the Study of Small Groups*. Cambridge, Mass.: Addison-Wesley Co., 1950.

Bradford, L. P., ed. *Group Development*. La Jolla, Calif.: University Associates, 1978.

Carroll, S., and W. Tosi. *Management by Objectives*. New York: Macmillan, 1973.

Cyert, R. M., and J. G. March. *A Behavioral Theory of the Firm*, Englewood Cliffs, N.J.: Prentice-Hall, 1963.

Davidson, W. H. "Small Group Activity at Musashi Semiconductor Works." *Sloan Management Review*, Spring 1982.

Davis, L. E. "Job Design and Productivity: A New Approach." *Personnel*, Vol. 33, 1957.

Davis, L. E. "Toward a New Theory of Job Design." *The Journal of Industrial Engineering*, Vol. 8, 1957.

Drucker, P. F. *The Practice of Management*, New York: Harper and Row, 1954.

Dunham, R. "Reactions to Job Characteristics: Moderating Effects of the Organization." *Academy of Management Journal*, Vol. 20, 1977.

Etzioni, A. *Modern Organization*. Englewood Cliffs, N.J.: Prentice-Hall, 1964.

Fein, M. "An Alternative to Traditional Managing" (unpublished), Hillsdale, N.J., 1977.

Fein, M. "Designing and Operating an Improshare Plan" (unpublished), Hillsdale, N.J., 1976.

Fein, M. *Improshare: An Alternative to Traditional Managing*. Norcross, Ga.: Industrial Engineering and Management Press, 1981.

Fein, M. "Improving Productivity by Improved Productivity Sharing." *The Conference Board Record*, Vol. 13, 1976.

Fein, M. "Motivation for Work," *Handbook of Work Organization and Society*, ed. R. Dubin. Chicago: Rand McNally, 1976.

Fein, M. *Rational Approaches to Raising Productivity*. Norcross, Ga.: Industrial Engineering and Management Press, 1974.

Ferratt, T. W., R. B. Dunham, and J. L. Pierce. "Self-Report Measures of Job Characteristics and Affective Responses: An Examination of Discriminant Validity." *Academy of Management Journal*, Vol. 24, No. 4, 1981.

Filley, A. C., R. J. House, and S. Kerr. *Managerial Process and Organizational Behavior*. Dallas: Scott, Foresman, 1976.

French, W. L., and R. W. Hollmann. "Management by Objectives: The Team Approach." *California Management Review*, Vol. XVII, No. 3, 1975.

French, J. R., E. Kay, and H. H. Meyer. "Participation and the Appraisal System." *Human Relations*, Vol. 19, 1966.

Graham-Moore, B. E., and T. L. Ross. *Productivity Gainsharing*. Englewood Cliffs, NJ: Prentice-Hall, 1983.

Griffin, R. W. *Task Design: An Integrated Approach*. Glenview, Ill.: Scott, Foresman, 1982.

Guetzkow, H., and W. R. Dill. "Factors in the Organizational Development of Task-Oriented Groups." *Sociometry*, Vol. 20, 1957.

Hackman, J. R., and E. E. Lawler. "Employee Reactions to Job Characteristics." *Journal of Applied Psychology Monograph*, Vol. 55, 1971.

Hackman, J. R., and G. R. Oldham. "Development of the Job Diagnostic Survey." *Journal of Applied Psychology*, Vol. 60, 1975.

Hackman, J. R., G. R. Oldham, R. Janson, and K. Purdy. "A New Strategy for Job Enrichment." *California Management Review*, Summer 1975.

Hackman, J. R., and G. R. Oldham. "Motivation through the Design of Work: Test of a Theory." *Organizational Behavior and Human Performance*, Vol. 16, 1976.

Hackman, J. R., and G. R. Oldham. *Work Redesign*. Reading, Mass.: Addison-Wesley, 1980.

Hare, A. P. "Small Group Discussions with Participatory and Supervisory Leadership." *Journal of Abnormal and Social Psychology*, Vol. 48, 1953.

Henderson, R. I. *Compensation Management* (4th Ed.) Reston, VA: Reston Pub. Co., 1984.

Hersey, P., and K. Blanchard. *Management of Organizational Behavior*, 4th ed. Englewood Cliffs, N.J.: Prentice-Hall, 1982.

Herzberg, F. *Work and the Nature of Man*. Cleveland: World, 1966.

Herzberg, F., B. Mausner, and B. Snyderman. *The Motivation to Work*, 2nd ed. New York: John Wiley, 1959.

Hirota, K. "Group Problem Solving and Communication." *Japanese Journal of Psychology*, Vol. 24, 1953.

Ivancevick, J. M., J. H. Donnelly, and H. L. Lyon. "A Study of the Impact of Management by Objectives on Perceived Need Satisfaction." *Personnel Psychology* 1970, 23, 139–151.

Kanter, R. M. "Dilemmas of Managing Participation." *Organizational Dynamics*, 1982.

Kanter, R. M. *The Changemasters*. New York: Simon and Schuster, 1983.

Kay, E., H. H. Meyer, and J. R. French. "The Effect of Threat in a Performance Appraisal Interview." *Journal of Applied Psychology*, Vol. 49, 1965.

Kerr, S. "On the Folly of Rewarding A, while Hoping for B." *Academy of Management Journal* 18, 1975.

Latham, G. P., and K. N. Wexley. *Increasing Productivity through Performance Appraisal*. Reading, Mass.: Addison-Wesley, 1981.

Lawrence, L. C., and P. C. Smith. "Group Decision and Employee Participation." *Journal of Applied Psychology*, Vol. 39, 1955.

Lawler, E. E. *Motivation in Work Organizations*, Monterey, Calif.: Brooks/Cole, 1973.

Lawler, E. E., and G. E. Ledford. "Productivity and Quality of Work Life." *National Productivity Review*, Vol. 1, No. 1, 1981.

Leavitt, H. J., and R. A. Mueller, "Some Effects of Feedback on Communication." *Human Relations*, Vol. 4, 1951.

Likert, R. *New Patterns of Management*, New York: McGraw-Hill, 1961.

Likert, R. and J. G. Likert. *New Ways of Managing Conflict*. New York: McGraw-Hill, 1976.

Locke, E. A. "Toward a Theory of Task Motivation and Incentives." *Organizational Behavior and Human Performance*, Vol. 3, 1968.

Locke, E. A., and J. F. Bryan. "Performance Goals as Determinants of Level of Performance and Boredom." *Journal of Applied Psychology*, Vol. 51, 1967.

Locke, E. A., et al. "The Relative Effectiveness of Four Methods of Motivating Employee Performance." *Changes in Working Life*. New York: John Wiley and Sons, 1980.

Luthans, F., and R. Kreitner. *Organizational Behavior Modification*, Glenview, Ill.. Scott, Foresman, 1975.

Luthans, F., and D. D. White. "Behavior Modification: An Application to Manpower Management." *Personnel Administration*, Vol. 34, No. 4, 1971.

Luthans, F., and R. Kreitner. "The Management of Behavioral Contingencies." *Personnel*, Vol. 51, July–August, 1974.

Maier, N. R. F. *Problem Solving and Creativity: In Individuals and Groups*, Belmont, Calif.: Brooks/Cole, 1970.

Maier, N. R. F. "The Quality of Group Decisions as Influenced by the Discussion Leader." *Human Relations*, Vol. 3, 1950.

Mali, P. *Improving Total Productivity: MBO Strategies for Business, Government, and Not-for-Profit Organizations*, New York: John Wiley, 1978.

Maslow, A. H. *Motivation and Personality*. New York: Harper and Row, 1954.

Maslow, A. H. *Motivation and Personality* (2nd ed.). New York: Harper and Row, 1970.

McClelland, D. C. *The Achieving Society*. Princeton, N.J.: Van Nostrand, 1961.

McGregor, D. *The Human Side of Enterprise*. New York: McGraw-Hill, 1960.

Mendleson, J. L. "Managerial Goal Setting: An Exploration into Meaning and Measurement." Unpublished doctoral dissertation, Michigan State University, Lansing, Mich., 1967.

Meyer, H. H., E. Kay, and J. R. French. "Split Roles in Performance Appraisal." *Harvard Business Review*, Vol. 43, 1965.

Miller, F. G., T. S. Dhaliwal, and L. J. Magas. "Job Rotation Raises Productivity." *Industrial Engineering*, Vol. 5, 1973.

Miller, L. *The Use of Knowledge of Results in Improving the Performance of Hourly Operators.* Behavioral Research Service, General Electric Company, 1965.

Moore, B. E., and T. L. Ross. *The Scanlon Way to Improved Productivity.* New York: John Wiley and Sons, 1978.

Morris, W. T. *Implementation Strategies for Industrial Engineers,* Columbus, Ohio: Grid, 1979.

Naisbitt, J. *Megatrends,* New York: Warner Books, 1982.

Nord, W. R. "Beyond the Teaching Machine: The Neglected Area of Operant Conditioning in the Theory and Practice of Management." *Organizational Behavior and Human Performance,* Vol. 4, 1969.

O'Dell, C. *Gainsharing: Involvement, Incentives and Productivity,* New York: AMACOM, 1981.

Oldham, G. R. "Job Characteristics and Internal Motivation: The Moderating Effect of Interpersonal and Individual Variables." *Human Relations,* Vol. 29, 1976.

Oldham, G. R., J. R. Hackman, and J. L. Pierce. "Conditions under Which Employees Respond Positively to Enriched Work." *Journal of Applied Psychology,* Vol. 61, 1976.

Ouchi, W. *Theory Z.* Philippines: Addison-Wesley, 1981.

Pierce, J. L., R. B. Dunham, and R. S. Blackburn. "Social Structure, Job Design, and Growth Need Strength: A Test of a Congruency Model." *Academy of Management Journal,* Vol. 22, 1979.

Preston, S. J. "J. Stone's Management by Objectives." *Personnel* (London), Vol. 1, 1968.

Raia, A. P. "Goal Setting and Self-Control." *Journal of Management Studies,* Vol. 2, 1965.

Roethlisberger, F. J., and W. J. Dickson. *Management and the Worker.* Cambridge, Mass.: Harvard University Press, 1956 (originally published in 1939).

Shaw, M. E. *Group Dynamics: The Psychology of Small Group Behavior,* New York: McGraw-Hill, 1976.

Simon, H. A. "On the Concept of Organizational Goal." *Administrative Science Quarterly,* Vol. 9, 1964.

Sims, H. P., and A. D. Szilagyi. "Job Characteristic Relationships: Individual and Structural Moderators." *Organizational Behavior and Human Performance,* Vol. 17, 1976.

Sims, H. P., A. D. Szilagyi, and R. T. Keller. "The Measurement of Job Characteristics." *Academy of Management Journal,* Vol. 19, 1976.

Skinner, B. F. *Science and Human Behavior,* New York: Macmillan, 1953.

Sloan, S., and D. Schreiber. *Hospital Management: An Evaluation.* Monograph No. 4, Bureau of Business Research and Service, University of Wisconsin, Madison, 1971.

Smith, A. *An Inquiry into the Nature and Causes of the Wealth of Nations.* New York: Modern Library, 1937 (originally published in 1776).

Smith, D. C., L. M. Kendall, and C. L. Hulin. "The Measurement of Satisfaction in Work and Retirement." Chicago: Rand McNally, 1969.

Smith, F. J. "The Index of Organization Reactions (IOR)." *JSAS Catalog of Selected Documents in Psychology.* Chicago: Rand McNally, 1976.

Smith, K. H. "Changes in Group Structure through Individual and Group Feedback." *Journal of Personality and Social Psychology,* Vol. 24, 1972.

Steers, R. M. and L. W. Porter. *Motivation and Work Behavior,* 3rd ed., New York: McGraw-Hill, 1983.

Stogdill, R. M. *Manual for Job Satisfaction and Job Expectations.* Bureau of Business Research, The Ohio State University, Columbus, Ohio, 1965.

Szilagyi, A. D. *Management and Performance.* Santa Monica, Calif.: Goodyear, 1981.

Taylor, F. *Scientific Management*. New York: Harper and Row, 1911.

Vroom, V. H. "A New Look at Managerial Decision Making." *Organizational Dynamics*, Vol. 1, No. 4, 1973.

Vroom, V. H. *Work and Motivation*. New York: John Wiley and Sons, 1964.

Vroom, V. H., and A. G. Yago. "Decision Making as a Social Process: Normative and Descriptive Models of Leader Behavior." *Decision Science*, Vol. V, 1974.

Vroom, V. H., and P. W. Yetton. *Leadership and Decision Making*. Pittsburgh, Pa.: University of Pittsburgh Press, 1973.

Walker, C. R. "The Problem of the Repetitive Job." *Harvard Business Review*, Vol. 28, 1950.

Walker, C. R., and R. Guest. *The Man and the Assembly Line*. Cambridge, Mass.: Harvard University Press, 1952.

Weiss, D. M., et al. *Manual for Minnesota Satisfaction Questionnaire*. Minnesota Studies in Vocational Rehabilitation: XXII, University of Minnesota, Minneapolis, 1967.

White, R. W. "Motivation Reconsidered: The Concept of Competence." *Psychological Review*, Vol. 66, 1959.

QUESTIONS AND APPLICATIONS

1. Investigate, compare, and evaluate the Scanlon, Rucker, and Improshare gainsharing plans.

2. What is the difference between a productivity gainsharing plan and a profit-sharing plan?

3. Based upon your knowledge of reinforcement theory and behavioral modification, generate a list of design criteria and performance specifications for a productivity gainsharing plan. Using these results, evaluate the plans presented in this chapter.

4. Based upon your knowledge of factors influencing individual, group, and organizational performance, develop a set of factors or variables that you feel must be managed effectively in order to positively influence productivity.

5. What are acceptable levels of performance? What are motivated levels of performance? How do we objectively define these? How does one develop motivated levels of performance?

6. Of the seven performance criteria discussed in Chapter 2, which do you think are most important? Develop a causal model for the seven criteria. (Which one causes the next?) Which of the seven criteria is most directly impacted by goal setting? Based on your answer, how important is goal setting as a performance management technique? *The One-Minute Manager* is a very popular book today. What are the three lessons given in the book? Are they consistent with the priorities you placed on the seven criteria? What additional lessons are gained from the follow-up book entitled *Putting the One-Minute Manager to Work*?

7. Administer the Job Diagnostic Survey or the Job Characteristic Inventory to yourself and then to a sample of people in various jobs. Analyze, evaluate, and discuss the results. Do you feel the instruments are valid? What kinds of validity can be tested for? Have these instruments been tested for various types of validity? If so, where would you find

the results? How would you interpret the results? Assuming the instruments are valid, can they be used to improve productivity and performance?

8. Investigate quality circles. Write a two-page summary of the technique. Focus on its operating characteristics, such as what it does; how and why it does what it does; desired or intended outcomes; potential dysfunctional consequences or outcomes; costs and potential benefits; and a typical action plan for program (not just a circle) implementation.

9. Investigate and discuss small-group behavior. You have some understanding now of individual behavior and motivation. What are the major differences between individual and group behavior and performance? Model and depict group behavior and performance.

10. Compare and contrast quality circles with performance/productivity action teams. Be systematic in your comparison.

11. Select several situations in which you have been involved where a decision could have been approached in a participatory fashion. Using the decision tree presented in Figure 12.16, analyze the decision situation and select the most preferred decision style. Analyze the following decision situations using this model:
  (a) The pilot of a 747 is faced with an emergency situation.
  (b) A manager is faced with a complex inventory problem that has significant financial implications to the plant.
  (c) The president of a fraternity faces planning decisions at the beginning of the school year for the entire academic year.
  (d) A supervisor is concerned about morale and discipline problems.
  (e) The president of a university has to develop a strategic (five- to ten-year) plan for the university.
  (f) The director of industrial engineering in a plant is faced with the dilemma of how to better posture industrial engineering in the company structure.
  (g) A plant superintendent is faced with a mandate to improve productivity in the plant in six months or else lose his job.
  (h) A city manager is faced with personnel problems, interpersonnel conflict, and performance problems in the police, fire, and service departments.
  (i) A supervisor has been asked to resolve a technical problem created within your department.

12. Discuss how the following factors might moderate your decisions on the situations in Question 11:
  (a) The maturity of your follower(s)
  (b) Time allotted for the decision
  (c) Managerial styles; psychological types
  (d) Critical nature of the decision
  (e) Ego or desire to look good by doing it yourself
  (f) Need for employee development in order to have someone ready to take your job when you get promoted

13. At the heart of the productivity improvement issue is the need for a process in American organizations that focuses on continually searching and striving to do things better than we did yesterday. Comment on this. What keeps us from accomplishing this? What are major roadblocks to productivity improvement? Have one of the students in class volunteer to facilitate an NGT session on the class. Another student will act as

assistant. Use the following question as a task statement:

Please identify roadblocks to productivity improvement in our country and its organizations today.

Discuss the results. If you mail them to me, I will return a copy of nationwide results of NGT sessions focusing on the same question so that you might compare your results with those received by other groups of students and managers.

CHAPTER

13

PERFORMANCE/PRODUCTIVITY ACTION TEAMS

HIGHLIGHTS

- Evolution of Participative Management
- Quality Circles
  Evolution
  The Process in Brief
- The Productivity Action Team Process
  Steps in the PAT Process
  Development of a Performance/Productivity Action Team Program
  Implementation of the PAT Process

OBJECTIVES

- To expose the reader to a process that incorporates many of the motivational and performance improvement principles, theories, and approaches presented in Chapters 11 and 12. In essence this chapter represents the operationalization of concepts presented in those chapters.
- To present a relatively simple process that has been tested and has the potential to spark innovation, improve communication, improve implementation effectiveness, improve coordination, reduce conflict, increase perceptions of quality of work life, improve the quality of management, enhance or accomplish team building, identify roadblocks to productivity and performance, and improve performance and productivity.
- To present a participative problem-solving technique/process that has been designed to work best in American organizations, with American employees and American management. The process has subtle features that make it superior to quality circles in most applications.

The productivity action team process originally evolved from the Ohio State productivity measurement studies discussed in Chapter 4. In 1975, the Productivity Research Group at Ohio State began their now classic development of the NPMM. Since the completion of that National Science Foundation-sponsored study in 1977, continued development on many aspects of that initial research has taken place in a variety of organizational systems.

For example, Stewart (1978) examined the refinement of the NPMM through application of the multiattribute utility theory. His early work prompted and pointed the direction for the objectives matrix (Riggs and Felix, 1983) and the NPMM (Sink, 1978). Morris (1977) further investigated applications of the Delphi technique and the hierarchical Delphi technique in developing productivity measurement systems. The American Productivity Center, Westinghouse, Honeywell, General Electric, and many other organizations have utilized the NGT in productivity measurement and improvement applications as a result of the Ohio State research. Of course, William T. Morris and George L. Smith are to be credited with the insight, creativity, and innovation necessary to bring together the social sciences, behavioral sciences, industrial engineering, and management into a highly effective, efficient, and unique approach to measuring and improving productivity.

The development of a participatory problem-solving and productivity improvement process modeled after the NPMM has continued at Oklahoma State University from 1978 to the present (Sink, 1978, 1982). This process has been implemented and tested in academic departments (as a planning process), with custodians in a school system (as an improvement and measurement process), with maintenance/service personnel for a small city government (as an improvement process), with process engineers in a process engineering division of a large petroleum company (as an improvement process), with plant managers in a variety of manufacturing plants (as a goal-setting and improvement process), in a heat exchanger manufacturing plant with welders (as an improvement process), in a fiberglass product manufacturing plant (as an improvement focus), in an agricultural chemicals processing plant (as a planning, supervisory development, communication, and productivity improvement program), and in a small steel mill (as a performance improvement process). Implementations and tests of this process have taken place in organizations ranging in size from 30 employees to 20,000 employees. The process has been implemented in a top-down as well as bottom-up fashion. Applied research results have led to improved designs and implementation strategies, which are presented in this chapter. The general approach has and can be expected to produce different outcomes, depending on the situation and how the process is designed, developed, and implemented.

Evolution of Participative Management

The evolution of participative management can be viewed largely as driven by attempts by academicians and practitioners to understand and develop improved processes for managing the performance of organizational systems. One can see that the evolution of thought on how managers should manage coincides interestingly with the conditions of the times. Szilagyi (1981) points out that the skills and attitudes needed for effective managerial performance depend, to some extent, on the state of the environment at any given time. The environment makes demands on managers, and in an attempt to respond appropriately, effective managers adapt in terms of (1) management style; (2) leadership style; (3) organizational design and structure; and (4) variables on which they focus.

Participative management, however, did not undergo intensive development until recently. In the sixties and seventies, participative management was largely a set of theories (Theories X and Y, McGregor; two-factor theory, Herzberg; need-based theories, Maslow and Alderfer; and so forth) with very few well-developed techniques. Most managers lacked the discipline or the skills to use these theories effectively. Furthermore, it is reasonable to suspect that many managers then and now doubt, if not reject, the premises on which participative management is based. In short, many managers simply do not believe the benefit/cost ratios for participative management techniques are greater than 1.00.

In general, the relationships between improved job performance and changes that occur in the areas of participation and job enrichment as a result of a quality circles program, orthodox job enrichment programs, productivity action team programs, or any such involvement strategy are not clear. Yet, there is still sufficient evidence to suggest that participation may, in the long term, be a valuable if not necessary precondition for productivity improvement.*

The dilemma surrounding participative management is primarily this: Participation in and of itself is probably not sufficient to significantly improve employee performance. However, some form of participation and/or involvement is likely an important if not necessary precondition for other productivity improvement techniques aimed at the human element. One might view participation as the foundation on which other improvement techniques are built. If it is a solid foundation, the structure will be strong. If it is a weak foundation, the structure will topple easily. Techniques like quality circles and productivity action teams have largely been responsible for bringing participative management out of the textbook and classrooms and into the organization.

Quality Circles

Involvement strategies and techniques are playing an increasingly important role in productivity improvement efforts in this country and in others. Management theory and practice have been evolving toward an increasing emphasis on expanded involvement of employees at all levels of the organization.

---

*Recall the particularly low median improvement (+ 0.5 percent) in performance that resulted in studies involving applications of participation techniques. The implications, however, are not as obvious as they might appear. For example, although the study involving job enrichment resulted in low median improvements in quantity performance, such improvements were actually much higher (+ 80 percent). Also, the research examined in this particular study measured only concrete, "hard" performance improvements. Improvement in such areas as cooperation, communication, and attitude, which might have an impact on long-term performance improvement, would not have shown up in these studies.

Participation in all elements of the problem-solving process (problem analysis, decision analysis, and implementation analysis) is now not uncommon even at the lowest level of an organization. Our increased knowledge of group behavior, an increasing "maturity" on the part of employees, rising expectations, declining productivity, inflation, and foreign competition have been major factors causing this most recent trend in management philosophy and practice.

A number of techniques have emerged as potentially viable methods for involving workers in the organizational problem-solving process in order to (1) "tap the human resource;" (2) improve quality of work life; (3) reduce costs; (4) reduce nonproductive time; and (5) improve the organizational effectiveness, efficiency, and productivity. Quality circles are one such technique and are becoming increasingly popular. There are currently some 2000 to 3000 quality circles operating in the United States. (In 1970, there were 20,000 operating in Japan.) Companies such as Lockheed, American Airlines, Westinghouse, Dresser Industries, Honeywell, General Motors, and General Electric are all currently experimenting widely with the technique. It is important to note that very little research on this technique has taken place. And, in comparison with Japan, there is relatively little experience with this technique in the United States. (Lockheed has the oldest program, which began in 1974.)

Evolution

Quality circles began in Japan after World War II when American management, statistical quality control, and other authorities were asked to assist Japan in a rebuilding process. Extensive statistically based quality-control and management training was provided to Japanese management and workers for about 15 years prior to the development of quality circles. In the early to middle fifties, Dr. Kaoru Ishikawa began to integrate certain behavioral and social science theories and techniques with statistically based quality-control concepts, approaches, and techniques. The result was quality control circles. He worked closely with the Japanese Union of Scientists and Engineers (JUSE) in developing training materials for the circles.

Quality circles are actually a very structured problem-solving process that teaches people to break down problems or processes into small components. Causes of problems can then be treated rather than chasing symptoms.

Quality circles were slow to catch on in Japan, with progress of the program initially limited to the success of well-designed pilot studies. However, in the late sixties and early seventies, quality control circles spread rapidly. There are presently over 100,000 quality circles registered with JUSE, involving over 1,000,000 Japanese employees. It is estimated that there are between 5 and 10 nonregistered circles for every registered circle in Japan (Patchin, 1983). If these programs are even nearly as successful as the Musashi semiconductor small-group activity program cited earlier, one can appreciate the significance of these numbers on Japan's dominance in world markets.

Japan's success with this structured employee participation and quality-focused technique prompted efforts in the early seventies to replicate this approach in the United States. Patchin (1983), director of Productivity Improvement Programs at Northrop's Aircraft Group, indicates that the first quality circle was established at Smith Kline Instruments of Palo Alto, California, in 1970. However, since the Japanese developed the quality circle concept based on United States' theories, research, innovation, and techniques, it is safe to assume that similar if not identical small-group problem-solving practices were already in existence in the United States for some time before 1970. "A rose by any other name is still a rose."

A few years later, Rieker, who was then operations manager at Lockheed's Missile and Space Division, first heard of quality circles. In 1973, he and a group from Lockheed visited plants in Japan and witnessed quality circles in action. Based on what they saw, they set up Lockheed's program, one of the most important in the application of quality circles in the United States. This program has developed a number of innovations with the process and provided necessary visibility for the process.

In 1978, Northrop Corporation of Los Angeles began a quality circles program in an attempt to deal with rising costs and quality of work life problems. It began a pilot project of six circles at a cost of $27,000 (circle members' time excluded). Northrop found that (1) there was no production loss despite the fact that the circles met on company time; (2) a net savings of $106,000 was documented; (3) improved communications was a hidden but significant benefit; and (4) worker morale and attitude toward the job and supervision increased substantially. Today, Northrop has over 60 circles, with 100 circle leaders. About 100 volunteer facilitators have been trained, and over 1000 circle members have participated in the process.

It is estimated that well over 3000 circles currently operate in the United States. The International Association of Quality Circles located in Midwest City, Oklahoma, had 3400 members in 1982. The technique appears to be spreading in terms of number and range of applications in the United States.

The Process in Brief

In simplest form, quality circles entail identifying a problem, gathering and analyzing data, developing solutions, evaluating the merit or cost/benefit of solutions, presenting recommendations, and deciding and implementing solutions. Anyone familiar with any structured problem-solving approach, such as the Kepner and Tregoe systematic approach to problem solving and decision making, would have difficulty detecting a difference between the two approaches aside from the quality focus and the participatory or group aspect. And, on close examination, quality circles really are not new. They are creative and innovative integrated combinations of theory, principles, and techniques developed in the United States during the fifties and sixties. In Drucker's words, "What we need to learn from the Japanese is not what to do, but to do it."

Strategies for planning and developing a quality circles program abound in the literature. Suggested readings on the subject are provided at the end of this chapter for interested readers. The reader must, however, keep in mind that implementation of quality circles in the United States has largely been accomplished by trial and error. American managers have *not* generally taken the time and care necessary to ensure effective implementation. Many mistakes have and will be made. Many quality circles programs have and will fail because they are not appropriate, they were rushed into, they were implemented bottom-up without ensuring continued commitment from upper and middle management, or they never had a chance because of poor training and development support. In short, the technique or program called quality circles is a good technique, but when implemented by inexperienced staff in adverse conditions, even the best program will likely fail.

Implementing quality circles will obviously be a different process and experience in each organization. However, there is a common, generic technique behind all the articles and books on how to successfully proceed. At the root of all versions is a structured, participative, small-group problem-solving process. As outlined for you in the last section, problem solving comprises at least 12 elements:

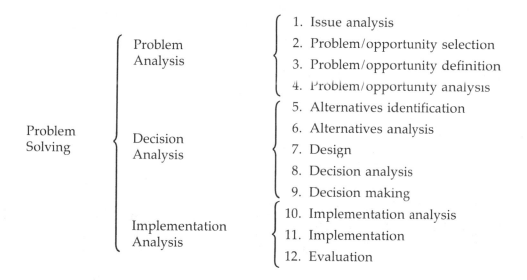

Most quality circles programs implemented at the lower levels in an organization involve employees in elements 2 through 6 and 8. Sometimes circles are asked to become involved implementation, but that depends highly on the problem addressed and solution arrived at.

So, the most distinguishing element of quality circles is that they introduce and involve a broader range of employees into the structured problem-solving process. They do this by training group or circle leaders (usually first-line supervisors) and facilitators (group-process and quality circle technique "experts" who facilitate the progress and development of circle programs) in the following areas:

1. Structured problem solving (Kepner- and Tregoe-type approaches to problem solving and decision making)

2. Group behavior

3. Elementary statistical quality control

4. Data collection and analysis

5. Data presentation, charting, developing graphs

6. Cause-and-effect diagrams

7. Making presentations

8. The quality circles process and program itself

These group leaders and facilitators then implement the process by actually involving work groups or circles during one-hour meetings that take place once a week. During these meetings, the circle, under the guidance of the group leader and a facilitator, actually work through a problem-solving process (elements 2 through 6, 8, and perhaps 10 and 11 from above).

Once the circle has proceeded to step or element 8, a proposal and presentation is made to a steering committee comprising appropriate management personnel responsible and accountable for decision making in the circle's work area. The steering committee, usually made up of 5 to 12 members, evaluates the proposal and decides whether to accept, modify, or reject it. Experience suggests that most proposals are so well developed by the time they reach the steering committee that most (80 to 90 percent or higher) are accepted. The leader and facilitator play a big role in the quality control.

The process then begins again. The circle selects another problem or opportunity and works through the problem-solving process. The technique is, in reality, no more complex than this.

Of course, the peripheral management activities associated with the establishment of such a program are significantly more complex and critical. Questions such as the following are essential to the successful development of a quality circles program.

1. What kind of training, how much training, and when is the training provided?

2. How should the program be structured, organized, and managed?

3. How should the appropriateness of the program for a given organizational system be assessed? How should the readiness of the management and employees for such a program be evaluated?

4. How should the stage be set or the foundation laid to enhance the program's success?

5. How should the success of the program over time be evaluated?

6. How should the evolution of the program over time be managed?

These questions are not easily answered and have not really even been seriously addressed by many managers attempting to implement a quality circles program. It appears that most programs are being implemented the way one buys a piece of business systems equipment. We evaluate alternatives, select one, buy it, install it, turn it on, and away we go. It is reasonably clear by now, after almost 10 years of experience with quality-circles-type programs in the United States, that we cannot effectively and efficiently implement a behavioral and social science program in this manner and expect the success rate to be acceptable.

Dr. Ishikawa has analyzed why some quality circle programs have failed in Japan and offers the following reasons:

1. Inadequate top management understanding, support, and continued legitimization

2. Restriction of the voluntary feature by goading people to join or enforcing joining by edict

3. Emphasis of the people-using instead of the people-building theme

4. Inadequate training of members in problem-solving methodology

5. Emphasis on individual effort rather than team effort (Schleicher, 1981).

The PAT process was initially developed independent of the quality circles program. Recently, though, from 1979 to the present, the PAT process has incorporated certain features of quality circles and, in particular, knowledge regarding successful program implementation. The performance/productivity action team process is a uniquely American employee involvement or participative program that many feel has significantly better design features than traditional quality circles. Some may view PATs as a second-generation quality-circles-type program. Regardless of how PATs are viewed, they appear to incorporate more appropriate design features for the typical American organization, manager, and employee and, as such, represent a potentially significant performance or productivity improvement technique. We will now examine this process.

Performance/Productivity Action Team Process

The actual PAT process began as an improvement of the NPMM. Instead of focusing on measurement, it was decided that the process should focus on ideas for productivity improvement. For example, in stage 1 of the NPMM, the NGT is utilized to identify and prioritize measures, ratios, and/or indexes of productivity for a given organizational system. Early in the development of the PAT process, the task statement for the NGT was simply reworded to solicit ideas for productivity improvement rather than measures of productivity (Sink, 1978). The result was an early form of the PAT process that developed quite independently of quality circles. Once the prioritized ideas for productivity improvement were generated, worker groups were asked to develop these ideas further so that management could evaluate their merit. The first successful application of this early form of the PAT process was in the Service and the Police departments of a small city north of Columbus, Ohio, in 1978. The city manager was quite pleased with the results, and the program survived until he vacated his position.

Since those early experiments, the PAT process has been refined and developed to further incorporate appropriate and successful design features of quality circles throughout the United States. The knowledge gained from quality circle experiments in various companies has also been incorporated into the process. Research results and technique development in the areas of goal setting, reinforcement theory (behavior modification and gainsharing systems), job characteristics theory and job enrichment, and, of course, participation strategies have been incorporated into the present design of the PAT process as well. Consideration for the long-term success of the process has led to a top-down implementation strategy. This alteration has caused, among other things, the process title to be expanded to performance action teams since upper-level managers must be concerned with the broader, integrated performance issue and not just with productivity. Implementation of the PAT program in unionized companies has also necessitated use of the word "performance" as opposed to "productivity." If you are after performance, though, you should tell people that, define it for them, help them understand what it is for their specific organizational system, then solicit their help in improving it. Quality circles, too, is a misnomer for most applications of that technique since circles typically devote less than 20 percent of their time to strictly quality-related topics. Efficiency, costs, morale, safety, and equipment are typical examples of high-priority topics addressed by circles.

The basic outline for the performance and/or productivity action team process is as follows:

SATISFACTION OF NECESSARY PRECONDITIONS
- Productivity management strategic planning
- Base period measurement and evaluation
- Extensive management briefings
- Extensive employee briefings
- Management style/productivity improvement technique compatability analysis
- Critical resistance points control
- Top management commitment to implement process top-down
- Upper and middle management modified PAT process execution

There are ten steps to this process, which we will now examine.

Steps in the PAT Process

Step 1: Performance/productivity action group (PAG)*

- Ideas for performance and/or productivity improvement: NGT ($1\frac{1}{2}$ to 2 hours).
- Output: Consensus list of prioritized ideas for productivity or performance improvement.

Step 2: Performance/productivity action group (PAG)

- Discussion and evaluation of NGT results, PAT formation, three or four of the top-priority ideas are assigned (participatively) to smaller teams of three or four persons each from the PAG. Each member of the PAG belongs to only one PAT. Issue analysis; process training.
- Output: PAT formation, process training and discussion, teams begin problem solving.

Step 3: Performance/productivity action team (PAT) problem solving

- Problem-solving process begins in meetings with line employees that are typically held on company time for one hour each week. Meetings with professional employees are also held with more flexible schedules. It will typically take one to two months for a complete problem-solving process. However, this time frame depends highly on the nature of the problem, decision, design, or implementation selected. Selected training on an as-needed basis is provided in this step.
- Output: A proposal and presentation from each PAT relating to each group's specific idea for performance or productivity improvement.

Step 4: PAT review sessions to PAG

- Results of problem-solving process are presented. The proposal is given in "dry-run" or dress rehearsal fashion to the PAG by each PAT. This gives each PAT a chance to experience giving a presentation and the other PAG members a chance to critique the proposals. This step takes approximately 2 hours, $\frac{1}{2}$ hour per PAT.
- Output/Outcome: Improved proposals, improved presentation skills, improved presentations.

Step 5: Management review session

- The PAT's present their proposals to a manager or management review committee. This step takes 1–2 hours, $\frac{1}{2}$ hour per PAT, if necessary.
- Output/Outcome: Management still controls resource allocation decisions. Opportunity to have management observe development of employees. Improved labor-management communication. Approval, if called for, can be made in this meeting.

*A performance action group (PAG) is 6–15 persons who are brought together to participate in the PAT process. It may be a natural work unit or it might be a heterogeneous collection of individuals brought together to address a specific topic.

Step 6: Management feedback session
- Proposal decisions and project management assignments made. Implementation issues discussed. Reinforcement theory invoked. Allow 1 hour for this step.
- Output/Outcome: $E \rightarrow P$ expectancies realized. Project management for implementation communicated.

Step 7: Implementation
- Highly variable depending on nature of idea.
- Output: Implemented proposals.

Step 8: Measurement, evaluation, feedback, charting
- Behavioral modification principles invoked. Base period measurement process continued.
- Output/Outcome: Improved motivation and performance, increased proactivity on part of PATs as to opportunities and problems.

Step 9: Reinforcement system design, development, implementation
- Once results are measured, some mechanism for confirming or establishing $P \rightarrow O$ linkages is needed. Management and/or analysts develop a reinforcement (performance gainsharing) system to integrate into the PAT process at this point.

Step 10: Process evaluation, modification, redesign, evolution, training and development analysis, performance tracking, recycling, and maintenance
- The PAG, if a work group, can recycle this process two to three times a year, or continually as must happen at Musashi semiconductor works. I recommend that the process cycle once a year to avoid burnout. The groups will obviously become more mature as the process continues and more efficient and effective in their execution of the steps. Management and program staff need to be open to evolutionary, positive changes as the participants mature. Training and development needs identified during the process can be handled prior to recycling. Performance tracking should continue. Continued development of feedback and charting is critical.

Figure 13.1 depicts the PAT process in a flow process fashion. Note that the process starts out with what Kepner and Tregoe (1965) call a situation, a mess, or fuzzy problems/opportunities. In Step 1, the NGT is utilized at the outset to give the group a sense of the structure for the process. No prior training is provided other than precondition briefings which give a general overview of the program and the process. Getting the group right into the NGT session in Step 1 gives them a sense of the action orientation for the program. The NGT also, of course, helps them shape consensus, which for many groups is a new experience. So, in Step 1 we start with the goal of developing a consensus list of priority ideas for performance or productivity improvement, and the NGT is utilized to accomplish this goal.

Keep in mind that the steps in this process assume that all necessary preconditions have been met. This includes upper and middle management exposure to a modified PAT process in the form of a strategic planning process. The preconditions will be discussed after the PAT process steps are detailed. Examples of the form and kind of output that can be expected from Step 1 were depicted in Figures 4.5, 4.6, and 4.11.

In Step 2, which usually occurs the following week, a discussion and evaluation of session 1 results takes place. Figure 13.2 depicts the kind of questions addressed for issue analysis during this session. From the PAG, which is typically a work group consisting of 10 to 12 persons (managers, supervisors, employees), PATs consisting of a subset (3 or 4) of the PAG are formed. Typically, the PATs share a common interest or have some degree of expertise regarding a given idea for performance or productivity improvement. They also are usually grouped on a voluntary basis, although sometimes a manager or group leader may wish to assign members to certain PATs to get the right combination of people addressing a given idea. Keep in mind, however, that the ideas selected from the NGT list for study by the PATs should be chosen by consensus. The choice is influenced by the votes received (priority) and certain issues, such as doability, chance of success, level of difficulty, cost, budget issues, resource availability, and expertise availability. (the SUG principle may also be used; seriousness, urgency, growth)

Any ideas that are designated as issues (fuzzy, ill-defined problems or op-

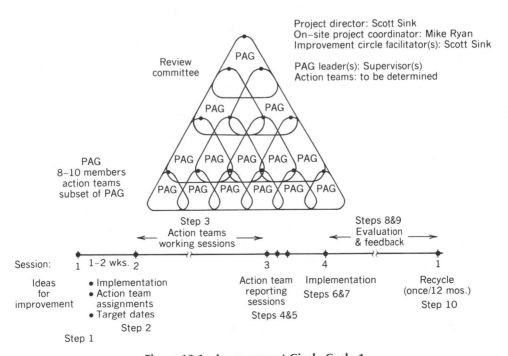

Figure 13.1 Improvement Circle Cycle 1

1.1 ISSUE ANALYSIS

- Something is wrong—we're not sure what.
- We have a vague idea of the situation.
- Emotions as well as facts are involved.

Goal: Turn issues into well-defined problems. Separate emotion from fact.

STEPS

I. ISSUE STATEMENT
- Identify in words what your concept of the issue is. Try to be as explicit as you can.

II. EXTENT OF THE ISSUE
- Think about how widespread the issue is. Who/what does it affect?
- Complete the "Extent of the Issue" worksheet.

ISSUE ANALYSIS WORKSHEET

DATE __ / __ / __

I. ISSUE STATEMENT (WHAT IS THE ISSUE?)   NAME OF THE PARTICIPANTS (PAG)

_____   _____

_____   _____

II. EXTENT OF THE ISSUE

Describe the present situation.   Describe the desired situation.

Specify: Specify:

1. What is happening?   1. What should be happening?

_____   _____

_____   _____

_____   _____

_____   _____

2. When is it noticed or observed?   2. When is it not observed?

_____   _____

_____   _____

_____   _____

_____   _____

III. POSSIBLE CAUSES OF THE ISSUE

1. Establish three of four major categories of causes.
2. Each member of the group silent generate all possible causes under each established major category.
3. Group coordinator round robin listing the causes under each category.

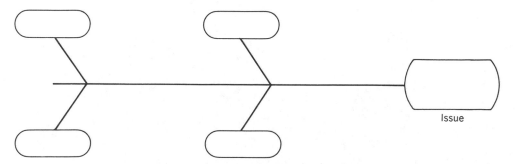

Issue

IV. FACT VS. FEELING

One of the goals of this exercise is to be able to define this issue and separate facts and feelings. It is extremely important to carefully review your perception of the facts as well as the feelings regarding this issue. Feelings can be what you feel as well as what you believe others feel. List as many specific facts regarding this issue as you can. Then list all the perceptions, feelings, etc., which you believe exist. Both lists are *equally important*.

FACTS   *FEELINGS*

_____   _____

_____   _____

_____   _____

_____   _____

_____   _____

_____   _____

_____   _____

_____   _____

V. RESTATEMENT OF THE ISSUE

Rewrite your issue statement below. Keep in mind the facts regarding the issue and try to only include the facts in this particular statement.

_____

_____

_____

_____

Are you comfortable with the statement above? If you considered the solution for the statement above, could you clearly define goals, steps by which you would solve it? If yes, go on to Step 5. If no, go back to Step 3 and rethink the facts vs. feelings. Then again restate the issue.

VI. FEELING STATEMENT

Does there seem to be a common thread regarding the feelings you have about the issue? If so, what do you perceive could be done to alleviate the problem?

_____

_____

_____

_____

Figure 13.2   Exemplary Structured Format for Issue Analysis. Similar Structure Can Be Provided and is Available for Problem Analysis, Decision Analysis, Design Analysis, and Implementation Analysis

portunities; concerns; complaints; criticisms) are clarified in this second step (See figure 13.2). Experience suggests that roughly 30 to 40 percent of the ideas are problems or opportunities, 20 to 30 percent simply require decision analysis, and 5 to 10 percent simply require implementation analysis. Kepner and Tregoe have developed a problem-solving model that depicts the relationship between issues, problems, decisions, and implementations. This structured approach assists clarification of the concept of entry in the problem-solving process at various stages. Figure 13.3, which is an adapted version of the Kepner–Tregoe model, depicts the problem-solving process. Note that a complete-cycle problem-solving process would involve entering at the issue stage, clarifying, defining, delimiting, and specifying the issue. Very often, an issue just needs clarifying so that employees will know what is going on.

During early stages of one PAT program in a small manufacturing plant, employees had ranked wage and salary administration as an idea for productivity improvement. The president of the firm did not wish to discuss this issue, arguing that salaries and any related decisions were solely the concern of management. But his employees were expressing their concerns about a topic they considered important. Denying them the right to at least communicate their

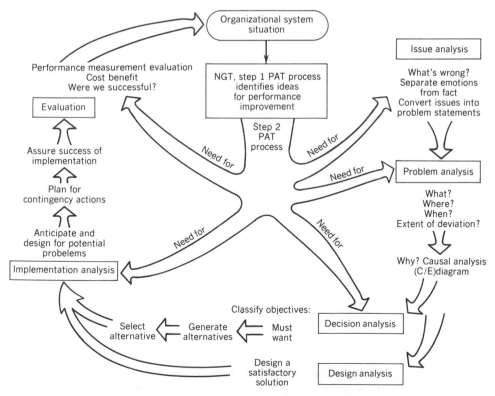

Figure 13.3   Adapted Kepner-Tregoe Problem-Solving Model

opinions is one of the reasons for poor labor-management communication in the United States. When the employees' concerns were treated as a genuine issue, it was clear that they simply wanted to find out if there was a systematic wage and salary administration program and, if so, how it worked. The president came to the next meeting, spent a half hour explaining how wages and salaries were administered, answered some questions, got to know some of his employees, and that was the end of the issue.

The point to be made from this story is that some ideas will turn into issues. Some of the issues can be dispensed with easily but others, once clarified, may turn into one or many problems that can then be analyzed. The full problem-solving process involves starting with a mess, evolving to issues, developing issues into problems and performing problem analysis, moving from problem analysis into decision analysis, and evolving from a decision to implementation analysis. Note, however, that many times we find ourselves entering the process somewhat downstream. Also, it is not unlikely for the problem-solving process to terminate prior to the next stage. The process is intended to be and should be very flexible. Figure 13.4 graphically depicts the process.

> Every goal man reaches provides a new starting point, and the sum of all man's days is just a beginning (Lewis Mumford).

In summary, Step 2 is designed primarily to establish the action teams (PATs), which, as mentioned, are a subset of the PAG. There will typically be 2 to 4 mutually exclusive PATs in each PAG, each formed either by assignment or by a mutual interest in a given idea that the group members wish to develop further. The second step also is utilized to discuss the NGT results and, in particular, to separate the ideas into issues, problems, decisions, designs, or implementations. The issues should be dealt with in this session and either immediately resolved, rescheduled for solution during the next several weeks, or converted into problem(s) or opportunities. In addition, some PAT process training takes

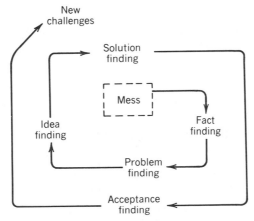

Figure 13.4 Creative Problem-Solving Process

place during this second session. Hand-out material on the process and on problem solving is usually provided, and overheads and/or slides are shown depicting the NGT results and the process steps. Specific technique training is done on an as need basis during the problem solving phase. PAG members and PATs are encouraged to ask questions about the process and, in particular, the next several steps.

Meetings for the PATs are scheduled once a week. We have found it appropriate for many groups to schedule the one-hour meeting at the same time each week. However, wherever managers, professional employees, and so forth are involved, it is often not possible to schedule a fixed meeting that everyone can attend. In this case, we let each PAT, based on the schedules of the members, determine a one-hour slot each week when everyone is available. As long as management continues to support the program and the PATs are forced to develop a project plan with milestones and deadlines, the PATs will be conscientious. The meetings must all be productive. An agenda is a must, beginning and ending on time is essential, and a group-process observer is helpful. Each PAT is encouraged to exercise good group-process management. Materials and training are made available to PATs on an as-needed basis. The PAT process does not invoke unnecessary training. Training is provided when a facilitator, group leader, or program coordinator sees a need for it or when a PAT specifically either requests it or is visibly floundering because of the lack of it.

Step 3, a structured group problem-solving process, is the heart of the problem solving component of the PAT process. It entails each PAT turning its specific idea into a proposal and presentation for management. At each stage of the problem-solving process, the PATs are encouraged to collect and analyze data to support the development of their idea. There are literally an infinite number of data collection and analysis techniques. Operations analysis, checklists, cause-and-effect diagrams, statistics, engineering economic analysis, methods engineering, value engineering, Pareto analysis, flow process charts, sequence charts, time study, and occurrence sampling are all various forms of data collection and analysis. This list doesn't even include the variety of informal or "ad hoc" data collection and analysis procedures available to a PAT.

During the first meeting of Step 3, which will typically be the third week of the process for a specific PAG and PAT, the PATs are asked to develop a projet plan. This project plan will necessarily include the following elements:

1. Statement of the idea and determination as to what the idea is (problem, design, decision, implementation).

2. Statement or list of desired outcomes for the PAT. What does the PAT intend to accomplish?

3. A list of fairly complete and detailed mutually exclusive steps or activities that the PAT feels must be accomplished in order to attain the desired outcomes.

4. A Gantt chart that sequences the activities over the expected duration of this particular project. On the Gantt chart (example depicted in Figure 13.5), mile-

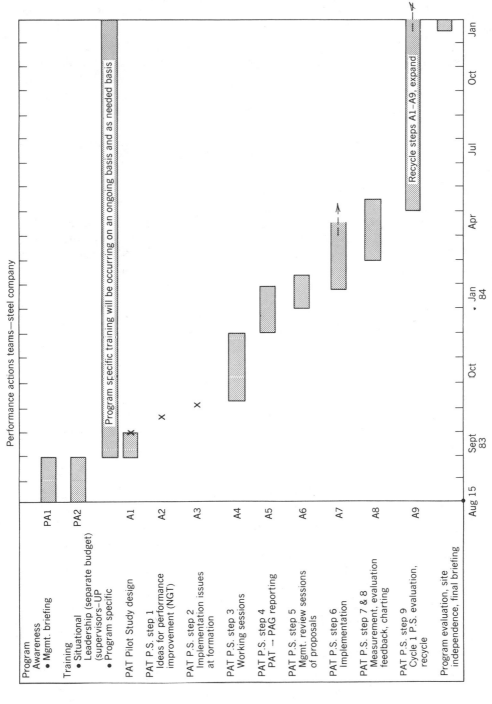

Figure 13.5   Gantt Chart Development. Example is for Project Planning for a Cycle of the PAT Process

stones or targets should be established for major activities and for the overall project completion date.

5. Resource allocation for the project should be considered for all major activities. Whether the activities are to be accomplished by an assigned individual between weekly PAT meetings or by the PAT itself must be decided on. When external resources (human or otherwise) for given activities are used, the program co-ordinator or facilitator will need to review and seek approval.

The Gantt chart development will take up most of the first PAT session. Subsequent PAT sessions begin the problem-solving process by executing the project plan. PATs usually find it necessary to designate a project leader. This person is responsible for ensuring that the project is on schedule and that the work is being done correctly.

The PAT program coordinator(s), the facilitators for the group meetings, and the group leader will need to be constantly assessing training and development needs. The range of problems, designs, decisions, and implementations that can be identified and addressed by PATs is immense. Therefore, it is virtually impossible to front-end load the training for the PAT process. As PATs proceed, specific technique development in terms of data collections and analysis and design development can be provided. We have found it best to train on an as-needed basis. Adults are much more receptive to training when they have a need to know.

There is not enough space, nor is it particularly appropriate, to present specific training modules that have been developed for the PAT process. However, Table 13.1 lists training modules that we have found useful and various sources for more detail on particular data collection and analysis, decision analysis, and design methodology techniques. The practitioner and student should keep in mind that many of these techniques are actually just structured ways of collecting, presenting, and evaluating data regarding specific types of problems. The key thing to remember is that regardless of the focus of the idea selected by an action team, some method for approaching the development of a solution, proposal, or recommendation is necessary. An inherent and critical early step in problem solving and design is data collection and analysis.

Progress in most disciplines is measured by the existence of techniques for analyzing the phenomena studied by that discipline. These techniques tell students and practitioners what data are important, how to capture or present the data, how to format data, how to analyze the data, how to interpret the results, and often what to do based on certain results. Think of well-developed fields, professions, or disciplines, such as medicine, engineering, quality control, work measurement, and operations research, and you will be able to envision the data collections, analysis, and evaluation systems. Structured, group problem-solving processes like quality circles and performance/productivity action teams are attempts to spin off or delegate downward certain fundamental, basic, generic, and specific techniques for analyzing problems and opportunities commonly faced in organizations.

Table 13.1 Performance/Productivity Action Team Training Module Sources

RESOURCE, REFERENCE/ TRAINING & DEVELOPMENT MODULE	STRUCTURED PROBLEM SOLVING	TRADITIONAL QUALITY CIRCLES	GENERAL DATA COLLECTION AND ANALYSIS TECHNIQUES	DESIGN METHODOLOGY AND PROCESSES	METHODS DESIGN & ENGINEERING, RAISE ENGINEERING, OPERATIONS ANALYSIS	STATISTICALLY BASED QUALITY CONTROL AND QUALITY MANAGEMENT	BASIC PROJECT MANAGEMENT	SMALL GROUP BEHAVIOR AND PROCESSES	STRATEGIC PLANNING	IMPLEMENTATION STRATEGIES
Kepner, C. H. and B. B. Tregoe. The Rational Manager, New York: McGraw-Hill, 1965.	✓									
Kepner-Tregoe, Inc. The Results Planning Manual, 1968.	✓									
Dewar, D. L. The Quality Circle Guide to Participation Management, Englewood Cliffs, N.J.: Prentice-Hall, 1980.	✓	✓	✓							
Gryna, F. M. Quality Circles: A Team Approach to Problem-Solving. New York: AMACOM, 1981.		✓								
Dewar, D. Quality Circle Leader Manual. Midwest City, OK.: IAQC.		✓								
Dewar, D. and Beardsley, J. F. Quality Circle Text. Midwest City, OK: IAQC.		✓								

411

Table 13.1 (Continued)

RESOURCE, REFERENCE/ TRAINING & DEVELOPMENT MODULE	STRUCTURED PROBLEM SOLVING	TRADITIONAL QUALITY CIRCLES	GENERAL DATA COLLECTION AND ANALYSIS TECHNIQUES	DESIGN METHODOLOGY AND PROCESSES	METHODS DESIGN & ENGINEERING, RAISE ENGINEERING, OPERATIONS ANALYSIS	STATISTICALLY BASED QUALITY CONTROL AND QUALITY MANAGEMENT	BASIC PROJECT MANAGEMENT	SMALL GROUP BEHAVIOR AND PROCESSES	STRATEGIC PLANNING	IMPLEMENTATION STRATEGIES
Anderson, D. R., D. J., Sweeney, and T. A. Williams, Introduction To Statistics: An Application Approach, St. Paul, Minn: West, 1981.			Chs. 1 & 2							
Tukey, J. W. Exploratory Data Analysis, Reading, Mass.: Addison-Wesley, 1977.			Chs. 1–6							
Konz, S. Work Design: Industrial Ergonomics, 2nd ed. Columbus, Ohio: Grid, 1983.	Ch. 5		Chs. 5–7, 9, 21 & 25, 26	Chs. 4–7, 26	Chs. 6, 7, 13–16					
Smith, G. L. Work Measurement: A Systems Approach. Columbus, Ohio: Grid, 1978.			Chs. 1 & 2							
Bailey, R. W. Human Performance Engineering: A Guide for System Designers, Englewood Cliffs, N.J.: Prentice-Hall, 1982.	Ch. 7		Chs. 23–25	Ch. 10–11						

Reference			
Barnes, R. M. Motion and Time Study Design and Measurement of Work. 7th ed. New York: John Wiley and Sons, 1980.	Ch. 4	Chs. 5, 6	Chs. 5–17
———. Motion and Time Study. Homewood, Ill.: Irwin, 1976.	Chs. 3–7, 11		Chs. 3–7, 9–11, 21–22
Juran, J. M. and F. M. Gryna, Quality Planning and Analysis, New York: McGraw-Hill, 1980.	Ch. 4		
Ishikawa, K. Guide To Quality Control, Asian Productivity Organization, Tokyo, Japan, 1976.	Chs. 1–6		Chs. 7–13
Besterfield, D. H., Quality Control, Englewood Cliffs, N.J.: Prentice-Hall, 1979.		✓	
Charbonneau, H. C. and G. L. Webster, Industrial Quality Control, Englewood Cliffs, N.J.: Prentice-Hall, 1978.		✓	

413

In a production environment, where quality circles first were initiated, quality and process control was a key issue. So much of the training focused on quality control that the program was, therefore, named quality control circles. In the United States, "quality circles" have been implemented to address broader issues. Studies have shown that in traditional quality circles applications, quality is the focus of attention less than 20 percent of the time, efficiency is the focus approximately 15 percent of the time, cost the focus 10 percent, technology or equipment the focus 10 percent, safety 5 percent, process control 10 percent, methods and procedures 10 percent, and other topics 10 percent. So, as one can easily see, the term "quality circles" is a misnomer. Yet what we find is that the same quality-related techniques and training are still provided to these expanded-focus participatory problem-solving programs. What we have attempted to do in the PAT process is to expand the range of training modules to meet the expanded needs of groups that are actually working to improve productivity or, even broader, performance.

What one finds as the concepts of participatory group problem solving are applied with managers, professional staff, clerical employees, as well as line personnel is that the traditional quality circles training is somewhat appropriate but that more often than not it is a case of a "hammer looking for a nail." PAT and for that matter quality circles program coordinators and facilitators are going to have to be more broadly trained and versed than at present. A bias of mine is that certain industrial engineers represent the closest ideal to PAT process coordinators and facilitators.

Training for the PATs needs to be highly pragmatic and focus only on what the group needs immediately or will need in the near future. The action orientation cannot be sacrificed at the expense of training and development. In particular, we have found that the following training modules are a necessary and rather predictable subset that will likely need to be presented during the course of the PAT process:

1. *PAT general process training:* outline of steps, rationale, purpose, group problem solving

2. *Project management:* how to construct a simple Gantt chart for the PAT

3. *Basic problem solving:* problem analysis, decision analysis, implementation analysis, check sheets, questions

4. *Basic design methodology:* how to describe, document, analyze, and present a system; how to systematically design a new, improved system

5. *Basic small-group behavior:* group process productivity, group-process observer, structuring for performance

Other training modules are utilized on a specific as-need basis. If there is a PAT with a methods-related idea, then basic methods training is provided to that team. The purpose is to help employees develop while solving problems, taking advantage of opportunities, and improving productivity and performance, not vice versa.

Step 3, then, is a structured group problem-solving process. The desired output is a proposal that summarizes the team's analysis and presents and justifies a specific recommendation. Each PAT is expected to prepare a presentation and a proposal less than or equal to five pages. Common elements for a proposal are (1) statement of the idea of performance improvement; (2) priority given the idea by the PAT (the NGT results); (3) discussion as to how the idea was developed (problem analysis, design analysis, decision analysis); (4) presentation analysis highlights; (5) proposed system, recommendation(s), solution(s), design(s), description; (6) justification, cost-benefit analysis, present versus proposed system comparison; and (7) implementation issues.

Step 4 of the PAT process entails each PAT presenting its proposal to the PAG. This provides the PAG with an opportunity to review and critique the work and to provide input prior to management review. It also provides each PAT with an opportunity to "dry-run" or "debug" its presentation in a low-risk, lower-pressure setting than the PAT might face in the management briefing. Each team is given 30 minutes to make its presentation and to respond to questions.

Step 5 is the management review session. Each PAT will present its proposal in written and oral format to a select management review committee. Again, the teams are given 30 minutes to present their proposals. Management will not necessarily evaluate the proposals in this meeting. In most cases, management will require a week to think through the proposals.

Step 6 is the management feedback session. After taking a short time to review the PAT proposals, the management review committee will reconvene the PAG and PATs to provide them with a decision and feedback. This is certainly, as in Step 5, a good opportunity for management to practice its positive reinforcement skills. Management will need to have thought through implementation questions, such as resource allocation, a timetable, and evaluation. It is preferable for management to have a Gantt chart developed for the implementation of each recommendation or proposal. In some cases, PAT and PAG members may actually implement their own proposal. In other situations, different persons or groups in the organization may be assigned implementation responsibilities.

Step 7 is the actual implementation phase of the PAT process. This is the payoff step and is perhaps the single most important step in the process. It is critical that management ensure that the ball is not dropped here. Implementation should occur in a very timely fashion. PAG and, in particular, PAT members must be kept informed (preferably with a posted and updated Gantt chart) as to implementation progress. Feedback at this stage is essential to the ability of management to capitalize on the motivational potential of this process.

Step 8 is a step that actually should be initiated, for the program at least, during the time that necessary preconditions are being ensured. One of the necessary preconditions, as you will recall, is base-period measurement and evaluation. During Step 8, specific measurements need to be developed relative to each proposal's potential impact on performance. In particular, behavior modification principles and techniques need to be invoked. Specific, explicit, and highly visible charts need to be developed that depict and monitor cause-and-effect relationships between the implemented proposal and organizational system performance.

Prior to, as well as throughout, the PAT process, management should be considering the integration of some sort of reinforcement system with the PAT process. As we discussed, participation in and of itself may have relatively little impact on performance if it is not effectively integrated with goal setting, feedback, and/or reinforcement. Step 9 specifically addresses the design of the reinforcement system and is intended to maintain efforts on the part of the group to follow through with results and continue the process. A fairly predictable human trait is to ask the question, ''What's in it for me?'' This step should be

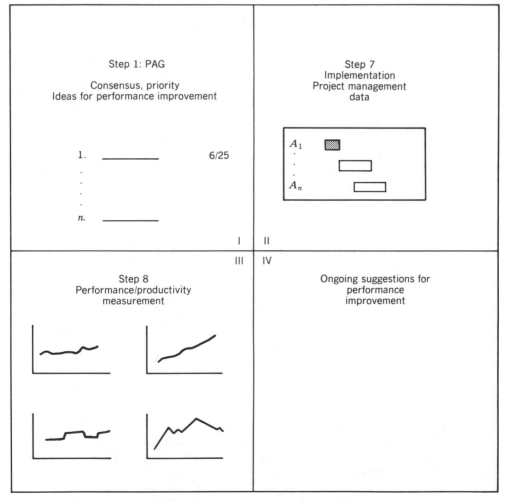

Figure 13.6   Performance Action Team Visibility System Component

an attempt on the part of management to answer that question. Note that management's ability to address the reinforcement issue depends directly on the quality with which it completed Step 8.

Step 10, the final step in the PAT process, involves stepping back and assessing each cycle (one completed problem-solving process for a PAG) from both a subjective as well as an objective perspective. Management and participants in the process should be involved. Recommendations for improving the program and process should be considered and incorporated for future cycles where appropriate.

Recycling the process for a given PAG can take place immediately after Step 7, or the PAG can be allowed to go into a maintenance phase. A maintenance phase is where a PAG, if a work group or unit, continues performing assigned responsibilities while monitoring the impact of the implementation of its proposals. We have found it useful during this phase, which typically lasts three to six months, to utilize a PAG feedback board, such as the one presented in Figure 13.6.* This board is divided into four areas. Quadrant I contains the Step 1 results or the NGT list of prioritized ideas for performance improvement. Quadrant II includes project management information (perhaps the Gantt charts) for implementation of proposals. Quadrant III contains elements of the performance measurement system that track the impact of the program, process, and proposals on various aspects of performance. And Quadrant IV contains room for a continuous input of new ideas for performance improvement.

During the maintenance phase, the PAG would gather around this feedback board at least once a week for a short time (5 to 15 minutes) with the group leader to discuss results. At certain points, new suggestions received might be prioritized and discussion might focus on possible PAT formation.

After three to six months of a maintenance period, the PAT process would be recycled from Step 1. Of course, preconditions are still as essential as before and must be ensured. If the PAT cycle is a pilot study, then a check for satisfied preconditions essentially is a program continuation and potential expansion indication.

Development of a Performance/Productivity Action Team Program

The discussion regarding PATs to this point has focused on describing the process itself. The larger question surrounding how to initiate, develop, and implement a PAT program will now be addressed. We have experimented with several approaches to introducing a program of this nature into an organization and have arrived at what appears to be an effective strategy for most organizations.

Figure 13.7 depicts the overall top-down implementation strategy. Research and experience suggest that the biggest sources of implementation failures for

*One of these boards is made up for each PAG in an organization; they are placed in a common visibility room.

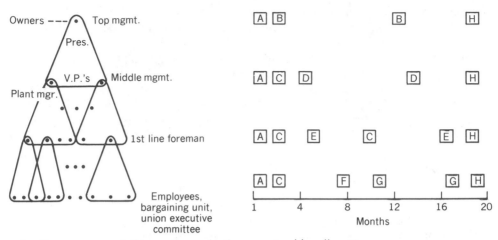

A—Program presentations, communication, approval by all parties
B—Top mgmt./union executive committee strategic planning retreat, steps 1–6
C—Situational leadership/leadership effectiveness/PAT process training, selected persons
D—Middle mgmt./union exec. committee action planning session, steps 4–7
E—1st line foreman/bargaining unit performance action group pilot, steps 6 & 7
F—Work group performance action group pilot, steps 6 & 7 (OPC facilitation)
G—Work group performance action group pilot, steps 6 & 7 (site facilitation)
H—Site independence

Figure 13.7   Top-down PAT Process Implementation

these types of programs are disregard for preconditions and the bottom-up approach. The bottom-up approach does not allow for the supportive climate and culture to establish itself, develop, or kill the program prior to raising, unnecessarily, employee expectations. Experience suggests that most employees are ready for structured participatory management programs while most management is not. The top-down implementation strategy, in an action research fashion, allows the program to gain support naturally as a result of a positive affective reaction on the part of management to the basic participatory characteristics of the process. Management actually experiences the NGT and structured group approaches to planning, problem solving, and decision making, and if enough managers at high levels don't feel comfortable with the process, the program dies an early death before it has a chance to elevate employee expectations.

STRATEGIC PLANNING RETREAT

A PAT program can be developed most effectively and successfully when it is fully understood and believed in by all levels of management. Top management is no exception. As a way to expose upper management to the basic characteristics of the PAT process in a meaningful and productive fashion, a structured, participative strategic planning process has been developed. The steps in the planning process are

Step 1: Day 1, A.M.
External Strategic Audit: What external strategic (two to five years varying from organization to organization) factors (competitors, trends, conditions, assumptions) should be considered in our planning process? Stress is on a creative search process. Silent generation, round-robin, much clarification and discussion, *no* closure (no voting and ranking). This step typically takes 2 hours. The unit of analysis must be specified.

Step 2: Day 1, P.M.
Internal Strategic Audit: What internal strategic (two to five years or other appropriate planning horizon) factors (strengths, weaknesses, conditions, trends, persons, positions, programs) should be considered during our planning process? Stress again is on a creative search process. Silent generation, round-robin, dicussion. Lasts 2 hours.

Step 3: Day 1, P.M.
External and internal factors are converted into planning promises and assumptions. These are then evaluated in terms of their importance to future plans (low importance to high importance) and in terms of the planner's certainty that the assumptions are valid. A matrix with four quadrants, such as the one shown below, is provided to each participant. Each assumption is labeled as external or internal and is numbered (for example, *E1*, *I5*, *E22*).

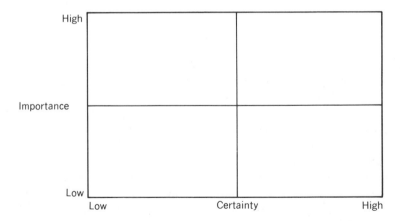

Each assumption is then placed on the grid, which is drawn on a piece of flip chart paper for all to see. The intent is to seek some level of consensus as to general placement and to carefully consider assumptions placed in the northeast and northwest quadrants during the next step. This step takes approximately 1 hour.

Step 4: Day 2, A.M.

Strategic Planning: The NGT is utilized to develop two-to-five-year (or other appropriate planning horizon) goals, objectives, desired outcomes. Note that all the flip chart sheets from Day 1 sessions should be left on the walls in full sight of all participants. This step will take 2 hours.

Step 5: Day 2, P.M.

Performance Objectives: The group is now asked to address performance measurement, evaluation, and control. The questions addressed are, How will we know if we succeed with our planning? How can we measure the performance of the organizational system we are examining? This step can utilize the NGT, which would mean you are essentially executing part of the NPMM. At this stage, however, just parts of the NGT are typically utilized. Silent generation, round-robin, clarification, discussion. This step typically takes 1 hour.

Step 6: Day 2, P.M.

Identification and prioritization of strategic, tactical, and operational action programs. What programs or plans will have to be budgeted for in the short term (one year) as well as the longer run (two to five years) in order to accomplish the goals and objectives from Step 4? A simultaneous operational focus (one year) with a strategic (innovation, growth, continued development and support) focus (two to five years) is required. NGT is utilized. Step 6 requires 2 hours.

Step 7:

Program planning and resource. This step typically takes place outside the retreat format.

Step 8:

Program evaluation, review, maintenance. Again, this is an activity that is executed after the retreat. Wrap-up, discussion of the process. PAT management briefing. Discuss common characteristics with the planning exercise. Intermix productivity basics. Lasts a maximum of 1 hour.

This simple and pragmatic strategic planning process has been tested in a variety of situations and applications. It has proven to be a very effective way of accomplishing planning while simultaneously exposing management to the nature of the participative problem-solving process its employees will become involved with. The planning process is best executed in a two-day retreat format. The process can be designed to accommodate anywhere from 5 to 30 managers (8 to 12 is ideal). The primary objective of the retreat is to develop strategic, tactical, and operational plans for the organization. However, intermixed with the planning process itself are the opportunities for PAT process familiarization and briefings, productivity management basics, as well as social activities. If designed and executed correctly, this retreat can acomplish many purposes, not the least of which would be team building.

TACTICAL AND OPERATIONAL PLANNING AND IMPROVEMENT

Assuming that top management receives the PAT process favorably as a result of the planning retreat, the plan then calls for moving exposure to the process downward in the organization. At the middle levels of the organization, we have found it useful to execute a collaborative or group management by objectives (MBO) process using the NGT as the device for shaping consensus on priority department, function, division, or plant objectives. The focus is on objective setting and action planning for a unit of analysis below the firm level, yet above the group level itself. The applications might be for a division, a plant, a function, or any other group break down for which a group of middle-level managers are responsible.

PAT process familiarization comes in two forms. The first is through the CMBO process, which incorporates certain characteristics or features of the PAT process itself. So, again, middle-level management has an opportunity to experience and feel the process firsthand. If too many feel threatened by or are uncomfortable with the process, it may not be implemented further or at least not without some sort of intervention to create improved support for the program. The second orientation to the PAT process itself comes in the form of an actual briefing that outlines the process. The CMBO process and PAT briefing can be done in one-to-two-day retreat formats; however, it is often done in-house.

In addition to the PAT process orientation, middle-level managers should be exposed to productivity management training. The Virginia Productivity Center utilizes a course entitled "The Essentials of Productivity Management: Planning, Measurement, Evaluation, Control, and Improvement" for this purpose. That short course is simply a three-day program covering elements of this book.

Implementation of the PAT Process

Moving exposure of the PAT process down to the operational levels in an organization could, depending on size, take anywhere from six months to two years or more. Pace of program implementation depends on many factors. Perhaps one of the most important, however, is the degree to which this type of program is congruent with management style. By the time that upper and middle levels of management have been exposed to the PAT process via planning and CMBO-type involvement, a program coordinator should have a good feel for level of support, areas of resistance, necessary program modifications, and so forth. At this stage, the program could be pilot tested in a few work groups. Of course, unions, office supervisors, and employees will need to be briefed on the program and provided with ample opportunity to raise questions and concerns.

When implementing the PAT process at the operational level of the firm, we have found it effective in the long run to comprise the first pilot study of all supervisors and perhaps some related support personnel, such as engineers, quality-control people, and maintenance personnel. We typically select 10 to 12 supervisors from a related area of the firm or plant. We then simply execute the ten-step PAT process. It should be stressed to employees, supervisors, unions, and management that such studies are simply dry runs or experiments with a new management process. No strings, no hidden agendas, no guarantees of success are attached to the pilot-study process.

The supervisory group(s) pilot study usually precedes other employee work-group pilot studies by three to four months. This gives program coordinator(s) and facilitator(s) the opportunity to learn from the supervisors' reactions to the process. After the first pilot study is well underway, another one or two PATs can be initiated. If the first pilot study consists of a group of supervisors, one or two of them should be selected as candidates for work-group PAT pilot studies. This ensures that these supervisors, as PAG leaders, will be knowledgeable and, hopefully, supportive of the process.

After the two or three pilot studies have completed Step 7 of the PAT process, management must ensure that Steps 8, 9, and 10 are executed. In particular, Step 10 takes on special meaning early in the program development. Appropriate management at all levels should step back and objectively assess the impact and worth of the program. It is recommended that management review various aspects of the program initiation, particularly the pilot studies, over the span of two to three months. This will give management time to decide whether to proceed with a maintenance phase experiment with the pilot study groups and expand the program to other areas of the firm.

The PAT process itself is deceptively simple. Successful execution of the program and the process requires considerable skill that can come only through much reading, planning, and experience. Persons starting a program of this nature should keep in mind, however, that "good decision making comes from wisdom, judgment, and experience. And, wisdom, judgment, and experience come from bad decision making." Program design, development, and implementation of a PAT process must be viewed as action research. Management, program coordinators, and facilitators must be well versed in the relevant behavioral and social science literature and, above all, must be patient, persistent, and consistent.

Executed correctly, the PAT process has the ability to unlock and unleash the kind of potential witnessed by Musashi semiconductor works. At Musashi, 370 worker groups submitted over 100,000 proposals of which more than 85 percent were implemented over the course of one year. American managers, wake up to the ways of your competition. We have taught them well. Let's hope the teacher can once again become a good student!

REFERENCES

Kepner, C. H., and B. B. Tregoe. *The Rational Manager: A Systematic Approach to Problem Solving and Decision Making.* New York: McGraw-Hill, 1965.

Kepner, C. H., and B. B. Tregoe. *The New Rational Manager.* Princeton, NJ: Princeton Research Press, 1981.

Morris, K. E. "The Application of the Delphi Technique to the Development of Productivity Measures." NSF Grant No. APR 75-20561, Ohio State University Research Foundation, Columbus, Ohio, 1977.

Morris, W. T. *Implementation Strategies for Industrial Engineers.* Columbus: Grid, 1979.

Patchin, R. I., and R. Cunningham, eds. *The Management and Maintenance of Quality Circles.* Homewood, Ill.: Dow Jones-Irwin, 1983.

Riggs, J. L., and G. H. Felix. *Productivity by Objectives.* Englewood Cliffs, N.J.: Prentice-Hall, 1983.

Schleicher, W. F. "Quality Circles: The Participative Team Approach." *Quality*, October 1981.

Sink, D. S. "The ABC's of Theories X, Y, and Z." *1982 Fall Industrial Engineering Conference Proceedings*, Norcross, Ga., 1982b.

Sink, D. S. "Development and Implementation of Productivity Measurement Systems with Emphasis on Interorganizational Relationships." Unpublished dissertation, The Ohio State University, Columbus, Ohio, 1978.

Sink, D. S. *The Essentials of Productivity Management: Planning, Measurement, Evaluation, Control, and Improvement.* Short Course Notebook, LINPRIM, Inc., Blacksburg, Va., 1982c.

Sink, D. S. "Productivity Action Teams: An Alternative Involvement Strategy to Quality Circles." *1982 Spring Institute of Industrial Engineering Conference Proceedings*, Norcross, Ga., 1982a.

Stewart, W. T. "The Facilitation of Productivity Measurement and Improvement in Manufacturing Organizations." Unpublished dissertation, The Ohio State University, Columbus, Ohio, 1978.

QUESTIONS AND APPLICATIONS

1. For the following situations and organizations, develop a strategic plan for a productivity improvement program or effort:

(a) You are the plant industrial engineer of a steel plant characterized by strong foreign competition, outdated and outmoded equipment, traditionally poor labor relations, and several years of running in the red. New corporate management sincerely wants to turn things around. There is a strong USW local, but the executive board is willing to cooperate. Markets for your product are good if you can improve quality, timeliness, and reduce costs and, hence, prices by 10 percent.

(b) You are the manager of a large grocery store. You own a local franchise of a national chain and have considerable autonomy and control over local operations.

(c) You are director of accounting services for a large corporation. Your vice president has asked you to develop a productivity improvement program for the group that might serve as a model for other directors of service-type functions. The program is to be comprehensive within your division.

(d) You are the general manager of an automobile assembly plant. You have decided to develop a program that might work at the assembly plant level and later integrate well with any programs that might come out of Detroit. You want the plan to consider all factors but focus especially on cooperation, communication, and coordination between areas of the plant. You have little control over major technology changes since the capital improvement decisions are spearheaded in Detroit.

(e) You are the president of a large state university that is a land grant college. You are convinced that the university must aggressively revitalize itself. Your concern is over a redefinition of what performance really means and to focus on the definition of improved performance, measurement, and improvement strategies and techniques at all levels of the organization. You realize any initial steps must come from your office.

(f) You are the productivity administrator for a large, rather decentralized and dispersed corporation. Products range from consumer goods to a wide variety of industrial products. Plants are located all over the United States as well as in many countries overseas. You have just been appointed to this position. The president of the company expects to see a strategic plan of your approach in six months.

2. Discuss how PATs might be developed to focus on:
(a) Productivity improvement
(b) Goal or objective setting
(c) Performance measurement and evaluation
(d) Quality improvement
(e) Efficiency enhancement
(f) Quality of work life enhancement
(g) Profitability improvement
(h) Innovation enhancement
(i) Effectiveness improvement

3. Investigate Quality Circles. Compare and contrast quality circles with PATs. What are the major differences? Evaluate the quality of the design of both techniques.

4. What are the critical factors determining participative management program successes?

5. How would you best characterize PATs? Where would you categorize them in the productivity improvement taxonomy presented in Chapter 9?

6. Why is it so important to implement PATs top-down in an organization?

7. What is the role of strategic planning in the PAT process?

8. Compare the PAT process with the concepts presented by Ouchi in his book *Theory Z* and by Kanter in her book *The Changemasters*.

9. Discuss the relative roles of skill versus knowledge in regards to effective implementation of the PAT process.

10. How will the concepts Morris presents in his book referenced below be useful to implementing a PAT process?

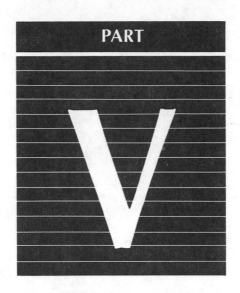

PART

V

PRODUCTIVITY MANAGEMENT PROGRAMS

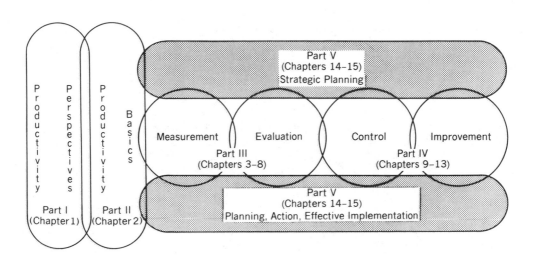

CHAPTER

14

DESIGNING, DEVELOPING, AND IMPLEMENTING SUCCESSFUL PRODUCTIVITY MANAGEMENT PROGRAMS/EFFORTS

HIGHLIGHTS

- The "Grand Strategy" for Designing, Developing, and Implementing Productivity Management Efforts
- Strategic Planning for Productivity Management Programs
    Contextual Issues
    Strategic Planning Process
    Synthesis

OBJECTIVES

- To introduce a generic "grand strategy" that seems to form the foundation of many excellent firms' approaches to productivity management program design, development, and implementation.
- To provide an example of how one major American corporation (a division within that company has incorporated the "grand strategy" into their productivity management approach.
- To present a pragmatic, effective, structured, group-process-oriented strategic planning plan. To describe how a two-to-five-year plan for a productivity management program can be easily developed.

To this point we have presented and discussed the kind of fundamental understanding regarding productivity that is essential to successfully execute a productivity management program. Chapter 1 presented perspectives on productivity and different views on the problems and solutions as well as data to indicate the magnitude of the dilemma. Chapter 2 presented basics so that managers and staff and students have a systematic, consistent understanding of performance concepts.

We have also examined approaches, theories, techniques, and models that can be utilized in different situations for different purposes to measure and evaluate productivity. After a recent exhaustive search of the literature and field of practice, we found primarily only those techniques presented in Part III.

Finally, we have looked at approaches, theories, and concepts that form the basics for controlling and improving productivity in an organizational system. Technology, capital, cost reduction, quality control, management by objectives, manufacturing requirements planning (MRP), production control, inventory control, work measurement, job enrichment, organizational development, employee involvement, and computer-integrated manufacturing are all mechanisms or approaches for controlling and improving productivity. Quality design and effective implementation of these systems are the key to tapping the productivity improvement potential of each approach.

Part IV focused on creating the willingness and ability on the part of the *people* in an organizational system to seek ways to become more effective, efficient, productive, innovative, and profitable/cost effective while maintaining and/or improving quality and quality of work life. Many organizations and managers interviewed during the preparation of this book over the past five years have talked about the primacy of the human component in developing innovative ways to improve productivity. Honeywell's Productivity and Quality Center in its Aerospace and Defense Group and Boeing's Industrial Engineering effort at the Renton plant come to mind, but there are many other examples. More and more managers are saying, "If we did not introduce one new technological innovation for the next five years but focused on more effective and efficient use of the technology we presently have, we could improve productivity dramatically and compete very favorably."

This sounds blasphemous in an era of robotics and computer-assisted manufacturing. However, such comments reflect a maturity and wisdom regarding the roles that certain resources play in overall, long-run performance and, of course, survival. This comment reflects a balanced perspective on the relative importance of technology versus people and systems. And it supports Naisbitt's (1982) perception that human resource management issues lag technological issues at present.

The managers we have spoken with over the past five years are all very sensitive to the "high tech/high touch" phenomenon. So, Part IV does not attempt to deemphasize the importance of such factors as capital, technology, and advanced processes to productivity improvement in the coming decades. Rather, it attempts to focus on an element that is a critical and complex parameter in the "equation."

In the three-day short course on the subjects covered in this book, we finish our discussion of Part IV topics on the third day at noon. Once managers have been exposed to the basic pieces of the productivity/performance puzzle, one question invariably arises: "How do we put these pieces of the puzzle together?"

The answer to this question is not an easy one. With 40 to 60 managers from different organizations, doing different things in different ways, it is extremely difficult to prescribe "one best way." Yet most of the current available resources adhere to this "one best way" mentality. The literature abounds with case studies of how many organizations have designed, developed, organized, and executed the productivity efforts. Conferences, seminars, and associations, are being held almost weekly on this topic. And, of course, the consultants available to help you design your program are numerous. All of these approaches suggest copying or at best modifying what others have done.

Managers should implement a strategic planning process that will develop their own unique plan for a productivity management effort. The techniques for measurement, evaluation, control, and improvement they use may be identical to those used by other organizations. However, the way in which managers use them and the way in which the techniques are developed in a system will be unique to their own organizations. The logic for the development of the productivity management effort will come from key people in their organizations. It will be *their* program. Managers will avoid the dysfunctional consequences of having a program where top management tells middle management what to do and so on down the line.

What follows is an application and extension of the eight-step strategic planning process presented in Chapter 13. It is a participative process designed to effectively and efficiently create a product (a strategic plan for a productivity management effort), and to shape consensus and commitment to the resulting plan. It may take longer to implement the process than to copy someone else's program, the long-run implementation effectiveness are likely to be significantly higher. The process will necessarily be pulled off differently from organization to organization depending on:

1. The scope of the program to be developed (companywide, plantwide, within a given function or department, and so forth);

2. The personalities and politics associated with the program and the key people involved and affected;

3. The skills of the people attempting to develop this program;

4. The actual desired outcomes for the program;

5. Peripherally related programs and efforts;

6. The maturity of the leadership and the followers with respect to participative approaches to innovation.

Prior to presenting the strategic planning process itself and its application to the development of a productivity management effort, a generic strategic plan or "road map" for developing a productivity program will be outlined and discussed. Although the strategy outlined here may not specifically represent any one firm's approach, it reflects the approach being taken by progressive and successful firms today.

On the surface, at least, a "grand strategy" appears to be driving or shaping the specific operational plans for productivity management. Case studies on productivity management efforts over the past seven years from both Japanese and American organizations indicate that there *is* a "least common denominator" to their efforts. The "grand strategy" has six reasonably distinct stages that will be discussed next.

The "Grand Strategy" for Designing, Developing, and Implementing Productivity Management Programs

Figure 14.1 depicts the six stages of this "grand strategy" and sequences them out over time. For a typical American organization, the planning horizon is five to ten years and the phasing in of the various stages is depicted. Note that your organization will need to develop its own program planning and schedules based on specific characteristics of your own organizational system.

STAGE 0: BASICS

- Basic management
- Product and service engineering
- Facilities management
- Quality management
- Vital function interface
- Industrial engineering
- Product and process innovation
- Productivity basics training

Effective productivity management is based on good, sound, disciplined general management. Effective and efficient planning, leading, coordinating, controlling, and adapting are the basics of management.

The art and science of management has changed dramatically in the past 15 years. Yet too many American managers view the management process in the same light as learning to ride a bicycle: "Once you learn, you never forget." However, employees, technology, societies, and the world are simply too dynamic for this analogy to be valid. Current, sound, basic, continuing management education may well be one of the most critical factors in productivity management. If the leadership function fails for lack of basic knowledge, the entire process is weakened significantly.

There are weaknesses at other levels as well. Robert Lynas, vice president and general manager, automotive worldwide, for TRW, recently presented the

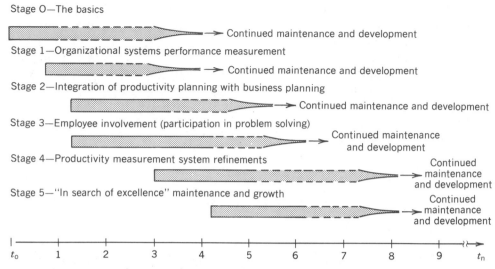

Figure 14.1   The "Grand Strategy" for Designing, Developing and Managing Successful Productivity Management Programs

comparisons of the relative effectiveness of certain critical functions in American and Japanese industry shown in Figure 14.2.

These comparisons are reinforced by a case example provided by Dennis Ossola, director of operations for Matsushita Industrial Company in Franklin Park, Illinois. The well-publicized case in which Matsushita acquired the Quasar Television plant from Motorola in the mid-seventies indicates not only that many successful American firms are struggling with Stage 0, but also that there appears to be a critical gap in effectiveness between American and Japanese firms for certain critical functions.

In 1974, the Quasar plant was experiencing a 140 percent in-process rejection rate, high warranty costs, absenteeism in the neighborhood of 10 to 12 percent, and test rejects of 2.6 percent. The plant was considered to be in dire trouble.

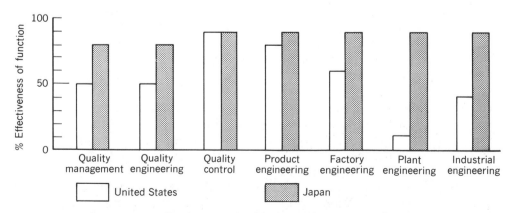

Figure 14.2   Effectiveness of Critical Functions, U.S. and Japan

By 1982, Matsushita Industrial Company had reduced in-process rejection rates to 5 to 7 percent, lowered warranty costs significantly, reduced absenteeism to 2.25 percent and test rejects to 0.32 percent, decreased field work by 90 percent, and increased labor productivity by 30 percent. All of this was done while essentially maintaining the same sales price on the finished product.

According to Ossola, this was accomplished with the same work force and the following changes in operations:

- Product engineering—26 percent fewer parts
- Automated assembly
- Cooperation between design and manufacturing
- Operator training
- Quality control/reliability
- Simplified management systems
- Management team attitude

Stage 0 took six to eight years of concentration on the basics, Ossola said. "It wasn't hard to go from a 140-percent in-process rejection rate to 10 to 12 percent. We just went back to basics." But, he added, "Going from 10 to 12 percent to 1 to 2 percent (the corporate goal) will require the full effective involvement and commitment of the work force."

In Stage 0 the organization needs to focus on the fundamental management process. Closing the gap between Japanese and American effectiveness at executing selected critical functions can have a significant impact on reversing the present productivity trends.

The message in calling this Stage 0 is that you aren't out of the starting gate until you audit and evaluate the success with which certain basic functions are being accomplished. Are you effective? Are you efficient? Are the primary functions in your organizational systems effective, efficient, and producing a quality product and/or service?

Do not leave Stage 0 unless management is willing to risk longer-term ineffectiveness and inefficiency in the productivity management effort. Don't get hung up here with too high a level of expectations and standards, but on the other hand, don't proceed to Stage 1 until an acceptable level of performance has been achieved. There is a lot of productivity improvement potential in every organization to be gained, as Matsushita proved, simply by concentrating on the basics.

STAGE 1: MEASURING PERFORMANCE

- Strategic planning for measurement system design
- Performance appraisal systems
- Group performance appraisal systems
- Management by objectives (MBO), collaborative MBO
- Work measurement systems
- Productivity measurement systems

Organizational systems (firms, corporations, departments, divisions, plants, functions, and work groups) measure and evaluate performance in a variety of ways. Most organizations have control systems for behaviors, costs, prices, information, decisions, financial performance, production, and quality.

There are many ways to classify control systems. For instance, we could categorize them with respect to the resource they are intended to manage (financial, production, quality) or with respect to a particular type of system performance (effectiveness, quality, efficiency, productivity, profitability, quality of work life).

Stage 1 involves examining systems in an integrated, strategic fashion and evaluating the effectiveness, comprehensiveness, and adequacy of an organization's performance measurement systems. Where deficiencies are detected, improvements will need to be made.

Performance appraisal, work measurement, management by objectives, quality, production, inventory, and all other measurement systems will need to be analyzed as an integrated system. Note that many of these measurement systems evaluate transformation efficiency in the organization.

Productivity measurement systems, if in existence, will also need to be evaluated. If there are none, this is when they would begin to be developed.

This is the stage at which the techniques for auditing your performance measurement, evaluation, and control systems can be implemented. What is performance? How do you communicate this to your management and employees? Do you reward what you say you want? Do you measure performance, operationally, according to what you feel performance should be? Do your performance appraisal (individual and group) systems work, or are they simply given lip service? Have MBO or comparable systems for goal setting become just a process of going through the motions and filling out the forms? Are your work measurement systems effective and efficient? Do you measure productivity? Are there gaps in your performance measurement systems? Do you get what you measure? Do you want what you measure? These questions and many more must be systematically examined in Stage 1 of this "grand strategy." *Do not* proceed to Stage 2 without giving serious attention to the issues raised in Stage 1.

STAGE 2: INTEGRATED PLANNING

- Productivity planning
- Integration with behavioral, financial, production, and quality control systems
- Improved strategic planning
- Improved effectiveness at linking planning to action and follow-through
- Specific productivity improvement strategies, programs, techniques

American managers have been criticized for their overemphasis on short-run, short-term performance. The characteristics of the planning process in the United States are under fire. Stage 2 is devoted to integrating specific productivity measurement and improvement goals and targets into the general business planning process. It is an attempt to balance productivity planning, strategic planning, and the traditional budget-oriented planning process and to refocus attention on managing productivities rather than only costs. Many of the issues Thurow (1984) discusses in his recent excellent article should be considered in this stage.

You must play an important catalytic role in

- Convincing management that productivity planning is important
- Providing an efficient and effective mechanism for soliciting productivity plans from the organization
- Linking specific productivity improvement programs and actions

Every organizational system has some form of business process, formal or informal, explicit or implicit. Few have a productivity component within that business planning process. There is typically a financial component, a market component, a product development component, an organizational structure/ design component, but very seldom a productivity management component. This will have to change in the coming decades if American organizations are going to successfully compete in an increasingly competitive world marketplace. Every supervisor, manager, and leader in every organizational system should have a productivity management plan. The alternative is that there won't be a need to have one. (Thurow, 1984)

STAGE 3: EMPLOYEE INVOLVEMENT

- Top management commitment and legitimization
- Design perspective
- Team building
- Structured group processes
- Supervisor/subordinate relationship focus
- Supervisor development
- Problem solving and effective implementation
- Effective and efficient techniques
- PAT process design, development and implementation

The Japanese spent approximately 16 or 17 years mastering the concepts and techniques of statistically based quality control, production control, and management and behavioral science. In the late sixties, a Tokyo University professor named Ishikawa developed the concepts and techniques for quality circles. By this time, Stage 0 for most major Japanese firms was well established. The groundwork had been laid, and the time was ripe to effectively and efficiently involve the Japanese work force in the ongoing problem-solving process.

During the sixties and seventies, Japanese organizations began to look different from their American counterparts. Figure 14.3 reveals this subtle, but significant, difference.

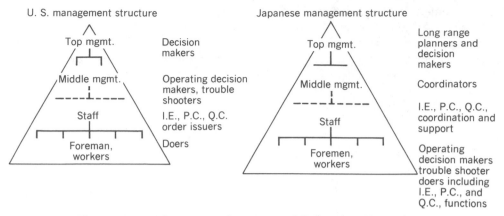

Figure 14.3   U.S./Japanese Structure and Delegation Comparisons

The following hypothesis reflects the motivation for efforts in this area (quality circles, productivity action teams, and so forth): If an organization could train and develop its work force on an ongoing basis and could effectively and efficiently involve that work force in a systematic and ongoing problem-solving process, it would have a significant long-run competitive advantage over a firm that did not employ this strategy.

The Musashi semiconductor works case study presented in Chapter 12 is a dramatic example. I present this example in my three-day seminar and in companies for which we are implementing the PAT process. Not one American company with which I have worked can come close to the level of proactive innovation on the part of the Japanese work force. "If the Japanese can do it, why can't we?" The answer is that we can, but it will take time, patience, persistence, and consistency. Most American organizational systems cannot hope to go from the characteristic organizational processes that are implicit in the structure on the left side of Figure 14.3 to processes implicit in the structure on the right side of the figure in less than five to ten years. You simply cannot change patterns of behavior any faster.

We are presently helping a steel comapny in Oklahoma to implement a PAT process. The strategic plan to implement pilot studies in a top-down fashion has a horizon of 18 months. When the plan was presented to management, it felt very comfortable with the timetable. The executive committee for the union also understood and agreed with the timetable. When the plan was presented to the bargaining unit, however, certain members criticized the pace of implementation. They wanted the program to involve all bargaining unit members immediately. It was explained that you can't change overnight a pattern of labor-management relations that took 75 years to establish. If a program or technique for employee involvement in participative problem solving is implemented without first trying to change a culture and patterns of behavior in the organization, there is bound to be failure. This does not mean that every employee, every supervisor, every manager has to buy into the philosophies of participative problem solving. One simply needs to minimize the pockets of resistance and increase the number of pockets of support.

Stages 0, 1, and 2 of the "grand strategy" are excellent opportunities to begin to build the foundation for what can take place in Stage 3. Unfortunately, most productivity improvement programs today start and end with quality circles. Management concludes it must do something about productivity and grabs onto the first "solution" that comes along, which in the late seventies and early eighties has been quality circles. We have attempted to implement a participative technique on characteristically autocratic organizational processes and leadership. And we have attempted to implement a participative problem-solving technique with "immature" (unable and often unwilling) employees. Employees know their jobs. What they are not particularly skilled at is how to systematically solve problems. Stage 3 has to be phased into an overall productivity management effort in an appropriate fashion and at the apropriate time. The PAT process or some alternative design would be initiated at this stage.

STAGE 4: PRODUCTIVITY AND PERFORMANCE MEASUREMENT SYSTEM REFINEMENT

• Multifactor measurement models
• Normative, partial-factor productivity measurement models
• Integration with other performance systems

We have learned by now, the hard way, that attempts to begin with productivity measurement as the focus of the productivity program tend to bog down, if not fail. The problem seems to be threefold.

First, there is a tendency to want to "act our way into a new way of thinking rather than to think our way into a new way of acting." Asking managers or employees to assist in the development of, or to actually develop, a productivity measurement system has been viewed as too passive.

Second, measurement and its use in control systems is a difficult subject. Productivity itself is misunderstood. The differences and relationships between performance measurement, profitability measurement, effectiveness measurement, efficiency measurement, and productivity measurement are not well understood.

Third, productivity measurement system development, motivation, and rationale have typically been driven from the top down. The logic behind the need for a productivity measurement system as a mechanism for controlling the organization at any level or for any function has not been clear to most managers.

An organizational system needs to be prepared to address productivity measurement issues. Certain productivity measurement system refinements are better off being delayed until Stage 4 in the interest of successful long-run implementation. In particular, attempts to develop normative productivity measurement systems through the use of structured group processes, such as the nominal group technique or the Delphi technique, are likely to be more effective at Stage 4.

The strategy is not to completely defer all forms of productivity measurement until Stage 4 but to begin measurement, evaluation, and the development of control systems in Stage 1 by auditing existing performance measurement systems and starting to develop simple yet effective productivity measurement systems. Throughout Stages 2 and 3, productivity measurement systems will be developed and maintained. By the time an organizational system successfully moves into Stage 3, the people in those systems will be psychologically more willing to support productivity measurement efforts, such as the ones presented in Part III.

In a sense, Stage 3 is a participative, proactive, innovative, action-oriented performance and productivity improvement effort. As you will recall from Chapter 13, the PAT process focuses on ideas for performance improvement in early steps and then in later steps directs attention to measurement, evaluation, and control of the results of PAT efforts. The reason for this is that we have found that it is much easier to motivate interest in developing decoupled performance/productivity measurement systems if the effort is phased after an improvement focus. Members of PATs will actually ask management for feedback in terms of the impact of their proposals on performance/productivity. This becomes a powerful driving force behind the type of Stage 4 developments that are possible.

STAGE 5: BEYOND "IN SEARCH OF EXCELLENCE"—MAINTENANCE AND CONTINUED GROWTH AND DEVELOPMENT

- Linking productivity improvement to measurement.
- Goal congruency (labor/management relations, knowledge of what is expected)
- Performance/productivity gainsharing (making rewards contingent on performance)
- Continued evolution of PATs at all levels of the organization
- Continued measurement, evaluation, and development of control system.
- Management development

An organization that successfully accomplishes Stages 0 through 4 would be in a position to maintain or exceed reasonably impressive levels of performance. This, of course, assumes that many other factors and conditions are being managed properly. It also assumes an "internal locus of control" for the firm and its management, or that the organization, through its management, is largely in control of its own destiny.

The major questions/issues to be addressed in Stage 5 are how to maintain

1. Motivated levels of performance on the part of all employees in the organizational system

2. High levels of proactivity with regard to innovation for improved performance and productivity

3. A sense of commitment to the kinds of issues and goals raised by Ouchi in

Theory Z, Kanter in *The Changemasters*, Peters and Waterman in *In Search of Excellence*, and Drucker in *Managing in Turbulent Times*

4. High levels of communication, coordination, and cooperation within and between organizational systems;

5. Progress toward improving the quality of management, the quality of work, and the quality of work life

One answer may lie in productivity gainsharing. As we have seen, money is still a powerful motivating factor. If the organizational system improves its performance, which in turn creates additional profitability, those portions of increased profits attributed to productivity improvement should be shared. The Multi-Factor Productivity Measurement Model is an excellent tool for measuring and determining what portion of profits should be shared. These profits should be shared in such a way as to reward those components of the system that created the gains. If technology was responsible, then capital is reinvested in technological innovation. If the people created the gains, then they share them proportionately to their contribution. We must be careful to differentiate increased profits created through productivity gains from profits created through other mechanisms. We are not simply sharing profits; they must be earned profits. And it is critical *how* we share the profits from productivity. The reward system must adhere to the basic design principles for good, positive reinforcement. This means that the productivity gainsharing should be timely and proportionate to the contribution made.

Another answer to the questions/issues raised in this stage may well be participative problem-solving approaches, such as the PAT process. The PAT process and similiar approaches have tremendous organizational system capacity building capability. We have found that when PATs are implemented effectively in an organizational system, communication, coordination, cooperation, and innovation increase in the short run. Attitudes about the organization improve. Problems/opportunities that had always fallen in the cracks, that had always drained the performance potential of the organizational system, that no one ever was willing to be responsible and accountable for are now being resolved or taken advantage of. And, in the long run, we find that quality increases, productivity imporves, effectiveness and efficiency increase, quality of work life is positively impacted. Moreover, all of these outcomes occur because the PAT process sparks and facilitates innovation in the organizational system.

There are, to be sure, many ways for an organization to reach Stage 5 and to accomplish the objectives of this stage. Like PATs, productivity gainsharing and processes are critical elements. The "grand strategy" outlined here for you may not resemble exactly the plan taken by the organization with which you are acquainted. However, if you stand back and examine objectively the strategies that successful organizations are taking regarding productivity management, you will observe patterns or trends. This "grand strategy" represents a sort of skeleton of those plans and efforts. Although this strategy may vary in practice in terms of the techniques and approaches utilized to execute each stage, the amount or degree of overlap between each stage, and the relative appropriateness and success of the strategy as a whole, the general components and sequences appear to be fairly typical.

What remains, therefore, is to present a methodology that might help you operationally develop your own specific "grand strategy." The next section will discuss the eight-step strategic planning process presented in Chapter 13 as applied to the design, development, and implementation of a productivity management "grand strategy." Keep in mind that depending on your particular situation, this planning effort may be focused on development of a plan for your entire organization or any subset organizational system. You must define the unit of analysis for the planning effort and make necessary application decisions regarding this strategic planning process.

Strategic Planning for Productivity Management Programs

Planning for productivity management efforts, regardless of the size or complexity of the organizational systems(s) involved, is a challenging task. In the eighties, more and more staff groups, industrial engineering departments, human resource departments, middle-level managers, and homeless upper-level managers have found themselves in the situation of having to develop and implement a productivity program. The order comes down from upper-level management to establish an exemplary productivity program. The recipients of this dictate often seek clarification but to no avail, for, as is often the case, even top management isn't exactly sure what a productivity management program is or should be. Top management hopes for creativity, innovation, and a quality, all-encompassing program. Unfortunately, all too often it gets a poorly thought through, band-aid approach to productivity improvement. Worse yet, it might receive a recommendation to start a quality circles program.

The quality of the planning process in early developmental stages is essential to the success of a long-term productivity management effort. The extent to which all critical viewpoints and perspectives are involved in this development process determines the long-run commitment to executing the eventual components of the program. The larger the scope of the program, the larger the organizational system, and the longer the planning effort will take to accommodate a broader range of interest.

Program planners do recognize the need for a well-developed plan, but they too often lack an approach or technique by which to achieve a quality planning *process*. As a result, there are futile attempts to develop well-thought-through strategic plans for a productivity management program. The final plans are often too narrow in focus, the product of too few key decision makers in the organizational system(s), too theoretical, unrealistic, unintegrated, and, of course, unsuccessful in the long run. Many productivity programs in even the largest corporations are products of a staff group reporting to upper management. They are detached from reality and not well supported or endorsed by the operations people. In fact, in some of the better-recognized productivity management programs in the United States, the operations people think the efforts are a waste of time and money and consider the productivity managers and their programs to be "window dressing." In some of the more visible productivity management efforts, the real productivity management successes are being developed and driven by the operations people entirely independent of the official "productivity management" program. Contacting the official company productivity administorator is not always the most effective or efficient way to really find out what a given company is doing in the way of productivity management.

Not all of the productivity management efforts in the United States are failures. But whenever you establish a productivity administrator in a staff capacity, you have one strike against you. The reason is that once an organization establishes a *staff* productivity function there is a tendency for line managers to breathe a sigh of relief and assume they now don't have to worry about the productivity issue. We have done this in this country with quality and are now in jeopardy of doing it with productivity. The line and operations people, the staff people, management, and everyone involved must view productivity and quality as an integral part of their job. You cannot delegate that responsibility or the accountability for those performance criteria to someone else.

Contextual Issues

Productivity management efforts are more prevalent today because the term "productivity" has become such a powerful buzz word. American management is highly subject to the influence of popular fads in much the same way as is the American consumer. Therefore, critical elements of the context within which productivity programs exist and are initiated are the causal factors and forces shaping the program development. We can assume that many programs will be initiated by persons who actually know very little about what they want. Many managers request such a program simply because it seems to be the thing to do. This contextual reality will lead to the use of program initiators to clarify and crystallize the desired outcomes for the program. Many developers of productivity programs rush off to begin their efforts without ever really clarifying program desires.

A second contextual factor to consider is the current economic and technological climate. In particular, we are entering an era that some have called "the information or computer revolution." This revolution is considered as challenging and disruptive as was the Industrial Revolution at the turn of the century. The implications of rapid technological innovation, complex sociotechnical issues, world economic instability, and extremely dynamic competitive forces and market conditions are paramount to this planning process. These factors will affect not only the specific operational features and elements of the program itself, but also the nature of the planning process. The uncertainty and risk involved in elements of a productivity management program and the dynamic aspects of the focus of this plan necessitate a planning process that is flexible and dynamic. The complexity of the productivity and overall performance issue for even the simplest organizational system requires involvement by a wide variety of persons to ensure quality and thoroughness in the eventual program.

A third and perhaps most critical contextual factor to consider is the political aspect of the organizational system for which the program is to be developed. Key actors in the system(s), points of resistance, managerial styles, and power bases are such factors that will need to be addressed as the strategic planning process is being developed.

The skills and relative credibility of the person or persons asked to design, develop, and execute the planning process and eventually the program itself are a fourth critical contextual factor. The skills can be acquired rather easily, although not everyone can perform the elements outlined. And even with a skilled leader, the subsequent planning process is not necessarily infallible. A skilled leader is necessary to carry out even the best process.

Finally, it must be assumed that if management requests a productivity management program, regardless of the level of understanding of what one looks like, management is truly committed to its development. However, this is not really a valid assumption. When a middle manager who has been given the responsibility to develop a productivity management program asks us what to do first, we always tell him or her to involve key upper management in a one-to-two-day strategic planning session for the effort. Typically, this middle manager informs us that he or she could not obtain even one or two hours from these people. We in turn ask if this program is of strategic importance to the organization. The manager says yes. We then ask why top management wouldn't participate in this type of activity. The manager typically responds by saying that they are too busy to plan for this type of program. Our recommendation is that the manager refuse to accept the assignment unless top management commits to participating early in the direction setting for the program.

Good, well-thought-through and solid requests, ideas, plans, and programs will almost always fall on receptive ears. Half-baked, ill-prepared ideas, plans, and programs will almost always be avoided. Competency that is aggressive and firm will be respected and rewarded. Develop a good plan for your strategic planning process and then demand that top management play its appropriate role. Don't settle for less. If the productivity management program is important enough to commit resources to it, then it is important enough for top management to spend one to two days in the strategic planning process described in the next section.

Strategic Planning Process

The basic eight-step strategic planning process to be developed here was initially presented, in part, in Chapter 13 in conjunction with discussion on implementation of the PAT process. This generic strategic planning process has been successfully applied to top management planning for redesign of a university's computer-support system needs; long-range planning for a college of engineering; strategic, tactical, and action planning for engineering departments; strategic, tactical, and action business planning between labor (organized) and management for a steel company; and, of course, design and development of a productivity management program. It is the specific application of this planning process to design and development of a productivity management program that is presented and discussed now.

The planning process presented is a structured group-planning activity. The process is designed to be executed with an appropriately selected group of 5 to 15 persons from the organizational system. The process could be executed a number of times with different groups of persons to achieve different purposes or to develop similar plans from different perspectives. If, in fact, there is need for cultural change and development of commitment to a productivity management program, repetition of all or part of this eight-step process with various critical groups of persons in the organizational system can facilitate this greatly. The sequence of strategic planning sessions, shown in Figure 14.4, is recommended.

Execution of this sequence at all levels in the organizational system and across all affected functions ensures appropriate input in the initial design of the productivity management program. More importantly, this total system design and development involvement strategy improves communication, builds teamwork, enhances coordination between units and functions, and establishes a strong foundation of commitment and consensus for future developmental steps. The costs are the time and effort involved in planning for and carrying out these

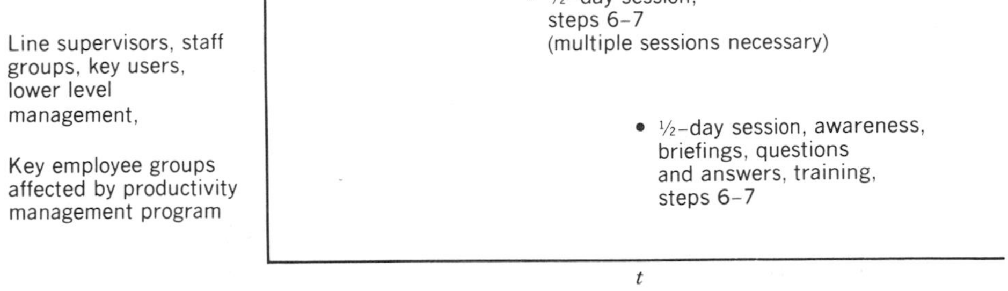

Top management
of focal organizational
system

- Two-day session,
  steps 1–6
  (multiple sessions possible)

Middle mgmt., mgmt.
of user groups,
mgmt. of staff groups,
key operations mgmt.

- 1- to 1½- day session
  steps 4–7
  (multiple sessions probable)

Line supervisors, staff
groups, key users,
lower level
management,

- ½-day session,
  steps 6–7
  (multiple sessions necessary)

Key employee groups
affected by productivity
management program

- ½-day session, awareness,
  briefings, questions
  and answers, training,
  steps 6–7

t

Figure 14.4   Sequence of Strategic Planning Sessions to Develop Consensus and Commitment to Productivity Management Program

sessions. The benefits are more effective implementation of the productivity program.

The eight-step planning process as applied to the design and development of a productivity management program is as follows:

STEP 1: INTERNAL STRATEGIC AUDIT (A CRITICAL LOOK WITHIN)

The first step of the eight-step strategic planning process focuses attention internally to the organizational system for which the program is being developed. The group is asked to respond to the following question:

> What internal factors (strengths, weaknesses, problems, opportunities, and other programs) should be considered in the design, development, and implementation of our productivity management effort?

Participants are given this question in written form and are asked to respond in a similar fashion. After a 5-to-15 minute silent generation period, the group is asked to present one factor at a time in round-robin fashion. These responses are placed on flip-chart paper and posted around the room. Discussion follows the conclusion of this round-robin process. Step 1 can be expected to take one hour. It is important for the facilitator to keep the group focused on factors within the organizational system(s) that will or should affect the design and development of the program.

STEP 2: EXTERNAL STRATEGIC AUDIT (A CRITICAL LOOK AROUND)

The second step focuses attention on factors external to the organizational system(s) that will/should influence the design and development of the program. The group is asked to respond to the following question:

> What external factors (trends, other organizational systems, problems, constraints, opportunities, vendors, technologies, competitors, people, etc.) should be considered or incorporated during the design and development of our productivity management effort?

Again, both the question and the responses are given in written form. The same process of round-robin presentation, posting, and discussion occurs. Step 2 usually takes a little less time than Step 1, because typically some Step 2 issues are inadvertently handled in the first step.

STEP 3: CONVERSION OF OUTPUT FROM STEPS 1 AND 2 TO PLANNING PREMISES AND/OR ASSUMPTIONS: DEVELOPMENT OF IMPORTANCE-CERTAINTY GRID

When the results of Steps 1 and 2 are posted around the room, they should be in the form of short (two-to-three-word) phrases regarding internal- and external-related factors influencing the design and development of this program. Step 3 focuses on converting these factors to specific, major planning premises and/or assumptions on which the program will be developed. We are attempting to focus attention on major planning premises and to crystallize perceptions of these premises. A list of 60 to 70 external and internal factors will typically reduce to 15 to 25 specific planning premises.

Once these major planning premises are developed, each premise is given a number. Each premise, by number, is then placed on the importance-certainty grid shown in Figure 14.5. Each participant individually and silently evaluates the importance of the premise to the eventual plan and the certainty surrounding the premise's impact on the plan. The group facilitator then simply asks for a show of hands as to the positioning of each premise on the matrix. Accuracy is not the most critical parameter in this phase. It is simply important to position the premise approximately where group consensus dictates. Discussion then focuses on the relationship between the premises, their relative importance and certainty, and specific plan development.

STEPS 1–3 LOGISTICS

For a day-and-one-half or a two-day session with upper management, Steps 1 through 3 will be completed on day one. These steps are typically not executed with middle- and lower-level management. A notebook with the planning proc-

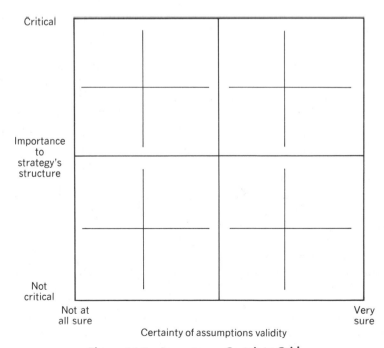

Figure 14.5   Importance-Certainty Grid

ess steps, task statement sheets, importance-certainty grids, and appropriate background material is provided.

In a one-and-one-half or two-day format, there will be time to allow for reasonable discussion sessions. It is particularly important to let the group have enough flexibility to wander away from the planning process, particularly if the discussion relates to the productivity program development. Task needs and social needs must be balanced. Luncheons and even receptions and dinners with the management groups can help to build support for and understanding of the program to be developed.

Appropriate background material might consist of articles on what others are doing, case studies, special-topic productivity-related articles, films, and outside speakers. Don't ignore the opportunity to create improved awareness and understanding of productivity basics.

STEP 4: STRATEGIC PLANNING

Step 4 involves the actual creation of consensus goals and/or objectives for the productivity management program. The nominal group technique is used to execute this step. The group is asked to respond to the following task statement:

> Please identify strategic (two-to-five-year horizon) goals and/or objectives that you feel our productivity management program should strive for and attain.

This step is executed in typical NGT session fashion. We have found, however, that it is perhaps most effective to perform the NGT without even telling the participants about the technique itself. This approach seems to take some of the "technique/gimmick being done to us" perception away.

The facilitator should concentrate on obtaining concise but clear goal statements. We have found it helpful during this step to separate superordinate or "must" goals and objectives from "want" goals and objectives. If you make this separation, you will rank the "want" goals/objectives in the typical fashion using the 3 × 5 index cards. You may also choose to rank the "must" goals/objectives if there are enough of them.

STEP 5: PERFORMANCE MEASURE DEVELOPMENT

Step 5 focuses on the development of specific criteria against which the productivity management program will be judged, evaluated, and assessed. The group is asked to respond to the following task statement:

> Please identify specific measures/criteria/standards against which we should assess the performance (effectiveness, efficiency, quality innovativeness, benefit/burden) of our productivity management effort.

The group is asked to silently generate responses and then present them in round-robin fashion. Measures/criteria/standards are posted and discussed but not prioritized at this point. This output will/can be used as a starting point for either an NPPMM or an MCP/PMT application later in the development of the program.

STEP 6: ACTION PLANNING

Step six focuses on the development of specific actions that the group feels should be budgeted for (time and money) during the next year relative to the design and development of the productivity management program. The facilitator should keep in mind that these do not necessarily have to be steps that can be initiated and completed within a year. They may be action items of a strategic character that will take two to five years to complete, but that should be started in the next year. The group is provided with the following task statement and asked to respond:

> Please identify and list specific programs, actions, and plans that should/will be budgeted for during the next year in order for our program to successfully get off the ground.

The NGT process is utilized to develop a prioritized list of action programs. Each participant is asked to rank eight action programs. Lengthy discussion follows the voting and ranking step of the NGT. This step will probably occur at about 2:30 to 4:00 P.M. on the second day of the top management session. If the facilitator senses an interest and commitment on the part of the top management group to go further, Step 7 could/should be executed. Typically, however, Step 7 (action program project planning) is too detailed for upper management to want to become involved in.

STEP 7: ACTION PROGRAM PROJECT PLANNING

At the completion of Step 6, each group will have developed a consensus set of initial (first-year) action items for the productivity management effort. Step 7 involves separating these action items or programs into categories and assigning specific items to subgroups. In this way, three to five of the specific action items can be detailed further. Each subgroup is expected to develop a project plan complete with activity sequencing, resource estimates, responsibility and accountability specification, project management issues, and cost/benefit analysis. At the completion of Step 7, we then have three to five major action items from this group fairly well planned out.

Assuming that Step 7 was completed with middle management groups down to certain key employee groups, significant productivity program planning would have been accomplished. What remains is bringing all these planning results together, analyzing the results, integrating them into one overall plan, and involving appropriate parties in specific implementation issues.

STEP 8: PROGRAM REVIEW AND EVALUATION

Step 8 essentially is an extension development of Step 5. The goal is to operationalize the results obtained from the various management groups in that step.

Synthesis

What is being suggested is a top-down participative strategic planning process. This approach is quite effective at generating high-quality information with which to develop a productivity management effort. Furthermore, it develops or accomplishes team building, improved communication, strong initial commitment to and support of the program, crystallized goals and objectives for the program, crystallized goals and objectives for the program, and a motivation to participate in the actual development and implementation of the program.

The process requires certain skills and takes time. Moreover, it is important to note that the process itself does not develop the plan. Rather, it simply provides a great deal of data and information that can be used. Our experience with this process suggests that it is very pragmatic, well received by managers and employees alike, highly effective, and reasonably efficient. It is vastly superior to traditional consultative, autocratic, or expert-imposed approaches to developing a productivity management program. It results in a unique program for each organizational system. And although the "grand strategy" may serve as a roadmap for your organization, you will likely find it necessary to take detours along the way.

REFERENCES

Sink, D. S. "Strategic Planning: A Crucial Step Toward a Successful Productivity Management Program" *Industrial Engineering*, January 1985.

Thurow, L. "Revitalizing American Industry: Managing in a Competitive World Economy" *California Management Review*, Vol. xxvii, No. 1 Fall 1984.

QUESTIONS AND APPLICATIONS

1. Investigate the recent literature on strategic planning. Do your findings support the process suggested in this Chapter?

2. Redo Question 1 from Chapter 13 but focus on development of a strategic plan for a productivity *management* program or effort.

3. What are the components of a productivity management effort?

4. What are the likely stages of evolution of a productivity management effort?

CASE EXAMPLES: HONEYWELL'S PRODUCTIVITY AND QUALITY CENTER; PERFORMANCE ACTION TEAM PROCESS TOP-DOWN PLANNING AND EXECUTION IN A UNIONIZED STEEL COMPANY

Honeywell's Productivity and Quality Center

Introduction

This brief case study is intended to present the general approach taken to productivity management and measurement at Honeywell and in particular within Aerospace and Defense (A&D). In a short, one-day visit to an organization as complex as Honeywell, it is impossible to come away with a complete understanding of everything they are doing in A&D with respect to productivity or for that matter, productivity measurement. The Productivity/Quality Center (P/Q Center) was established in A&D in 1981 to support A&D operations in their efforts to increase productivity and improve quality of work. Since that time, under the general direction of Mr. G. E. Peters, Vice President and Staff Executive, there have been tremendous strides made to focus the entire organization's attention on productivity and quality. A balanced effort has been made to create awareness; educate; create priority of focus; improve proactivity with respect to productivity improvement project planning and implementation; document exemplary productivity improvement practices and projects; spark innovation; and promote and facilitate measurement and evaluation. Their accomplishments are no less impressive than the sophistication and maturity of the overall plan for the P/Q Center. Organizational development considerations and effective change strategies are considered equally important to technological innovation. The program has top-down support but is implemented bottom-up. There exists a nice blend of top-down interventions as well as decentralized, decoupled bottom-up driven/facilitated innovations.

Rosabeth Moss Kanter uses Honeywell as an example organization in her new book, *The Changemasters* (1983). The focus of her book is on innovation for productivity in the American corporation. She suggests that the "era" of strategic planning (control) may be over and that we are entering an era of tactical planning (response). Of the major functions of the executive—planning, organizing, leading, controlling, and adapting—adaptation will be the most focused upon in the coming decades. Successful, excellent firms will need to be externally and internally flexible; they will need to bring together resources quickly, in response to short-term recognition of opportunities, constraints, requirements, problems, and so forth. Additional support for these views is found in Naisbitt's *Megatrends* and Peters' and Waterman's *In Search of Excellence*. The challenge that the Honeywell Productivity/Quality Center has adopted is to address how to drive obvious and necessary productivity and quality improvement innovations in an effective and efficient top-down fashion while simultaneously facilitating and promoting proactivity with regard to productivity and quality improvement at the "grass-roots" level.

Honeywell A&D Background (Taken from a published interview with Dr. Mary Ann Donahue by Quality Circle Digest)

Honeywell Aerospace and Defense is a $1.5 billion dollar business for Honeywell and has 12 divisions and centers around the United States. About half of those are based in Minneapolis. There are approximately 19,000 people in the A&D business representing about one fourth of Honeywell. The basic structure of the A&D organization is depicted in Figure 15.1. Major explicit goals and objectives for Honeywell A&D are as follows:

Mission:	The mission of A&D is to provide quality components, subsystems, and systems to enhance the strength and effectiveness of military capabilities of the United States, our allies, and related commercial markets.
Strategic Purpose:	To provide the corporation with a stable and steadily growing profit stream, a source of technology development, and high-caliber management talent.
Financial:	Growth: 13%/Year Profit: 8.5% Operating Profit Assets: 2.7 Turns Return: 18% ROI
Engineering and Production:	All programs on or ahead of schedule. All work done with best achievable costs. All programs equal to or better than technical specifications. All work done right the first time: Quality is first. Consistent application of best available technology, tools, and practices. Fair and effective relationships, with quality vendors and suppliers.
Human:	Positive climate and relationships: "Our Way" philosophy and style. Maximum employment stability and personal development opportunity.
Performance:	Continuously improving quality of work life, quality of management.

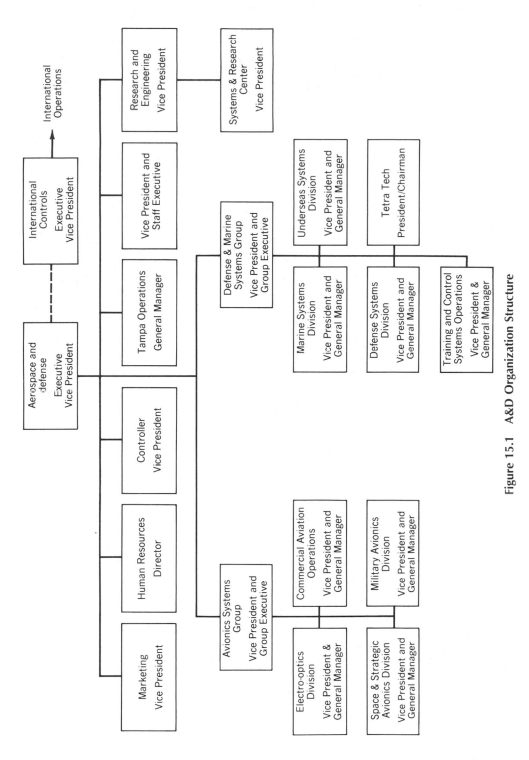

Figure 15.1 A&D Organization Structure

449

Marketing: Effective and constructive relationships with all customers. Respect for work quality by all customers. Significant position in markets served.

Honeywell: Open, constructive, professional relationships with all Honeywell people everywhere.

Community: Positive, quality, sharing citizenship.

Honeywell A&D Productivity/Quality Center Approach

The Productivity/Quality Center leadership determined that there are three critical quality issues that should be the themes upon which the center's efforts are directed:

Quality of work: managing productivity/quality; the degree to which the work output meets requirements

Quality of work life: managing culture; the degree to which work environment fosters employee "ownership" and contribution

Quality of management: managing leadership; the degree to which management accomplishes performance improvement through practices and commitment to quality of work and quality of work life

At the A&D (Division) level, there are three specific measurement criteria identified that are utilized to monitor, evaluate, and control specific performance areas: (1) scrap/rework in dollars and as a percent of sales; (2) gross inventory, average balance in dollars and as a percent of sales; and, (3) value-added sales/employee in dollars. Each of these three criteria has goals established for it and is tracked at all lower levels in the organization. These criteria are generic, top-down specifications that lower levels are expected to manage.

The first productivity measurement ratio employed by Honeywell was sales/employee, which was later changed to sales/pay in 1975 (*Honeywell* 1982). It continues to be used by the corporation along with a sales/employee ratio adjusted for inflation. Within A&D, the primary ratio used is value-added sales/employee, where values-added sales is stated in current year dollars. Value-added sales equals sales minus material costs with adjustments for any labor and burden inventory changes. As one can see, this is a partial-factor (primarily labor) productivity measurement approach using static ratios. The unit of analysis is the division, and the scope of the measurement period is annual.

Measurement at the department level (profit center, group, program/contract, product/product family, productivity improvement project, individual) is encouraged and facilitated with support, awareness, education, examples, and so forth. Selection of a productivity measure for a particular department is guided by the following suggestions:

• Select measurements that the department employees and management agree are representative of measuring their productivity. These people know their job best and can define good ways to measure their output.
• A family of measures with some understanding of the limitations and value of each is more useful than trying to have one all-inclusive measurement that has limitations.
• The set of measures for a department or group should reflect the most important resources of that department and the group's ability to manage that resource.
• Wherever possible, department-level measurements should utilize existing data available from the accounting system or other information systems in the organization.

A select sampling of measures used by various departments in Honeywell A&D is provided in Table 15.1.

Quality of work improvement projects are the responsibility of each department; however, the Productivity/Quality Center is made available to assist in the selection, planning, implementation, and measurement of results. Figure 15.2 (page 458) depicts the conceptual sequence of activities in a typical quality of work improvement/productivity improvement project.

The P/Q Center promotes the use of the nominal group technique (NGT) as a mechanism for prioritizing improvement ideas and for creating a sense of ownership, responsibility, and accountability for the projects themselves. The NGT is used to identify and prioritize opportunities/potential projects for productivity improvement as well as counterproductive practices.

The productivity management program at Honeywell A&D appears to be an appropriate blend of top-down planning and intervention and bottom-up encouraged innovation. Staffing for the P/Q Center comes directly from the operations groups. New staff joins the P/Q Center for a two-year assignment. Each staff member know he or she will return to operations at the end of his or her assignment. This accomplishes several things: (1) it promotes the potential for P/Q Center credibility since the staff are from operations and operations oriented; (2) it avoids the development of staff/support orientation stagnation since there is a constant flow of personnel from operations to the center and from the center to operations; and (3) it allows for operations personnel to take a sort of "sabbatical" from their particular work area and to develop innovative ideas they may not have had time to develop while in the often hectic world of operations. In less than three years, the P/Q Center has developed some very exciting and innovative products, programs, and services.

Honeywell A&D Productivity Measurement Approach Summary

The overall philosophy on productivity measurement at Honeywell A&D is to create decoupled measurement systems at the department level. Elements of the normative productivity measurement methodology exist in their approach. At the more micro level, there are a few overall measures of productivity that are to be used consistently within the A&D. Beyond that, however, the focus is on promoting departmentally generated measures that are viewed as useful for evaluating the success of productivity and quality of work improvement projects. Honeywell A&D is an excellent operational example of the NPMM in use.

There are obviously a very large number of examples of measurement and evaluation systems that have been and are being developed within the A&D. An excellent example is what has been developed in the Defense Systems Division (DSD) in the Engineering area. Figure 15.3 (pages 460–461) depicts the multidimensional performance measurement and evaluation system that is currently being developed. Note the 4 criteria that are common to all DSD, the 14 quality-of-work-related criteria, and the 4 financially related criteria. It is critical to understand that this philosophical approach to productivity/performance measurement recognizes the importance of the process of creating measures and using them to manage by and evaluate improvement. The process is as critical as the content (the measures themselves). As successively lower levels of management, eventually all the way to first-line foreman and supervisor, become more acquainted with the process of focused project planning, project implementation, and measurement and evaluation of results, the organization will be able to build the capacity necessary to be more proactively innovative.

Table 15.1 Examples of Productivity/QOW Measurements

PRODUCTION/PRODUCTION PROFIT CENTER MEASUREMENTS

$$\frac{\text{Direct Hours}}{\text{Standard Hours}} \qquad \frac{\text{Scrap Costs}}{\text{L, B, M Additions}} \qquad \frac{\text{Inventory Shortage}}{\text{Inventory Additions}}$$

$$\frac{\text{Earned Hours}}{\text{Direct Hours}} \qquad \frac{\text{L, B, M \& Support Costs}}{\text{No. of Units Produced}} \qquad \frac{\text{Total Production Hours}}{\text{Direct Earned Hours}}$$

$$\frac{\text{Indirect Hours}}{\text{Direct Hours}} \qquad \frac{\text{Fixed Price Cost of Sales}}{\text{Gross Net Inventory}} \qquad \frac{\text{Delinquent Units} \times \text{Selling Price}}{\text{Average Daily Sales}}$$

$$\frac{\text{Salvage Hours}}{\text{Direct Hours}} \qquad \frac{\text{Sales/VA Sales}}{\text{Direct Headcount}} \qquad \frac{\text{Production Support Costs}}{\text{Production L, B, M Costs}}$$

$$\frac{\text{Set-Up Hours}}{\text{Earned Hours}} \qquad \frac{\text{Sales/VA Sales}}{\text{Total Headcount}} \qquad \frac{\text{Indirect Headcount}}{\text{Direct Headcount}}$$

$$\frac{\text{Direct Labor \$}}{\text{Standard Hours}} \qquad \frac{\text{Product Build \& Support Hours}}{\text{Equivalent Units Produced}} \qquad \frac{\text{Production Hourly Headcount}}{\text{Production Control Headcount}}$$

$$\frac{\text{No. of Units Accepted}}{\text{No. of Units Inspected}} \qquad \frac{\text{Customer Accepted Lots}}{\text{Lots Submitted}} \qquad \frac{\text{Production Hourly Headcount}}{\text{Production Engineering Headcount}}$$

$$\frac{\text{Wait Time Hours}}{\text{Direct Labor Hours}} \qquad \frac{\text{Warranty Repair Costs}}{\text{Sales}} \qquad \frac{\text{No. of Defects}}{\text{No. of Units Inspected}}$$

$$\frac{\text{Units Scheduled}}{\text{Units Produced}} \qquad \frac{\text{Sales/VA Sales}}{\text{Indirect Headcount}} \qquad \frac{\text{Hrs. on Labor Ticket Rejects}}{\text{Total Hours Reported}}$$

$$\frac{\text{Complete Kits Issued}}{\text{Total Kits Issued}} \qquad \frac{\text{Cost of Quality}}{\text{Cost of Sales}} \qquad \frac{\text{Actual Burden Rate}}{\text{Planned Burden Rate}}$$

$$\frac{\text{PBIT}}{\text{Employees}}$$

$$\frac{\text{Direct Labor}}{\text{Total Time Reporting Labor}} \qquad \frac{\text{Projected Unit Build Cost}}{\text{Target Unit Build Cost}} \qquad \frac{\text{No. of Key Performance Specs Met}}{\text{Total No. of Key Performance Specs}}$$

$$\frac{\text{Sales/VA Sales}}{\text{Time Reporting Headcount}} \qquad \frac{\text{Production Support Costs}}{\text{Production L, B, M Costs}} \qquad \frac{\text{No. of Programs Where PVWA} > \text{Actual}}{\text{No. of Programs}}$$

$$\frac{\text{Sales/VA Sales}}{\text{Indirect Headcount}} \qquad \frac{\text{No. of ECO's}}{\text{No. of Drawings}} \qquad \frac{\text{Counter Productive Hours}}{\text{Total Engineering Hours}}$$

$$\frac{\text{Sales/VA Sales}}{\text{Total Headcount}} \qquad \frac{\text{Hrs. on Rejected Time Reports}}{\text{Total Hrs. Reported}} \qquad \frac{\text{No. of Drawings}}{\text{Drafting Headcount}}$$

$$\frac{\text{No. of Software Instructions}}{\text{No. of Software Engineers}} \qquad \frac{\text{Projects with Plans}}{\text{Total Projects}} \qquad \frac{\text{No. of ECO's}}{\text{No. of Engineers}}$$

$$\frac{\text{Cost to Prepare Drawings}}{\text{No. of Drawings Produced}} \qquad \frac{\text{Projects Overrun}}{\text{Total Projects}} \qquad \frac{\text{Bid Hours}}{\text{Estimated Hours}}$$

$$\frac{\text{Prod. Build Hrs. on Layouts}}{\text{Prod. Build Hrs.}} \qquad \frac{\text{Projects Overrun \$}}{\text{Total Project \$}} \qquad \frac{\text{Negotiated Hours}}{\text{Bid Hours}}$$

$$\frac{\text{PBIT}}{\text{Employees}} \qquad \frac{\text{CAD Hours Usage}}{\text{CAD Hours Available}} \qquad \frac{\text{Planned Cost All Programs}}{\text{Actual Cost All Programs}}$$

$$\text{BCWP BCWP} \qquad \frac{\text{Actual Burden Rate}}{\text{Planned Burden Rate}} \qquad \frac{\text{Factory Costs}}{\text{Production Engineering Costs}}$$

$$\text{BSWS ACWP} \qquad \frac{\text{Milestones Completed I-T-D}}{\text{Milestones Scheduled I-T-D}}$$

453

Table 15.1 (*Continued*)

QUALITY DEPARTMENT MEASUREMENTS

$\dfrac{\text{Quality Dept. Hours}}{\text{Production Hours}}$	$\dfrac{\text{Material Lots Inspected}}{\text{Receiving Inspection Headcount}}$
$\dfrac{\text{Quality Indirect Hours}}{\text{Total Quality Hours}}$	$\dfrac{\text{Total Operating Headcount}}{\text{Quality Dept. Headcount}}$
	$\dfrac{\text{Errors in Data Collection}}{\text{Volume of Data Collected}}$
	$\dfrac{\text{Actual Burden Rate}}{\text{Planned Burden Rate}}$
$\dfrac{\text{Earned Hours}}{\text{Direct Hours}}$	$\dfrac{\text{Operations Budget}}{\text{Quality Dept. Budget}}$
	$\dfrac{\text{Prevention Costs}}{\text{Cost of Quality}}$
$\dfrac{\text{Cost of Quality}}{\text{Cost of Sales}}$	$\dfrac{\text{Production Earned Hours}}{\text{Quality Eng. Support Hours}}$
	$\dfrac{\text{Appraisal Costs}}{\text{Cost of Quality}}$
$\dfrac{\text{Sales/VA Sales}}{\text{Product Assurance Headcount}}$	$\dfrac{\text{QE Support Costs}}{\text{Production L, B, M Costs}}$
	$\dfrac{\text{Failure Costs}}{\text{Cost of Quality}}$
$\dfrac{\text{Total Receiving Insp. Hours}}{\text{Lots Received}}$	$\dfrac{\text{Errors on Inspection Procedures}}{\text{Inspection Procedures Issued}}$

PROCUREMENT DEPARTMENT MEASUREMENTS

$\dfrac{\text{Purchase Order Errors}}{\text{Purchase Orders Audited}}$	$\dfrac{\text{Purchasing Dept. Budget}}{\text{No. of PO's Placed}}$
	$\dfrac{\text{Total Operations Headcount}}{\text{Purchasing Dept. Headcount}}$
$\dfrac{\text{Estimated Savings on Orders Placed}}{\text{Dollar Value of Orders Placed}}$	$\dfrac{\text{\$ Amount of Purchases}}{\text{Purchasing Dept. Headcount}}$
	$\dfrac{\text{Sales/VA Sales}}{\text{Procurement Dept. Headcount}}$
$\dfrac{\text{Material Proposal Records Received}}{\text{Material Proposal Records Completed}}$	$\dfrac{\text{Lots Received on Time}}{\text{Total Lots Received}}$
	$\dfrac{\text{\$ Amount of Purchases}}{\text{Purchasing Dept. Budget}}$
$\dfrac{\text{Incoming Material Lots Accepted}}{\text{Incoming Material Lots}}$	$\dfrac{\text{No. of PO's Placed}}{\text{Purchasing Dept. Headcount}}$
	$\dfrac{\text{No. of MPR's Returned on Time}}{\text{No. of MPR's Returned}}$

FINANCE DEPARTMENT MEASUREMENTS

$$\frac{\text{Trade Billed Receivable \$}}{\text{Avg. Trade Billed Sales/Day}}$$

$$\frac{\text{Invoices Processed} \times \text{Standard}}{\text{Invoicing Hours}}$$

$$\frac{\text{Disbursements Audit Functions} \times \text{Standard}}{\text{Disbursement Audit Hours}}$$

$$\frac{\text{Total Operations Personnel}}{\text{Finance Personnel}}$$

$$\frac{\text{No. of Pricing Proposals}}{\text{No. of Pricing People}}$$

$$\frac{\text{Operations Budget}}{\text{Finance Dept. Budget}}$$

$$\frac{\text{Sales/VA Sales}}{\text{Finance Personnel}}$$

$$\frac{\text{No. of DD250 Errors}}{\text{Total DD250's Processed}}$$

$$\frac{\text{Receivables Over 60 Days}}{\text{Total Receivables}}$$

$$\frac{\text{Incomplete Cost Standard}}{\text{Total Cost Standards}}$$

$$\frac{\text{Finance Dept. Budget}}{\text{Sales}}$$

$$\frac{\text{\$ Value of Pricing Proposals}}{\text{No. of Pricing People}}$$

$$\frac{\text{Invoicing Errors}}{\text{Invoices Processed}}$$

SERVICE ENGINEERING DEPARTMENT MEASUREMENTS

$$\frac{\text{Proposals Won}}{\text{Proposals Submitted}}$$

$$\frac{\text{\$ Orders Received Y-T-D}}{\text{\$ Orders Planned Y-T-D}}$$

$$\frac{\text{\$ Orders Received—Month/Year}}{\text{No. of Marketeers/Contract Admin.}}$$

$$\frac{\text{Total Operations Personnel}}{\text{Service Engineering Personnel}}$$

$$\frac{\text{Sales/VA Sales}}{\text{Service Engineering Headcount}}$$

$$\frac{\text{Operations Budget}}{\text{Service Eng. Budget}}$$

$$\frac{\text{\$ Delinquent Deliveries}}{\text{Average Daily Sales}}$$

$$\frac{\text{No. o Proposals}}{\text{No. of Marketing Reps.}}$$

$$\frac{\text{\$ Orders Received}}{\text{Service Eng. Budget}}$$

$$\frac{\text{No. of DD250 Errors}}{\text{Total DD250's Processed}}$$

$$\frac{\text{No. of Active Contracts}}{\text{No. of Contract Administrators}}$$

$$\frac{\text{FP Orders with Progress Payments}}{\text{Total No. of FP Orders}}$$

$$\frac{\text{Sales Proposals \$}}{\text{\$ Orders Received}}$$

$$\frac{\text{Service Eng. Budget}}{\text{Operations Sales}}$$

Table 15.1 (Continued)

EMPLOYEE RELATIONS DEPARTMENT

$$\frac{\text{Change Notices Processed}}{\text{No. of Compensation Clericals}} \qquad \frac{\text{Workers Compensation Costs}}{\text{Total Hours Worked}}$$

$$\frac{\text{Recruitment Costs}}{\text{No. of People Hired}} \qquad \frac{\text{No. of People Interviewed \& Hired}}{\text{No. of People Interviewed}} \qquad \frac{\text{Offers Made}}{\text{Offers Accepted}}$$

$$\frac{\text{Sales/VA Sales}}{\text{Employee Relations Headcount}} \qquad \frac{\text{Operations Support}}{\text{Employee Relations Budget}}$$

$$\frac{\text{Total Operations Headcount}}{\text{Employee Relations Headcount}} \qquad \frac{\text{Elapsed Time of Unprocessed ECR's}}{\text{No. of Unprocessed ECR's}} \qquad \frac{\text{Employees Terminating}}{\text{Total Employees}}$$

$$\frac{\text{No. of Change Notice Errors}}{\text{Total Change Notices}} \qquad \frac{\text{Insurance Claims Processed}}{\text{No. of Insurance Claim Clerks}}$$

$$\frac{\text{Lost Time for Injuries}}{\text{Total Hours Worked}}$$

INFORMATION SYSTEMS DEPARTMENT

$$\frac{\text{Output Distributed On Time}}{\text{Total Output Distributed}} \qquad \frac{\text{Keypunch Earned Hours}}{\text{Keypunch Actual Hours}} \qquad \frac{\text{Operations Budget}}{\text{IS Budget}}$$

$$\frac{\text{Hardware Up Time}}{\text{Total Hardware Time}} \qquad \frac{\text{Jobs Completed}}{\text{Jobs Scheduled}} \qquad \frac{\text{User Complaints}}{\text{Hours of Usage}}$$

$$\frac{\text{Out-of-Service Terminals}}{\text{Total No. of Terminals}} \qquad \frac{\text{Sales/VA Sales}}{\text{IS Headcount}} \qquad \frac{\text{Proj. Estimated Development Cost}}{\text{Proj. Actual Development Cost}}$$

$$\frac{\text{Trouble Calls Received}}{\text{Unit of Time (Week, Mo., Etc.)}} \qquad \frac{\text{Total Operations Headcount}}{\text{IS Headcount}} \qquad \frac{\text{MRP/HMS Performance/Usage}}{\text{Various MRP/HMS Criteria}}$$

COMMUNICATIONS DEPARTMENT MEASUREMENTS

$$\frac{\text{Reproduction Costs}}{\text{No. of Pages Produced}} \qquad \frac{\text{Sales/Va Sales}}{\text{Communications Dept. Headcount}} \qquad \frac{\text{Cost of Viewgraph Changes}}{\text{Total Graphics Cost}}$$

$$\frac{\text{Viewgraphs Redone}}{\text{Total Viewgraphs Produced}} \qquad \frac{\text{Operations Headcount}}{\text{Communications Dept. Headcount}}$$

$$\frac{\text{No. of Programs where PVWA} > \text{Actuals}}{\text{No. of Programs}}$$

$$\frac{\text{Average Maintenance Down Time of Gyros}}{75 \text{ Days}}$$

$$\frac{\text{Qty. of Spares Delivered}}{\text{Qty. of Spares to be Delivered Per Contract}}$$

$$\frac{\text{Maintenance Costs/Flt. Hr.}}{\$22} + \frac{\text{Units In-House for Repair}}{\text{Units Installed}} \quad 0.333$$

$$\frac{\text{Orders for Logistics Services}}{\text{Total Orders}}$$

$$\frac{\text{Average Grade Level of Fld. Engrs.}}{\text{Average Grade Level of Ideal Work Force}}$$

$$\frac{\text{Sales/VA Sales}}{\text{Logistics Headcount}}$$

$$\frac{\text{Specific Prog. Logistic Orders}}{\text{Specific Prog. Nonlogistics Orders}} \quad 0.20$$

GENERAL/MISCELLANEOUS MEASUREMENTS

$$\frac{\text{Actual Hrs./\$}}{\text{Estimated Hrs./\$}}$$

$$\frac{\text{Direct Headcount}}{\text{Indirect Headcount}}$$

$$\frac{\text{Operations Headcount}}{\text{Department Headcount}}$$

$$\frac{\text{Operations Sales/VA Sales}}{\text{Department Headcount}}$$

$$\frac{\text{Building Sq. Footage}}{\text{Maintenance Cleaning Personnel}}$$

$$\frac{\text{Maintenance Orders Within Estimate}}{\text{Total Maintenance Orders}}$$

$$\frac{\text{Unplanned Absent Hours}}{\text{Total Hours}}$$

$$\frac{\text{Backlog Hrs. on Maintenance Work Orders}}{\text{Maintenance Headcount}}$$

$$\frac{\text{Non-Productive Time}}{\text{Total Time Available}}$$

$$\frac{\text{Department Costs}}{\text{Department Budgeted Costs}}$$

$$\frac{\text{Headcount}}{\text{No. of Secretaries}}$$

$$\frac{\text{No. of People in QC Teams}}{\text{Total Employees}}$$

$$\frac{\text{Sales}}{\text{Assets}}$$

$$\frac{\text{Profit}}{\text{Employees}}$$

$$\frac{\text{Assets}}{\text{Employees}}$$

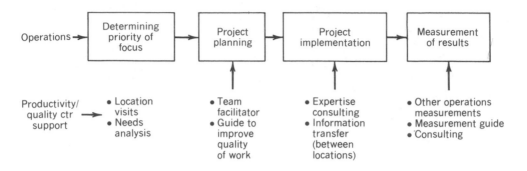

Figure 15.2   Productivity and Quality of Work Project Flow

Cooperative Labor-Management Strategic Planning in a Unionized Steel Plant (This section is a paper presented at the 1984 Spring Institute of Industrial Engineering Conference in Chicago by L. K. Swim and D. S. Sink, May 1984)

Abstract

Participative management techniques have been receiving widespread attention in America since the middle 1970s, facilitated largely by the sucess of quality circles in Japan. American organizations, American managers, and American employees are unique. They share a culture quite different from that of the Japanese. The success of a participative management program in an American organization can be highly influenced by the degree of flexibility the program allows and by how well the program is tailored for that specific organization. The performance/productivity action team (PAT) process is an integrated group-participative problem-solving technique, comprising a variety of programs, processes, and methodologies, and tailored specifically for American organizations. The PAT process encompasses the necessary structure, as well as the required flexibility, for successful employee participation in decision making and/or problem solving.

This paper will discuss the PAT program design used in the development of a labor-management participation program for a unionized steel plant in Oklahoma. A strategic planning process was integrated into the PAT program designed for this plant. The first major activity of the program was an off-site strategic planning retreat, attended by both union and management—an event certainly not typical for this company and one not typical for most American organizations. Six steps of an eight-step strategic planning process were implemented over the course of the two-and-one-half-day retreat to establish strategic, tactical, and operational action programs. A broad discussion of the PAT program design used, in addition to a focused discussion of the results of this company's strategic plan, will be presented.

Introduction

Faced with the impact of increased competition, and an era of adversarial labor management relations, the top management of a unionized steel company in Oklahoma surveyed available approaches for developing a labor-management participation program. The importance of pursuing this program was heightened by the requirement for such a program in the current bargaining unit contract. A discussion between the company president and the vice president of human resources regarding the offerings of the Oklahoma Productivity Center in this area established a potential match between the company's criteria and the characteristics of the Oklahoma Productivity Center's performance/productivity action team process. Prior to contract approval, on-site briefings were held first with the company's key management personnel, then with members of the Union Executive Board, and finally with both of these groups jointly. The purpose of the briefings was for the Oklahoma Productivity Center to communicate to these organizational members the design of the PAT program and to address questions and concerns. These briefings also served to involve key organizational members in the labor-management participation program selection, thus strengthening their commitment to the selected program.

Upon contract approval, the Oklahoma Productivity Center began the chalenging and interesting task of uniting labor and management in an effort to improve the company's performance and thus increase its chances of successfully competing in the dynamic environment it faces. The steel company produces approximately 200,000 tons of rebar and fence posts per year and employees 350 to 400 people. It has recently been divested from one of the country's largest steel companies and is struggling, to say the least, to survive. Adversarial relations between labor and management have developed and heightened through the years. Clearly, the company was in need of labor-management participation program tailored to meet its unique needs. This paper addresses the approach adopted by the Oklahoma Productivity Center in implementing a participatory management program capable of meeting the demands of this particular company. Focus is placed on the integration of strategic planning in the PAT program design.

In order to establish a foundation of knowledge on which to build, the discussion to follow will first address participative management in American organizations. Then, the PAT process will be presented, followed by a discussion of the role and description of strategic planning in the PAT process. Finally, focus will be placed on the steel company's strategic planning retreat to include a presentation of the strategic planning session results/outputs and outcomes.

Participative Management in American Organizations

Participation in decision making and/or problem solving does, in fact, require a decentralization of power and control (Sink, 1982). The risk of such an approach to management, as well as its contrast to what has become the "comfortable" manner of management, is highly visible to American managers. However, few would argue with the suggestion that organizations of today cannot perform at a competitive level if they lack the ability to innovate and change. Innovation and change appear to require a mixture of leadership and consensus, autonomy and participation—a balance of control (Kanter, 1983). Participative management provides an outlet for expression of the vital human resources of the organization. More democratic processes can guarantee that creative new ideas can develop in a synergistic manner, and that they will have an outlet for expression.

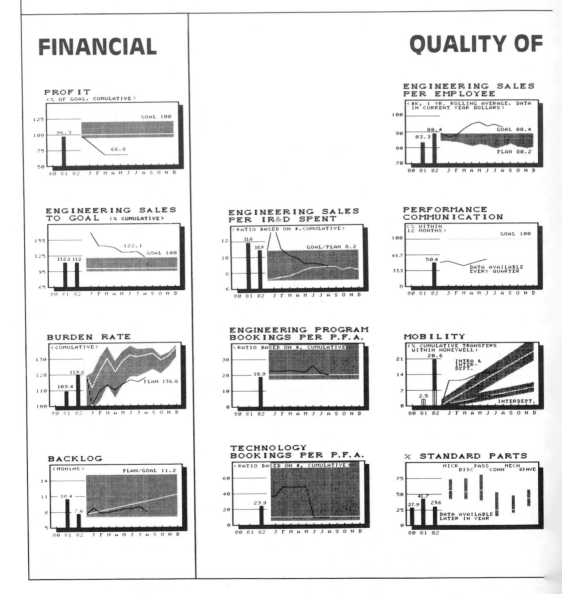

Figure 15.3   Example of an Application of NPMM

Division
PERFORMANCE

WORK

FOLLOW-ON PRGM (>$.1M) HIT RATE (CUMULATIVE)

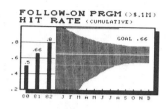

GOAL .66

.8 / .66 / .5 / .8

80 81 82 J F M A M J J A S O N D

TRAINING & TOOLS PER EMPLOYEE
($ THOUSANDS, TOOLS: GROSS UNDEPRECIATED, CUMULATIVE)

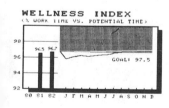

15.9

80 81 82 J F M A M J J A S O N D

PATENT DISCLOSURES
(# CUMULATIVE)

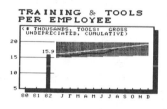

29 25

GOAL 28

80 81 82 J F M A M J J A S O N D

WELLNESS INDEX
(% WORK TIME VS. POTENTIAL TIME)

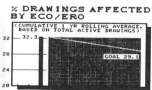

96.5 96.7

GOAL: 97.5

80 81 82 J F M A M J J A S O N D

INTERNAL COST REDUCTIONS
($ THOUSANDS CUMULATIVE)

467 525

GOAL 300K

80 81 82 J F M A M J J A S O N D

% DRAWINGS AFFECTED BY ECO/ERO
(CUMULATIVE 1 YR ROLLING AVERAGE, BASED ON TOTAL ACTIVE DRAWINGS)

32.3

GOAL 29.1

80 81 82 J F M A M J J A S O N D

SUPPORT ACTION COST REDUCTIONS
($ MILLIONS, CUMULATIVE)

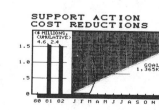

4.6 2.4

GOAL 1,365K

80 81 82 J F M A M J J A S O N D

ALL DSD

DSD VALUE ADDED SALES PER EMPLOYEE

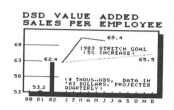

69.4

1983 STRETCH GOAL (5% INCREASE)

65.5

62.4

53.2

($ THOUSANDS, DATA IN '83 DOLLARS, PROJECTED QUARTERLY)

68 63 58 53

80 81 82 J F M A M J J A S O N D

DSD SCRAP & REWORK PER SALES
(CUMULATIVE, PROJECTED QUARTERLY)

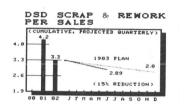

4.2 3.3

1983 PLAN 2.8

2.89

(15% REDUCTION)

4.0 3.3 2.6 1.9

80 81 82 J F M A M J J A S O N D

DSD GROSS INVENTORY TURNS
(CUMULATIVE, PROJECTED QUARTERLY)

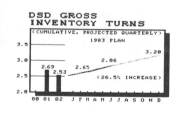

1983 PLAN 3.20

2.86

2.69 2.53 2.65

(26.5% INCREASE)

3.5 3.0 2.5 2.0

80 81 82 J F M A M J J A S O N D

DSD MOBILITY
(% CUMULATIVE TOTAL TRANSFERS)

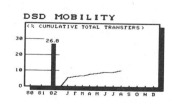

26.8

30 20 10 0

80 81 82 J F M A M J J A S O N D

Decoupled Productivity Measurement System

American organizations, American managers, and American employees are unique. In order for participative management to take hold and survive in America, the demands placed by American organizations upon this nontraditional approach to management must be recognized and acknowledged. In a very broad sense, American organizations appear to place the following demands on participative management programs:

1. The program should have the full support of and be understood by top management.

2. The program should be implemented in a pervasive manner, starting with the indoctrination of top-level managers and proceeding in a downward fashion through the layers of the organization.

3. The program should be linked tightly with current organizational objectives.

4. The program should allow for a range of organizational performance issues to be addressed to include effectiveness, efficiency, quality, quality of work life, productivity, profitability, innovation, and communication.

The Oklahoma Productivity Center at Oklahoma State University has been further developing and enhancing a process that was initiated in 1975 by the Productivity Research Group at The Ohio State University (Sink, 1978). This process began as an independent investigation into innovative methods for effectively and efficiently measuring and improving productivity and has evolved, through the efforts of the Oklahoma Productivity Center, to include aspects of a variety of techniques, programs, processes, and methodologies. The performance/productivity action team (PAT) process is truly a hybrid design that has integrated the necessary requirements for successful group participative problem solving in American organizations.

Performance/Productivity Action Team Process

The PAT methodology incorporates the use of the nominal group technique, a structured group process designed to achieve group consensus. The establishment of group consensus can serve as a catalyst to enhance organizational effectiveness. In addition, the methodology used to attain this consensus creates an environment that encourages structured, open communication. Group consensus in the determination of broad, as well as specific, organizational goals, objectives, and activities (determining what constitutes the "right" activities for which to commit such resources as time and dollars) makes what comprises effectiveness for an organizational system more understandable and visible to the group.

Although the PAT process is highly flexible, another characteristic that tailors it for American organizations, it follows a sequential ten-step procedure. The ten-step PAT process is outlined in Figure 15.4. The PAT process appears to be fairly simplistic, conceptually; however, it is an extremely robust procedure, rich in potential for providing the organization with a systematic way in which to (1) better utilize employees' talents; (2) create improved goal congruity between management and employees; (3) enrich the jobs of employees; (4) improve communication, cooperation, and coordination between functions; (5) identify and attack roadblocks to performance that have been "hiding in the cracks"; (6) create decision making and motivate action based on group consensus rather than on an autocratic or consultative style of decision making; (7) develop employees by allowing them the opportunity to become active participants in organizational problem solving; and (8) improve organizational performance by creating a motivation to improve as a result of involvement and commitment (Sink, 1982).

Step 1: Form productivity/performance action group (PAG)
Action: Nominal group technique (NGT) to obtain list of ideas for improvement (1.5–2 hours)
Outcome: Concensus, prioritization of ideas

Step 2: PAG continues discussion of NGT results
Action: Form PAT's (smaller teams) and assign high priority items from NGT (1.5–2 hours)
Outcome: PAT formation

Step 3: Individual PAT problem solving (issue analysis, problem analysis, decision analysis, design analysis, implementation analysis)
Action: Develop project plan, activities, proposal (3–4 months)
Outcome: PAT develops proposed solution

Step 4: PAT reports to PAG
Action: Proposal is presented to PAG to obtain input (2–4 hours)
Outcome: PAT practices presentations; skill in proposal presentations is developed

Step 5: Management review session
Action: PAT presents proposal to management review team (2 hours)
Outcome: Management is able to view progress of problem solution approach, as well as development of employees

Step 6: Management "feedback" session
Action: Management communicates its decision on PAT proposals; implementation issues are discussed (1 hour)
Outcome: Knowledge of where proposal stands; project plan for implementation detailed

Step 7: Implementation
Action: Follow project plan for implementation
Outcome: Proposals are implemented

Step 8: Measurement, Evaluation, Feedback, Charting
Action: Measurement process developed, success of proposal evaluated, feedback provided regarding proposal

Step 9: Reinforcement system design, development, and implementation
Action: Management develops reinforcement system
Outcome: Employees continue to feel pride in their work (proposals); motivation continues

Step 10: Process evaluation, modification, recycle
Action: PAT process improved by learning from previous cycle; PAT process repeated
Outcome: A second PAT process

Figure 15.4   The Ten-Step PAT Process

The PAT process is clearly tailored to the American way of thinking and acting. It provides the necessary structure for the successful adoption of the participative management approach. The PAT process, if introduced and implemented properly, can facilitate effective and efficient participative problem solving and, as a result, improvements in long-term organizational performance. A very broad interpretation of the role the PAT process can play in long-term organizational system performance is depicted in Figure 15.5. The illustration conveys that six of at least seven key organizational system performance measures or criteria can be positively impacted by the PAT process, with innovation serving as a moderating variable.

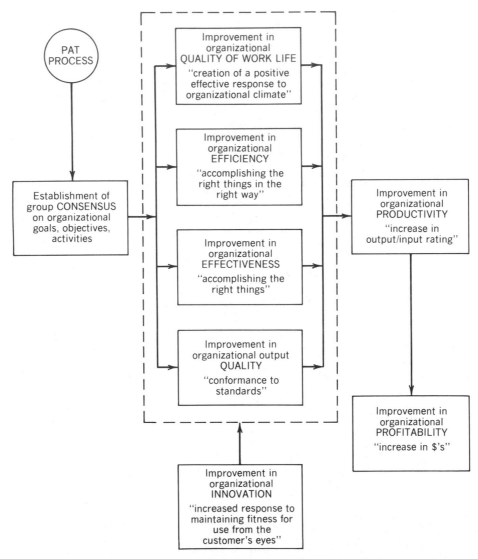

Figure 15.5   PAT Program's Role in Organizational System Performance Improvement

The Role and Description of Strategic Planning in the PAT Process

Strategic planning has traditionally been an activity or function executed by top management. The typical planning cycle and process in most businesses tends to be very lock-step and bureaucratic in character. We simply go through the motions and fill in the necessary forms. It is, most frequently, a very individualistic process and does not effectively tap "group wisdom." It is rarely a group-oriented process and the various perceptions and knowledge of the medium-to-long-term picture for the organization is seldom acquired from a large number of organizational participants. Planning has become what top management tells middle management to get the line personnel to do.

In reality, personnel at all levels of the organization have strategic (long-range), tactical (medium-range), and operational (shorter-range) views of the organizational systems within which they work. Figure 15.6 depicts this. There are, of course, primary, secondary, and tertiary views and perspectives at each level based on position and focus. However, the point is that all three views exist at all levels, and the tactical and strategic views should be tapped at lower levels in the organization than is presently the case. In order for organizations to be successful in the eighties and nineties (effective and innovative, productive and, in the longer term, profitable), they must develop and implement strategic planning processes that are more flexible, adaptive, effective, and, in particular, more participative in character. The trends we see and hear being described in *Megatrends* (Naisbitt), *Managing in Turbulent Times* (Drucker), *The Third Wave* (Toffler) all support

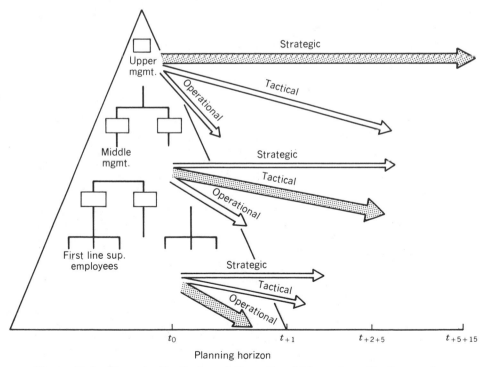

Figure 15.6   Strategic, Tactical, and Operational Views from the Perspective of Various Organizational Levels

this observation. And case studies that have been written regarding strategic planning process development in excellent companies also support this prescription (Fombrun, 1982; Barrett, 1981). We simply have to find ways to involve more of the organization in various elements of strategic planning. And we have to find ways to ensure the linkage of strategic planning to action planning and effective implementation.

A PRAGMATIC EIGHT-STEP STRATEGIC PLANNING PROCESS

After years of reading about strategic planning and listening to renowned speakers on the topic, we decided to convert theory and concepts into technique. The eight-step process that resulted incorporates developments by a management consultant named Maurice Mascarenhas and work done in the area of structured, group processes. The resulting process is a structured group-oriented and participative strategic planning approach. The goal of the process is to achieve consensus among a group of managers with respect to strategic goals and objectives, specific action plans/programs, and implementation steps. The assumption underlying the process is that effective movement toward goals and objectives springs from positions of consensus and the ensuing commitment to follow through. This assumption has been tested in a variety of organizations, and with the right amount of leadership intervention at appropriate times we have found the assumption to be very valid.

The eight-step process can be directed toward any number of desired focuses. For instance, the application of the process could be directed toward

1. Overall business planning

2. Productivity management program planning

3. Departmental planning

4. Quality management program planning

5. Cost reduction program planning

Figure 15.7 presents the basic eight-step process as it might be applied to the development of a plan for a productivity management program.

Note that steps 1–3 are an attempt to get the persons involved to critically consider factors (both external and internal) that will/should influence the strategic plan. We simply have the participants in the process silently respond to Steps 1 and 2 and then, in round-robin fashion, solicit their responses one at a time. The result is a list of factors and planning premises posted on flip-chart paper around the walls of the meeting room. The process of discussing these factors and premises creates a group gestalt for trends, issues, considerations, other viewpoints and perceptions, and so forth. We have found this process to be an uncomfortable and difficult one for most managers, albeit a very critical and important one. No one has doubted the importance of these first three steps, they simply discover the difficulty of thinking about these issues.

Step 4 utilizes the nominal group technique (Delbecq, Van de Ven, and Gustafson, 1975; Sink, 1983) to generate a prioritized and consensus list of strategic goals and objectives for the program or organizational system that is being focused on. The simplest way to think about this step is to realize that we are attempting to develop a clear perception of where the program or the organizational system is headed during the next planning horizon (often 2–5 years).

Pre-Condition/Step: Strategic and tactical planning awareness for the corporation and affected groups.

Step 1: Internal strategic audit (looking within). What internal factors (strengths, weaknesses, conditions, trends, persons, programs, assumptions, etc.) should be considered during the design, development, and potential implementation of our productivity management program?

Step 2: External strategic audit (looking around). What external factors (competitors, strengths, weaknesses, trends, conditions, organizations, assumptions, etc.) should be considered during the design, development, and potential implementation of our productivity management program?

Step 3: Planning premises, assumptions. Importance–certainty grid.

Step 4: Strategic Planning—2–5 year goals and objectives (desired outcomes) for the productivity programs.

Step 5: Prioritization and consensus of performance objectives in key results areas relative to the productivity program.

Step 6: Identification, prioritization, and concensus for/of strategic, tactical, and operational action programs.

What programs, plans, resources, etc. will have to be budgeted for in both the short run (1 yr.) as well as longer run (2–5 yrs.) in order for this program to succeed.

Simultaneous operational focus (1 yr.) with a strategic focus (innovation, growth, continued development and support) (2–5 yrs.)

Note: some programs will be doable in one year; others will take 2–5 years or more. Therefore, some subprograms are to be budgeted for in year 1 but are building for longer-term/payoff programs.

Step 7: Program planning and resource allocation.

Step 8: Program evaluation, review, maintenance.

Figure 15.7 A Practical Eight-Step Productivity Management Program Planning Process

Step 5 focuses on developing a list of performance objectives or measures for the program or the organizational system. The focus is on obtaining from the group their perception of how success should be evaluated. How will we know we have succeeded or are succeeding in accomplishing our goals and objectives?

Step 6 develops consensus and priority action programs that should be budgeted for and worked on in the next year. In other words, what has to occur in the next year in order for us to move successfully toward our priority goals and objectives? Step 7 then provides the opportunity for a team (2–3 persons) of the planning group (6–15 persons) to develop program planning in terms of specific implementation steps and resources to be allocated. The final step, step 8, is executed after the planning session itself and is the evaluation process. Have we carried through with the action program implementation? Are we moving toward our goals and objectives? What deviations/adaptations have occurred?

This is a brief presentation of the strategic planning process we have developed and tested. Next we will discuss the role this specific planning process plays in the establishment and implementation of the PAT process.

STRATEGIC PLANNING PROCESS AND THE PAT PROCESS IMPLEMENTATION

Most employee involvement/participative problem-solving programs fail as a result of management reactions and attitudes rather than as a result of the inability or unwillingness of employees to participate. Development of participative problem-solving programs requires significant attention to preparing the entire organization for this new managerial process. The organizational culture needs to be supportive of such processes in order for these types of programs to have a chance of success. All levels of management as well as the employees themselves must have realistic expectations of what the program will and will not do (Kanter, 1983).

Musashi Semiconductor Works, a division of Hitachi Corporation, began implementing a "small group activity program" like PATs in 1971. They spent from 1971 through 1975 developing an educational and cultural support system for their 3700 employee plant. The 2700 workers are organized into 360 groups of 8–10 people each. There is a reasonably complex infrastructure for this program that is described in detail in Davidson's article (1982). The first formal improvement proposal was filed by a worker group in 1977. In the second half of 1980, 112,022 proposals were submitted by the 360 groups. Of that total, 98,347 improvements were implemented by the end of the year. No company in the United States that we know of can boast anywhere near this level of accomplishment with their employee involvement effort. We believe the key lies in the foundation that was laid for the effort and the top-down implementation procedure.

Based on our knowledge and experience of factors correlated with participative problem-solving program implementation successes as well as failures, we proposed a top-down implementation of the eight-step strategic planning process as depicted in Figure 15.8. This implementation plan can accomplish several things:

1. It exposes participants to a participative goal setting and action planning process similar to what the employees and supervisors will experience later in the program. This provides management with an opportunity to affectively respond and react to this process.

2. The process exposes pockets of resistance to the program before the employees are involved.

3. It manages/controls expectations for the program. Implementation is slowed down so that the "culture" can adapt to this new way of doing business.

4. It provides an opportunity to execute a participative planning process while simultaneously exposing management at various levels to structured participative processes.

5. It allows for the educative component of this change process to be effectively integrated with a planning and action mode. The process moves toward performance improvement, while educating management and labor as to participative goal setting, problem solving, leadership effectiveness, the PAT process itself, and so forth.

Examples of output from various stages of this implementation plan are presented in the next section.

The Steel Company's Strategic Planning Retreat

The strategic planning retreat was held at an off-site resort location. The site was chosen to provide an environment conducive to both work and recreation. Top management, including the company president, vice presidents, and superintendents, and the Union

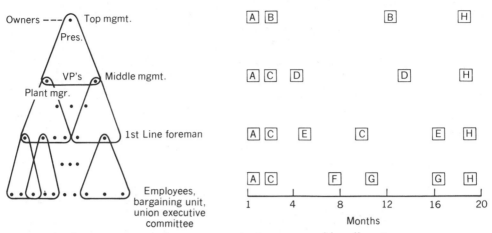

A—Program presentations, communication, approval by all parties
B—Top mgmt./union executive committee strategic planning retreat, steps 1–6
C—Situational leadership/leadership effectiveness/PAT process training, selected persons
D—Middle mgmt./union executive committee action planning session, steps 4–7
E—First-line foreman/bargaining unit performance action group pilot, steps 6 and 7
F—Work group performance action group pilot, steps 6 and 7 (OPC facilitation)
H—Site independence

Figure 15.8   Top-Down Implementation of Strategic Planning Process

Executive Board participated in the two-and-one-half-day session. The strategic planning retreat agenda, which served as a guide, is outlined in Figure 15.9. The discussion to follow focuses on the strategic planning session results/outputs and outcomes.

SESSION RESULTS/OUTPUTS

Partial outcomes of the six implemented strategic planning steps are shown in Figures 15.10, 15.11, and 15.12. Although the complete set of results for each step is not provided in this article in an effort to conserve space, the outcomes listed are a representative sample of the total set. The items listed for Steps 4 and 6 are, in fact, the top-ranked items for each of these steps. The linkage from one step to the next is made visible by the relationship between the actual outcomes of each step. Further, the narrowing focus enacted by the sequential progression through the steps is also highly visible in the outcomes. The action programs developed in Step 6 are much more action oriented than are the goals, objectives, or desired outcomes listed for Step 4.

The accomplishments of the session are quite impressive for several reasons. Fifteen individuals synergistically combined their knowledge and experiences to develop a strategic plan for their company—and did so during two eight-hour sessions. An examination of the session outputs reveals a very strong thread of logic that weaves the six steps together

---

15 Key Participants from Top Management and Union Executive Board

Day 1 (Evening): Introduction
Day 2 (Morning): Strategic planning
  Step 1: Internal strategic audit
      "Looking Within"
  Step 2: External strategic audit
      "Looking Around"
  Step 3: Planning premises and assumptions
Day 2 (Afternoon): Strategic planning
  Step 4: Strategic planning (NGT facilitated)
      "Identification of 2–5 Year Goals/Objectives and Desired Outcomes"
  Step 5: Development of priority and concensus performance objectives
      "Measures of Goal Attainment"
Day 3 (Morning): Strategic planning
  Step 6: Action planning (NGT facilitated)
      "Identification of Specific Action Programs"

Figure 15.9 Strategic Planning Retreat Agenda

---

in a unique pattern, governed by the unique characteristics of this particular company. Further, the process was performed in an efficient and effective manner made possible by the sheer structure of the process—a structure that allows for the necessary control over the process by the process facilitator but that affords the flexibility required for unsanctioned and cooperative participation. Perhaps the most important aspect of the strategic planning session results is the fact that they represent consensus priority results as determined by the group of participants. It's interesting to note that the list of top-priority action items developed by the company president for Step 6 were almost identical to those determined to be consensus priority action items by the group. However, the ease with which such items will be implemented will be greater since they have been formed through group consensus as opposed to being mandated by top management.

SESSION OUTCOMES

As previously discussed, the strategic planning session was integrated into the performance action team program design in order to accomplish those things correlated with participative problem-solving program implementation successes. The session allowed current organizational objectives (the development of a strategic plan) to be linked tightly with the PAT program and vice versa. The session also indoctrinated top management and the Union Executive Board to the elements inherent to the PAT process itself (the group-consensus-seeking technique). Both of these outcomes should serve to deepen the commitment to the PAT program and to broaden the understanding of the mechanics of the PAT program techniques.

Internal Strategic Audit (Step 1)

Strengths
- Quality of employees
- Good market place next 2–3 years
- Low cost billet with excess capacity
- Union concessions
- Attitude toward improvement

Weaknesses:
- Not enough participative management
- Apathy among employees
- Lack of diversified product line
- Assets in certain areas need modernization
- Union/management relations at front line level

Trends:
- Going from big steel companies to small
- Change from status quo to growth type industry
- Toward more teamwork
- Union/management cooperaiton

Problems:
- Mill quality
- Layoffs
- Employee rights, social responsibility

Opportunities:
- New technology Is available
- Cheap equipment available now—bargains

External Strategic Audit (Step 2)

Competitor's activities:
- Rebar not their main product
- More product lines
- New competitors
- More computer utilization

Trends:
- Mill closings, mergers, consolidation, etc. (equipment, labor availability)
- Construction market increase—new highways, bridges, paving, etc.
- Shipments by truck vs. rail
- EEOC and EPA

Strategic Planning Premises and Assumptions (Step 3)

- Expand product line
- Even though times get better, we will learn from the past and not fall back into old habits
- Will attain team attitude (labor-management)
- Quality will improve
- Better than average source of human resources
- Will find new sources of capital
- Open climate with respect to communication—management willing to share information
- No need for additonal concessions

Figure 15.10 Strategic Planning Process Outputs (Steps 1–3)

Strategic Planning 2–5 Year Goals (Step 4)

GOALS/OBJECTIVES OR DESIRED OUTCOMES	INDIVIDUAL VOTES RECEIVED (8 = MOST IMPORTANT, 1 = LEAST IMPORTANT)	NUMBER OF VOTES/TOTAL VOTE SCORE
1. Reduce debt-equity ratio	8–8–8–8–7–7–7–7–6–6 6–4–2	13/84
2. Improve labor/management relations • Enhance credibility between management and bargaining unit (trust and confidence) • Get management more union oriented (better understanding of employee needs, etc.) • Better relations between company (front-line) and union.	8–7–7–6–6–5–5–4–4–4 3–2–2–1	14/64
3. Update present rolling mill	8–8–8–8–6–6–6–5–3–2	10/60
4. Become best in existing product line	8–7–7–5–4–3–3–3–3–3–3	11/49
5. Develop other products to offset downturns	7–6–6–5–5–5–4–4–3–2	10/47
6. Have an effective management/ workforce team that truly understands what we are trying to accomplish	8–7–7–7–6–6–3–3	8/47

Measures of Goal Attainment (Step 5)

- Profits
- Attitudes
- Quality of product
- Success of PAT program
- Severity and frequency of labor/management confrontations
- Production costs
- Layoffs
- Equipment update status

Figure 15.11   Strategic Planning Process Outputs (Steps 4–5)

Specific Action Programs (Step 6)

ACTION PROGRAMS	INDIVIDUAL VOTES RECEIVED (8 = MOST IMPORTANT, 1 = LEAST IMPORTANT)	NUMBER OF VOTES/TOTAL VOTE SCORE
1. "Mill Revamp Program" • Repower the 10" mill and retire the MG Set • Re-do hot bed carry over system • Change 10" from fabric bearings to roller bearings • Rebuild mill furnace—lower hearth	8–8–8–8–8–8–8–8 8–8–8–8–6–5–4	15/111
2. Pursue PAT Program • Total coverage of PAT Process	8–8–7–7–6–6–6–5–5–5	13/73
3. Establish a Quality Control Committee • Significant improvement in quality, reduce present rejects and complaints • Random steel reduced or eliminated	8–7–6–6–3–2–2–1–1–1	10/37
4. Get a new product	7–7–6–6–5–4–2–2–1	9/40
5. Improve production marketing, planning, scheduling system in present product	7–7–7–6–6–3–2–2	8/40

Figure 15.12   Strategic Planning Process Output (Step 6)

A secondary, but extremely important, outcome of the strategic planning session was the attainment of an enhanced level of communication between the union representatives and top management. The joining of efforts by these two groups in establishing a "company" strategic plan was a new experience for both groups. The issue of improving labor-management communication, cooperation, and coordination was voiced repeatedly by members of both groups throughout the session. Providing a structured process that allowed such voices to be not only "heard" but "listened to" can serve only to initiate progression toward the desired improvements.

Conclusion

The integration of the strategic planning process with the PAT program development has proven to be an extremely effective approach for implementing a labor-management participation program in this particular company. Much progress has been made in the areas of communication, coordination, and cooperation since the initiation of the PAT program. Despite the deeply rooted adversarial relations between labor and management, the two groups have begun the change process to unite their efforts to strengthen their company's performance and competitive position. The top-priority consensus item of the action program (Step 6) phase of the strategic planning process, "Mill Revamp Program," has been acted on and when implemented, will make significant contributions to company performance improvement, specifically product quality and production effectiveness and efficiency. The action program of Step 6, "Total Coverage of PAT Process," which was rated second, falls primarily under the responsibility of the OPC. At this time, a middle management and Union Executive Board group, and first-line supervisor and bargaining unit representative group have been exposed to the PAT process. Both of these groups have progressed to Step 5 of the PAT process. A total of 11 ideas for performance improvement have been investigated by the action teams formed from both groups to include such items as "solicit employee input before installing new equipment or modifying old equipment," "better coordination between electric furnace and merchant mill," "training for craft people," and "gradual update on equipment and machinery."

As stated earlier, few would argue with the suggestion that organizations of today cannot perform at a competitive level if they lack the ability to innovate and change. A balanced mixture of leadership, consensus, autonomy, and participation with respect to control is a necessary element to enact innovation and change. The PAT process provides a structured approach with which to develop the required balance and offers a mechanism for fulfilling the demands placed on participative management programs by American organizations, managers, and employees.

REFERENCES

Barrett, E. M. "OST at Texas Instruments" Harvard Business School Case Study, HBS Case Services, HBS: Boston, 1981.

Davidson, W. H. "Small Group Activity at Musashi Semiconductor Works." *Sloan Management Review*, Spring 1982.

Delbecq, A. L., A. H. Van de Ven, and D. H. Gustafson. *Group Techniques for Program Planning: A Guide to Nominal Group and Delphi Processes.* Glenview, Ill.: Scott, Foresman, 1975.

Drucker, P. F. *Managing in Turbulent Times.* New York: Harper and Row, 1980.

Fombrun, C. "Conversation with Reginald H. Jones and Frank Doyle." *Organizational Dynamics*, Winter 1982.

Honeywell Corp. *A&D Guide to Improved Productivity and Quality of Work.* Minneapolis: Honeywell Corp, 1982.

Honywell Corp. *A&D Productivity/Quality Guide for Performance Improvement.* Minneapolis: 1984.

Kanter, R. M. "Dilemmas of Managing Participation." *Organizational Dynamics*, Summer 1982.

Kanter, R. M. *The Changemasters.* New York: Simon and Schuster, 1983.

Naisbitt, J. *Megatrends.* New York: Warner Books, 1982.

Peters and Waterman. *In Search of Excellence.* New York: Harper and Row, 1982.

Sink, D. S. *Development and Implementation of Productivity Measurement Systems with Emphasis on Interorganizational Relationships.* Unpublished Dissertation, Columbus: 1978.

Sink, D. S. "The ABC's of Theories X, Y, and Z," *1982 Fall Annual Institute of Industrial Engineers Conference Proceedings*, IIE: Norcross, Ga., November 1982.

Toffler, A. *The Third Wave.* New York: Bantom Books, 1981.

QUESTIONS AND APPLICATIONS

1. Choose an organizational system and develop a one-to-five-year plan for a productivity management program for that system. Be as specific and detailed as possible.

2. Discuss the respective characteristics, features, benefits, costs, and risks of a consultative, expert, staff-developed productivity management program as well as a participatively developed productivity management program. Defend each approach. When is each appropriate? Discuss moderating variables or factors that will cause each to succeed or fail. Discuss the skills necessary to execute each approach. Is this factor a critical constraint for either approach? If so, what can be done to meet this constraint? Develop a plan to execute a productivity management program using each approach.

3. What factors will influence the specific features or elements of a productivity management program?

4. What are the generic features, characteristics, or elements of a productivity management program?

5. Interview two or three managers. Ask them how they specifically manage productivity. Ask them how they manage performance.

6. Identify and discuss the critical research questions you envision as being associated with the design, development, and implementation of a productivity management program.

7. Develop a research proposal in the area of productivity management.

8. Select three current high-quality articles on strategic planning. Critique them. Discuss the implications of the articles on the process outlined in this chapter. Discuss the implications of the process on the articles.

9. Read an article or a chapter in a text on the following:
   (a) Quality management
   (b) Production and operations management
   (c) Engineering management
   (d) Energy management
   (e) Financial management
   (f) Management in general
   How do these forms or subsets of general management differ from productivity management? In what ways are they similar?

10. Prepare a 30-minute briefing, for a management audience, on productivity management. Include handout material, audiovisual support materials, and outline of agenda.

BIBLIOGRAPHY

Part I

Abernathy, W. J., K. B. Clark, and A. M. Kantrow. *Industrial Rennaissance: Producing a Competitive Future for America.* New York: Basic Books, Inc., 1983.

Anderson, C. A., and C. H. Kimzey. "National Productivity: The Responsibility of Engineering Educators." *Engineering Education*, November 1978.

Anthanosopoulos, C. N. *Corporate Productivity Atlas: Gross Corporate Product Measures 1982–1978.* Lincoln, Mass.: Delphi Research Center, 1983.

Behind the Productivity Headlines. Special collection of papers. *Harvard Business Review*, 1981.

Block, K. "International Business Strategies for the Eighties." Beta Gamma Sigma, St. Louis, Mo., 1982.

Bolino, A. C. *Productivity: Vocational Education's Role*. National Center for Research in Vocational Education, Columbus, Ohio, January 1981.

"Boosting Productivity: Crucial Task for the 1980's." *Civil Engineering*, Special Issue, October 1980.

Bueche, A. M. "Taking a Look at Fundamentals." *American Society for Metals*, April 1980.

Cowing, T. G., and R. E. Stevenson, eds. *Productivity Measurement in Regulated Industries*. New York: Academic Press, 1981.

Directory of U.S. Productivity and Innovation Centers. U.S. Chamber of Commerce, 1980.

Dogramaci, A., and N. R. Adam, eds. *Aggregate and Industry-Level Productivity Analysis* (Studies in Productivity Analysis, Volume II). Boston: Martimus Nijhoff, 1981.

"The End of the Industrial Society." *Business Week*, 50th Anniversary Issue, September 3, 1979.

"Engineering Management: The Key to Recharging U.S. Productivity." *Professional Engineer*, June 1982.

Entrepreneurship: The Japanese Experience. [An interesting monthly publication that began in 1982.] PHP Institute, Inc., Minami, Kyoto 601, Japan.

Filer, R. J. "Foreign Productivity Centers—What Can We Import from Them?" *Management Review*, January 1975.

"The Future of Productivity." The National Center for Productivity and Quality of Working Life, Washington, D.C., Winter 1977.

Grayson, C. J. "Productivity Slowdown—What Does It Mean to Management?" *Ideas*, Herman Miller, Inc., 1979.

Grossman, E. S., and G. E. Sadler. *Comparative Productivity Dynamics*. American Productivity Center, Houston, Texas, November 1982.

Hatvany, N., and V. Pucik. "Japanese Management Practices and Productivity." *Organizational Dynamics*, Spring 1981.

Hayes, R. H. and S. C. Wheelwright. *Restoring our Competitive Edge: Competing Through Manufacturing*. New York: John Wiley and Sons, 1984.

Henrici, S. B. "How Deadly is the Productivity Disease?" *Harvard Business Review*, November–December 1981.

Judson, A. S. "The Awkward Truth about Productivity." *Harvard Business Review*, September–October 1982.

Katzell, R. A., et al. *Work, Productivity, and Job Satisfaction: An Evaluation of Policy-Related Research*. The Psychological Corporation, 1975.

Kendrick, J. W. *Improving Company Productivity: Handbook with Case Studies*. Baltimore: The Johns Hopkins University Press, 1984.

Kendrick, J. W. *Understanding Productivity: An Introduction to the Dynamics of Productivity Change*. Baltimore: The Johns Hopkins University Press, 1977.

Lane, D. G. "Improving U.S. Productivity: Part 1—Organizing for Increasing Productivity." *Mechanical Engineering*, September 1978.

Leonard, L. "Productivity for a Price." *Production Engineering*, August 1979.

Marsland, S., and M. Beer. "The Evolution of Japanese Management: Lessons for U.S. Managers." *Organizational Dynamics*, Winter 1983.

McQuay Group. "Measuring Productivity." *Ideas*, Herman Miller, Inc., 1979.

"Meeting Japan's Challenge: The Need for Leadership." Study conducted for Motorola, Inc., 1982.

Mize, J. H. "Perspectives on Productivity." Republic of China Workshop on Productivity, August 1981.

Morris, W. T. *Work and Your Future*. Reston, VA: Reston Publishing Company, 1975.

Newland, K. "Productivity: The New Economic Context." Worldwatch Paper 49, June 1982.

Perspectives on Productivity: A Global View. A Sentry Study, Louis Harris & Associates, Inc., 1982.

Peters, T. J. "Putting Excellence into Management." *Business Week*, July 21, 1980.

"Productivity." *Executive*, Graduate School of Business and Public Administration, Cornell University, Fall 1980.

"The Productivity Crisis—A Special Section." *World*, Peat, Marwick, Mitchell & Company, New York, Winter 1981.

"Productivity Crunch." Survey conducted by the American Institute of Industrial Engineers.

Productivity Digest: January 1983. Guide to Productivity Resources. American Productivity Center, Houston, Texas, 1983.

"Productivity in the Changing World of the 1980's." The Final Report of the National Center for Productivity and Quality of Working Life, Washington, D.C., 1978.

"Productivity: Our Biggest Undeveloped Resource." *Business Week,* Special, September 9, 1972.

Productivity, People, and Public Policy. Economic Policy Division, U.S. Chamber of Commerce, June 1981.

"Productivity: The Difficulty Even Defining the Problem." *Business Week,* June 9, 1980.

"Productivity: The Link to Economic and Social Progress—A Swedish-American Exchange of Views." C. Prendergast, ed., Work in America Institute, 1976.

"Productivity: The Pieces That We Must Put Together." *Modern Materials Handling,* January 20, 1982.

"Pushing Productivity." *Dun's Review,* July 1978.

"Reaching a Higher Standard of Living." *The Exchange,* Office of Economic Research, January 1979.

Reed, W. J. and R. House. "Idea Management Makes Problem Solving More Productive." *Machine and Tool Blue Book,* February 1980.

Rowan, M. "Productivity Growth—What Can We Learn from Japan?" *Modern Materials Handling,* May 6, 1980.

Schoenblum, E. "Overview and Perspective on Productivity." *Management Focus,* November–December 1978.

Schroeder, P. "The Politics of Productivity." *Public Personnel Management Journal.*

Scott, B. R. "Can Industry Survive the Welfare State?" *Harvard Business Review,* September–October 1982.

Shaffer, J. D., and V. L. Sorenson. "Policies on Productivity—Some Lessons from Agriculture?" SAE Technical Paper Series, SAE, Warrendale, PA., 1983.

"Solving the Productivity Puzzle." *American Machinist,* June 1981.

Steele, C. G. "Measuring Productivity." *DH&S Reports,* No. 1, Deloitte, Haskins and Sells, 1982.

Sumanth, D. J. *Productivity Engineering and Management.* New York: McGraw Hill, 1984.

Sumanth, D. J., and N. G. Einspruch. "Productivity Awareness in the U.S.: A Survey of Some Major Corporations." *Industrial Engineering,* October 1980.

Suttermeister, R. A. *People and Productivity,* 3rd ed. New York: McGraw-Hill, 1976.

Takeuchi, H. "Productivity: Learning from the Japanese." *California Management Review,* Vol. XXIII, No. 4, Summer 1981.

Tribus, M., and J. H. Hollomon. "Policies and Strategies for Productivity Improvement." AAES National Meeting, May 1982.

Utt, R. D. "Statement of the Chamber of Commerce of the U.S. on Productivity." Subcommittee on Employment and Productivity of the Senate Committee on Labor and Human Resources, April 2, 1982.

White, R. B. "Productivity and Effectiveness in the Service Sector." *Ideas,* Herman Miller, Inc., 1979.

Williams, R. "Reindustrialization: Past and Present." *Technology Review,* MIT, Cambridge, Mass., November–December 1982.

Workers' Attitudes Toward Productivity: A New Survey, U.S. Chamber of Commerce, 1980.

Zenger, J. "Increasing Productivity How Behavioral Scientists Can Help." *Ideas,* Herman Miller, Inc., 1979.

Zussman, Y. M. "Learning from the Japanese: Management in a Resource-Scarce World." *Organizational Dynamics,* Winter 1983.

Part II

Anderson, C. A. and C. H. Kimzey. "National Productivity: The Responsibility of Engineering Educators." *Engineering Education,* November 1978.

Anderson, D. R., D. J. Sweeney, and T. A. Williams. *Introduction to Statistics: An Applications Approach.* St. Paul, Minn.: West, 1981.

Bailey, R. W. *Human Performance Engineering: A Guide for System Designers*. Englewood Cliffs, N.J.: Prentice-Hall, 1982.

Bain, D. *The Productivity Prescription: The Manager's Guide to Improving Productivity and Profits*. New York: McGraw-Hill, 1982.

Barnes, R. M. *Motion and Time Study Design and Measurement of Work*, 7th ed. New York: John Wiley, 1980.

Beard, D. W., and G. G. Dess. "Corporate Level Strategy, Business Level Strategy, and Firm Performance." *Academy of Management Journal*, Vol. 24, No. 4, 1981.

"Behind the Productivity Headlines." *Harvard Business Review* (Special Monograph), 1981.

Dewar, D. L. *The Quality Circle Guide to Participation Management*. Englewood Cliffs, N.J.: Prentice-Hall, 1980.

Dewar, D. L. *Quality Circle Leader Manual*. Midwest City, Okla.: IAQC.

Dewar, D. L., and J. F. Beardsley. *Quality Circle Text*. Midwest City, Okla.: IAQC.

Hanlon, M. D., and J. C. Williams. "In Jamestown: Labor–Management Committee at Work." *QWL Review*, Vol. 1, No. 3, 1982.

Heaton, H. *Productivity in Service Organizations*. New York: McGraw-Hill, 1977.

Henrici, S. B. "How Deadly is the Productivity Disease?" *Harvard Business Review*, November–December, 1981.

Judson, A. S. "The Awkward Truth about Productivity." Harvard Business Review, September–October, 1982.

Kepner, C. H., and B. B. Tregoe. *The Rational Manager*. New York: McGraw-Hill, 1965.

Kepner-Tregoe, Inc. *The Results Planning Manual*, 1968.

Lehrer, R. M., ed. *White Collar Productivity*, New York: McGraw-Hill, 1983.

MacMillan, J. C., D. C. Hambrick, and D. L. Day. "The Product Portfolio and Profitability— A PIMS Based Analysis of Industrial-Product Business." *Academy of Management Journal*, Vol. 25, No. 4, 1982.

Mali, P. *Improving Total Productivity: MBO Strategies for Business, Government, and Not-for-Profit Organizations*. New York: John Wiley and Sons, 1978.

Moss, S. "A Systems Approach to Productivity." *National Productivity Review*, Vol. 1, No. 3, 1982.

Peters, T. J., and R. H. Waterman. *In Search of Excellence: Lessons from America's Best-Run Companies*. New York: Harper and Row, 1982.

Productivity Through Work Innovations. Work in America Institute. New York: Pergamon Press, 1982.

Wacker, G., and G. Nadler. "Seven Myths About Quality of Working Life." *California Management Review*, Vol. XXII, No. 2, 1980.

Working Smarter (eds. of *Fortune*). New York: Viking Press, 1982.

Zager, R., and M. Rosow, eds. *The Innovative Organization: Productivity Programs in Action*. New York: Pergamon Press, 1982.

Part III

Adam, E. E., J. C. Hershauer, and W. A. Ruch. *Productivity and Quality: Measurement as a Basis for Improvement*. N.J.: Prentice-Hall, 1981.

Adam, N. R., and A. Dogramaci, eds. *Productivity Analysis at the Organizational Level*. Boston: Martimus Nijhoff, 1981.

Bailey, D., and T. Hubert. *Productivity Measurement*. Westmead, England: Gower Publication Company, 1980.

Bain, D. F. *The Productivity Prescription*. New York: McGraw-Hill, 1982.

Balk, W. L. "Technological Trends in Productivity Measurement." *Public Personnel Management*, March–April, 1975.

Bernolak, I. "Conventional Wisdom is Not Always Wise." Study Meeting on Productivity Measurement and Analysis, Asian Productivity Organization, February 1982.

Bernolak, I. "New Productivity Thrust from Effective Measurement." World Productivity Congress, Detroit 1981.

"A Better Way to Measure Productivity?" *Industry Week*, November 13, 1978.

Brayton, G. N. "Simplified Method of Measuring Productivity and Opportunities for Increasing It." *Industrial Engineering*, February 1983.

Burgess, R. W. "Integration of Productivity Studies with the Operating Accounting and Statistics of Industry." Paper presented at the Productivity Conference Panel, November 30, 1951.

Cameron, K. S., and D. A. Whetten. "Perceptions of Organizational Effectiveness over Organizational Life Cycles." *Administrative Science Quarterly*, December 1981.

Clegg, S. "Organization and Control." *Administrative Science Quarterly*, Vol. 26, 1981.

Cocks, D. L. "Company Total Factor Productivity: Refinements, Production Functions, and Certain Effects of Regulation." *Business Economics*, May 1981.

Cowing, T. G., and R. E. Stevenson, eds. *Productivity Measurement in Regulated Industries*. New York: Academic Press, 1981.

Christopher, W. F. "How to Develop Productivity Measures that Can Improve Productivity Performance." *Commentary*, Vol. 3, No. 3.

Craig, C. E., and R. C. Harris. "Total Productivity Measurement at the Firm Level." *Sloan Management Review*, Vol. 14, No. 3, Spring 1973.

Davis, H. S. *Productivity Accounting*, Industrial Research Unit, The Wharton School, University of Pennsylvania, Philadelphia, 1978 (originally published in 1955).

Diebold, J. "The Significance of Productivity Data." *Harvard Business Review*, July–August 1952.

Dogramaci, A., ed. *Developments in Econometric Analyses of Productivity: Measurement and Modeling Issues*. Boston: Martimus Nijhoff, 1983.

Dogramaci, A. *Productivity Analysis: A Range of Perspectives*. Boston: Maritmus Nijhoff, 1981.

Dogramaci, A., and N. R. Adam, eds. *Aggregate and Industry Level Productivity Analyses*. Boston: Martimus Nijhoff, 1981.

Evans, W. D., and I. H. Siegel. "The Meaning of Productivity Indexes." *Journal of the American Statistical Association*, March 1942.

Felix, G. *Productivity Measurement with the Objectives Matrix*, Oregon Productivity Center Corvallis, Or., 1983.

Flamholtz, E. "Organizational Control Systems as a Managerial Tool." *California Management Review*, Vol. XXII, No. 2, 1979.

Ford, J. D., and D. A. Schellenberg. "Conceptual Issues of Linkage in the Assessment of Organizational Performance." *Academy of Management Review*, Vol. 7, No. 1, 1982.

"Gauging Growth: Productivity Debate Is Clouded by Problem of Measuring Its Lag." *The Wall Street Journal*, October 14, 1980.

Gold, B. "Practical Productivity Analysis for Management Accounts." *Management Accounting*, May 1980.

Gold, B. "Practical Productivity Analysis for Management: Part 1: Analytical Framework." Institute of Industrial Engineers, *Transactions*, December 1982.

Greenburg, L. *A Practical Guide to Productivity Measurement*. Bureau of National Affairs, Washington, D.C. 1973.

Hitt, M. A., and R. D. Ireland. "Industrial Firms' Grand Strategy and Functional Importance: Moderating Effects of Technology and Uncertainty." *Academy of Management Journal*, Vol. 25, No. 2, 1982.

Honeywell ADG Guide to Improved Productivity and Qualtiy of Work, Aerospace and Defense Group, Productivity/Quality Center, Edina, Minn., 1982.

"How to Measure Programmer Productivity." *Computerworld*, April 6, 1981.

Hoy, F., and D. Hellriegel. "The Kilmann and Herden Model of Organizational Effectiveness Criteria for Small Business Managers." *Academy of Management Journal*, Vol. 25, No. 2, 1982.

Huber, G. P. "Multi-Attribute Utility Models: A Review of Field and Field-like Studies." *Management Science*, Vol. 20, 1974.

Huber, G. P., V. Sahney, and D. Ford. "A Study of Subjective Evaluation Models." *Behavioral Science*, Vol. 14, 1969.

"Improving Productivity through Industry and Company Measurement." The National Center for Productivity and Quality of Working Life, Washington, D.C., October 1976.

Kendrick, J. W. *Understanding Productivity: An Introduction to the Dynamics of Productivity Change*. Baltimore: The Johns Hopkins Press, 1977.

Kendrick, J. W., and E. S. Grossman. *Productivity in the United States*. Baltimore: The Johns Hopkins Press, 1980.

Kendrick, J. W., and staff of the American Productivity Center. *Measuring and Promoting Company Productivity: Handbook with Case Studies*. Baltimore: The Johns Hopkins Press, 1983.

Langenberg, W. "An Experiment in Productivity Measurement." *N.A.C.A. Bulletin*, January 1952.

Lehrer, R., ed. *White-Collar Productivity*. New York: McGraw-Hill, 1982.

Lipsey, R. C., and P. O. Steiner. *Micro-Economics*, 5th ed. New York: Harper and Row, 1979.

MacCrimmon, K. R., and M. Toda. "The Experimental Determination of Indifference Curves." *The Review of Economic Studies*, Vol. 36, 1969.

Mammone, J. L. "Productivity Measurement: A Conceptual Overview." *Management Accounting*, June 1980.

Meaning and Measurement of Productivity. U.S. Department of Labor, Bureau of Labor Statistics, Bulletin 1714, SN. 1972-0-457-061, 1972.

Measurement and Interpretation of Productivity. National Research Council, National Academy of Sciences, Washington, D.C., 1979.

Miller, J. R. *Professional Decision Making*. New York: Praeger, 1970.

Olson, V. *White Collar Waste*. Englewood Cliffs, N.J.: Prentice-Hall, 1983.

Sadler, G. E., and E. S. Grossman. "Measuring Your Productivity." *Plastics Technology*, Vol. 28, No. 1, 1982.

Schainblatt, A. H. "How Companies Measure the Productivity of Engineers and Scientists." *Research Management*, Vol. XXV, No. 3, May 1982.

Schaifer, R. O. *Analysis of Decisions under Uncertainty*. New York: McGraw-Hill, 1969.

Siegel, I. H. *Company Productivity: Measurement for Improvement*. The W. E. Upjohn Institute, Kalamazoo, Mich., 1980.

Siegel, I. H. *Productivity Measurement: An Evolving Art*. Scarsdale, N.Y.: Work in America Institute, 1980.

Sink, D. S. "The Essentials of Productivity Management: Measurement, Evaluation, Control, and Improvement." Short Course Notebook, LINPRIM, Inc., Blacksburg, Va., 1980, 1983, 1984.

Sink, D. S. "Multi-Criteria Performance/Productivity Measurement Technique." *Productivity Management*, Oklahoma Productivity Center, Oklahoma State University, Stillwater, Okla., 1983.

Sink, D. S., S. J. DeVries, and T. Tuttle. "An In-depth Study and Review of Existing Productivity Measurement Techniques." *Proceedings*, World Productivity Congress, Oslo, Norway, May 1984.

Sink, D. S., and J. B. Keats. "Quality Control Applications in Multi-Factor Productivity Measurement Models." Institute of Industrial Engineering Annual Conference *Proceedings*, May 1983 (Louisville Conference).

Sink, D. S., and J. B. Keats. "Using Quality Costs in Productivity Management." 1983 ASQC Quality Congress *Transactions*, Boston, May 1983.

Sink, D. S. "Initiating Productivity Measurement Systems in MIS." 1978 ASQC Technical Conference *Transactions*, Chicago, 1978.

Sink, D. S. "Organizational System Performance: Is Productivity a Critical Component?" Institute of Industrial Engineers 1983 Annual Conference *Proceedings*, Norcross, Ga., 1983 (Louisville Conference).

Sink, D. S. "Productivity Action Teams: An Alternative Involvement Strategy to Quality Circles." Institute of Industrial Engineers 1982 Annual Conference *Proceedings*, Norcross, Ga., 1982 (New Orleans Conference).

Smith, I. G. *The Measurement of Productivity*. Essex, England: Gower Press, Ltd., 1973.

Steele, C. G. "Measuring Productivity: A New Challenge." *Deloitte, Haskins and Sells Reports*, Vol. 19, No.1, 1982.

Steers, R. M. "Problems in the Measurement of Organizational Effectiveness." *Administrative Science Quarterly*, Vol. 20, December 1975.

Steffy, W., N. Ahmad, and T. Reyes. *Productivity and Cost Control for the Small and Medium-Sized Firm*. Industrial Development Division Institute of Science and Technology, The University of Michigan, Ann Arbor, Mich., 1980.

Stewart, W. T. "Performance Measurement and Improvements in Common Carriers." Institute of Interdisciplinary Engineering Studies, Purdue University, West Lafayette, Ind.

Strom, J. S. "Bank Profitability: A Function of Productivity and Price Recovery." *American Banker*, February 29, 1980.

Sumanth, D. J. *Productivity Engineering and Management*. New York: McGraw-Hill, 1984.

Thor, C. "Banking Industry Productivity." American Productivity Center, Houston, Texas, April 1983.

Thor, C. "Productivity Measurement." Supplement to *Corporate Controller's Manual*, Warren, Gorham and Lamont, 1982.

Total Bank Productivity Measurement: A Conceptual Framework. Bank Administration Institute, Rolling Meadows, Ill., 1982.

Tuttle, T. C. *Productivity Measurement Methods: Classification, Critique, and Implications for the Air Force*. Air Force Systems Command, Brooks Air Force Base, Texas, 1981.

Tuttle, T. C., et al. "Measuring and Enhancing Organizational Productivity: An Annotated Bibliography." Air Force Human Resources Laboratory, Manpower and Personnel Division, Brooks Air Force Base, Texas, 1981.

Tuttle, T. C., R. E. Wilkinson, and M. D. Matthews. *Field Test of a Methodology for Generating Efficiency and Effectiveness (MGEEM)*. Maryland Center Publication Series MC-6, AF Contract No. F33615-79-C-0019, University of Maryland Center for Productivity and Quality of Work Life, 1982.

Wooton, P. M., and J. L. Tarter. "The Productivity Audit: A Key Tool for Executives." *MSU Business Topics*, Spring 1976.

"The Uses and Abuses of Measuring Productivity." *Monthly Review*, June 1980.

Part IV

Arai, J. "U.S. and Japanese Technology: a Comparison." *National Productivity Review*, Vol. 1, No. 3, 1982.

Barnard, C. I. *The Functions of the Executive*. Cambridge, Mass.: Harvard University Press, 1938.

Blake, R. R., and J. S. Mouton. *Productivity: The Human Side*. New York: AMACOM, 1981.

Blanchard, K., and R. Lorber. *Putting the One Minute Manager to Work*. New York: William Morrow, 1984.

Blanchard, K., and P. Hersey. *Management of Organizational Behavior: Utilizing Human Resources*, 4th ed. Englewood Cliffs, N.J.: Prentice-Hall, 1982.

Bradford, L. P., ed. *Group Development*, La Jolla, Calif.: University Associates, 1978.

Brousseau, K. R. "Toward a Dynamic Model of Job-Person Relationships: Findings, Research Questions, and Implications for Work-System Design." *Academy of Management Review*, Vol. 8, No. 1, 1983.

Bryan, J. F., and E. A. Locke. "Goal Setting as a Means of Increasing Motivation." *Journal of Applied Psychology*, Vol. 43, 1959.

Buehler, V. M., and Y. K. Shetty. *Productivity Improvement: Cast Studies of Proven Practice*. New York: AMACOM, 1981.

Bullock, R. J., and T. L. Ross. *The Scanlon Way to Improved Productivity*. New York: John Wiley, 1978.

Crosby, P. B. *Quality Without Tears: The Art of Hassle-Free Management*. New York: McGraw-Hill, 1984.

Delbacq, A. L., A. H. Van de Ven, and D. H. Gustafson. *Group Techniques for Program Planning: A Guide to Nominal Group and Delphi Processes*. Glenview, Ill.: Scott, Foresman, 1975.

Demmings, W. E. "Improvement of Quality and Productivity through Action by Management." *National Productivity Review*, Vol. 1, No. 1, 1981.

Dessler, G. *Improving Productivity at Work*. Reston, Va.: Reston Publishing Company, 1983.

Fein, M. "An Alternative to Traditional Managing" (unpublished). Hillsdale, N.J., 1977.

Fein, M. "Designing and Operating an Improshare Plan" (unpublished). Hillsdale, N.J., 1976.

Fein, M. *Improshare: An Alternative to Traditional Managing.* Norcross, Ga.: Industrial Engineering and Management Press, 1981.

Fein, M. "Improving Productivity by Improved Productivity Sharing." *The Conference Board Record,* Vol. 13, 1976.

Fein, M. "Motivation for Work," in *Handbook of Work Organization and Society* (ed. R. Dubin). Chicago: Rand McNally, 1976.

Fein, M. *Rational Approaches to Raising Productivity.* Norcross, Ga.: Industrial Engineering and Management Press, 1974.

Gainsharing: A Collection of Papers. Norcross, Ga.: Industrial Engineering and Management Press, 1983.

Galbraith, J., and L. L. Cummins. "An Empirical Investigation of the Motivational Determinants of Task Performance: Interactive Efforts between Instrumentality-Valence and Motivation Ability." *Organizational Behavior and Human Performance,* Vol. 2, 1967.

Geneen, H. w/ Alvin Moscow, *Managing.* New York: Doubleday and Co. 1984.

Georgopoulos, B. S., G. M. Mahoney, and M. N. W. Jones. "A Path-Goal Approach to Productivity." *Journal of Applied Psychology,* Vol. 41, 1957.

Graham-Moore, B. E., T. L. Ross. *Productivity Gainsharing.* Englewood Cliffs, NJ: 1983.

Groover, M. P. "Meeting the Challenge of CAD/CAM." *National Productivity Review,* Vol. 2, No. 1, 1982.

Hellriegel, D., and J. W. Slocum, "Managerial Problem—Solving Styles." *Business Horizons,* December 1975.

Henderson, R. J. *Compensation Management: Rewarding Performance,* 3rd ed. Reston, Va.: Reston Publishing Company, 1983.

Herzberg, F. *Work and the Nature of Man.* Cleveland: World, 1966.

Herzberg, F., B. Mausner, and B. Snyderman. *The Motivation to Work,* 2nd ed. New York: John Wiley, 1959.

House, R. J., and E. Kay. "Improving Employee Productivity through Work Planning," in *The Personnel Job in a Changing World* (J. Blood ed.). American Management Association, New York, 1964.

Jehring, J. J. "A Contrast between Two Approaches to Total Systems Incentives." *California Management Review,* Vol. 10, Winter 1967.

Kahn, R. L. "Productivity and Job Satisfaction." *Personnel Psychology,* Vol. 13, 1960.

Kerr, S. "On the Folly of Rewarding A, While Hoping for B." *Academy of Management Journal,* 1975.

Kerr, S. "Overcoming the Dysfunctions of MBO." Management by Objectives, Vol. 5, No. 1, 1971.

Kimzey, C. H. "The Integrated Manufacturing Plant: A Practical Systems Approach." *National Productivity Review,* Vol. 2, No. 2, 1983.

Lawler, E. E. *Pay and Organizational Effectiveness: A Psychological View.* New York: McGraw-Hill, 1971.

Lawler, E. E., and G. E. Ledford. "Productivity and Quality of Work Life." *National Productivity Review,* Vol. 1, No. 1, 1981.

Lehrer, R. N. *Participative Productivity and Quality of Work Life.* Englewood Cliffs, N.J.: Prentice-Hall, 1982.

Lincoln, J. F. *Incentive Management.* The Lincoln Electric Company, Cleveland, Ohio, 1951.

March, J. G., and H. A. Simon. *Organizations.* New York: John Wiley, 1958.

Marsh, W. A. "Management Theories: Comparing X, Y, and Z." *Quality Progress,* December 1982.

Maslow, A. H. *Motivation and Personality,* 2nd ed. New York: Harper and Row, 1970.

McAfee, R. B., and W. Poffenberger. *Productivity Strategies: Enhancing Employee Job Performance.* Englewood Cliffs, N.J.: Prentice-Hall, 1982.

McCaulley, M. "Psychological Types in Engineering: Implications for Teaching." *Engineering Education,* April 1976.

McGregor, D. "An Uneasy Look at Performance Appraisal." *Harvard Business Review*, Vol. 35, No. 3, 1957.

Mikami, T. *Management and Productivity Improvement in Japan*, JMA Consultants, Tokyo, Japan, 1982.

Mitchell, T. R. "Motivation: New Directions for Theory, Research and Practice." *Academy of Management Review*, Vol. 7, No. 1, 1982.

Mohr, W. L., and H. Mohr. *Quality Circles*. Reading, Mass.: Addison-Wesley, 1983.

Monden, Y. *Toyota Production System*. Industrial Engineering and Management Press, Norcross, Ga., 1983.

Norsbitt, J. *Megatrends*. New York: Warner Books, 1982.

Odiorne, G. *Management by Objectives*. New York: Pittman, 1965.

The Ohio State University Productivity Research Group. "Productivity Measurement Systems for Administrative Computing and Information Services: An Executive Summary." NSF Grant No. APR 75-20561, Ohio State University Research Foundation, Columbus, Ohio, 1977.

The Ohio State University Productivity Research Group. "Productivity Measurement Systems for Administrative Computing and Information Services: A Manual of Structured Participation Methods." NSF Grant No. APR 75-20561, Ohio State University Research Foundation, Columbus, Ohio, 1977.

Otis, I. "Total Involvement for Increased Productivity." *Industrial Management*, November–December 1981.

Passmore, W. A. "Overcoming the Roadblocks in Work-Restructuring Efforts." *Organizational Dynamics*, Spring 1982.

Passmore, W., and F. Friedlander. "An Action-Research Program for Increasing Employee Involvement in Problem Solving." *Administrative Science Quarterly*, Vol. 27, No. 3, 1982.

Patton, J. A. *Patton's Complete Guide to Productivity Improvement*. New York: AMACOM, 1982.

Pava, C. H. P. "Designing Managerial and Professional Work for High Performance: A Sociotechnical Approach." *National Productivity Review*, Vol. 2, No. 2, 1983.

Pierce, J. L., R. B. Dunham, and R. S. Blackburn. "Social Structure, Job Design, and Growth Need Strength: A Test of a Congruency Model." *Academy of Management Journal*, Vol. 22, 1979.

Rolland, I., and R. Janson. "Total Involvement as a Productivity Strategy," *California Management Review*, Vol. XXIV, No. 3, 1981.

Sandman, W. E., and J. P. Hayes. *How To Win Productivity in Manufacturing*. Dresher, Pa.: Yellow Book of Pennsylvania, 1980.

Scanlon, J. N. "Adamson and His Profit-Sharing Plan." *Production*. New York: American Management Association, 1947.

Sink, D. S. "The ABC's of Theories X, Y, and Z." 1982 Fall Industrial Engineering Conference, *Proceedings*, Norcross, Ga., 1982.

Sink, D. S. *The Essentials of Productivity Management: Planning, Measurement, Evaluation, Control, and Improvement*. Short Course Notebook, LINPRIM, Inc., Blacksburg, Va., 1984.

Skinner, B. F. *Science and Human Behavior*. New York: Macmillan, 1953.

Skinner, B. F. *The Behavior of Organizations*. New York: Appleton Century-Crofts, 1938.

Stein, C. I. "Objective Management Systems: Two to Five Years After Implementation." *Personnel Journal*, October 1975.

Sutermeister, R. A. *People and Productivity*. New York: McGraw-Hill, 1969.

Szylagyi, A. D. *Management and Performance*. Santa Monica, CA: Goodyear Publishing Co., 1981.

Thor, C. Baking Goods Industry Productivity Study." Houston, TX: American Productivity Center, 1983.

Tosi, H. L., and S. J. Carroll. "Management Reaction to Management by Objectives." *Academy of Management Journal*, Vol. 11, 1968.

Townsend, R. *Further up the Organization: How to Stop Management from Stifling People and Strangling Productivity*. New York: Alfred A. Knopf, 1984.

Von Bertalanffy, L. "General System Theory—A Critical Review." *General Systems*, Vol. VII, 1962.

Vroom, V. H. *Work and Motivation*. New York: John Wiley, 1964.

Wallace, M. H., and C. H. Fay. *Compensation Theory and Practice*, Belmont, Calif.: Kent, 1983.

Part V

APC. *Case Study Series*. Houston, Tex.: American Productivity Center, 1981, 1982, 1983.

Buehler, V. M., and Y. K. Shetty, eds. *Productivity Improvement: Case Studies of Proven Practice*. New York: AMACOM, 1981.

Goodman, D. S. "Why Productivity Programs Fail: Reasons and Solutions." *National Productivity Review*, Vol. 1, No. 4, Autumn 1982.

Judson, A. S., and F. J. Cirillo. "Building on Success: Managing Productivity Improvement by Strategy." *National Productivity Review*, Vol. 2, No. 4, Autumn 1983.

Mooney, M. "Organizing for Productivity Management." *National Productivity Review*, Vol. 1, No. 2, 1982.

Morris, W. T. *Implementation Strategies for Industrial Engineers*. Columbus, Ohio: Grid, 1979.

Murray, T. J. "The Rise of the Productivity Manager." *Dun's Review*, January 1981.

Quinlan, J., ed. "How to Prevent Your Productivity Program from Fizzling." *Material Handling Engineering*, May 1980.

Ruch, W. A., and W. B. Werther, "Productivity Strategies at TRW." *National Productivity Review*, Vol. 2, No. 2, Spring 1983.

Sink, D. S. "Strategic Planning: A Crucial Step Toward a Successful Productivity Management Program." *Industrial Engineering*, January 1985.

PRODUCTIVITY MANAGEMENT

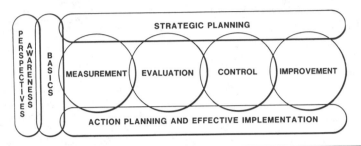

| Winter 1983 | Number Five | EDITOR: Camille Frye |

TABLE OF CONTENTS

• VPC PERFORMS!

VPI and State University
Blacksburg, VA 24061

FEATURE ARTICLE

"CAPITAL PRODUCTIVITY AND THE NEW TAX LAW"
by Dr. Kenneth E. Case, Ph.D. and Camille Frye, OPC Associate

Productivity is an issue that concerns every manager today. It has become the "new hope" for today's struggling economy. Politicians, consultants, educators and others view increased productivity as a solution to our economy's problem.

Increased productivity has long been associated with a strong economy. Following World War I, the U.S. experienced a high level of per capita production which can be linked directly to improved productivity. This long term growth of productivity from 1947-1967 has been followed by a drastic reduction in the growth rate from 1967 to the present.[1] The result is that the U.S. still leads all other nations in absolute productivity levels, but is near the bottom of the list in the rate of productivity growth.[2]

The question being raised by many in the 70's and 80's is "how to stimulate productivity growth and thus the economy."[3] There are many potential answers to the question. This article will focus on the role that capital investment might play.

The relationship between capital investment and productivity growth is fairly well established. Managers have long recognized this important relationship. A recent survey conducted by Gray-Judson, a Boston-based consulting firm, asked 236 top managers to indicate the major causes for their success/disappointment in improving productivity. Seventy-two percent indicated capital investment in plant, equipment and process as the primary reason for their increased productivity rate.[4] Other surveys indicate similar results.

The causal relationship between productivity growth and capital investment can be debated, but

485

historically, when capital investment in new technology, etc., increased, so have productivity rates. Thus, a reasonable hypothesis is that if capital investment is somehow increased, theoretically, productivity growth should occur, and thus stimulate the economy.

A potential solution to the problem is therefore clear: find ways to spark increases in capital investment. However, in today's struggling economy, it is sometimes difficult to find enough capital to "break even", let alone to invest in the future.

One answer to the dilemma is to make more capital available. Availability of capital depends upon several factors, including: 1) level of earnings and 2) amount of earnings retained for investment as opposed to being paid out in taxes, dividends, etc.

The new tax law provides some help in increasing investment capital. President Reagan's Economic Recovery Act of 1981 is designed to stimulate investment through "supply-side" economic policies. He hopes to achieve this in part by freeing more capital for investment through revisions in depreciations strategies and investment tax credit rules.

This article will address the changes in the tax laws which affect capital investment. Specifically, the new Accelerated Cost Recovery System (ACRS) and the new Investment Tax Credit rules will be examined. For further information, consult appropriate tax publications such as the Prentice-Hall Federal Tax Reports, Commerce Clearinghouse publications, etc., available at the public library.

ACRS

Under the ACRS, there are several significant changes from traditional depreciation methods. It is now common to speak of depreciation allowance as "recovery". Capital is recovered over the "recovery period" which is considerably shorter in many cases than the useful life. Also, it is no longer necessary to estimate a salvage value, since it is not considered when calculating recovery.

The Economic Recovery Act of 1981 classifies all recovery property according to recovery class as follows:

3 Year	Personal property with a short useful life such as automobiles, light trucks, and certain special tools for manufacturing.
5 Year	A wide range of personal property such as equipment, tooling, large trucks, ships, office equipment, and fixtures.
10 Year	Certain real property such as theme and amusement park structures and certain public utility property. Also, manufactured and mobile homes.
15 Year Real	Included as all real property such as buildings, other than those designated as 10 year property.
15 Year Public Utility	Long lived public utility property.

Most personal and real assets other than public utility property are either 5 year, or 15 year real property, respectively.

If a property was placed in service before 1981, the taxpayer must continue depreciating the asset using the same method as elected at that time. However, for property placed into service after 1980, the taxpayer must use the new Accelerated Cost Recovery System. (Minor exceptions to this rule exist, but we will not address these. For further information consult IRS publication 334.)

Unlike past methods of depreciation, ACRS does not consider expected salvage value when computing the allowable deductions. Thus, the entire investment is recovered. In essence, there is no longer a need to estimate salvage value. Likewise, since fixed recovery periods have replaced useful lives, there is no need to estimate and "haggle over" useful life.

Investment Tax Credit (IC)

The new tax law continues to allow the 10% IC for investment in personal property (i.e., equipment, machinery, computers), but now allows the full credit for assets with an ACRS recovery class greater than 3 years. In addition, it allows a 6% IC to be taken on assets with a recovery class of 3 years. If the IC is taken, the cost basis is reduced by half the IC for recovery purposes. The IC is of particular interest because it results in a full tax credit as opposed to merely a tax deduction.

The IRS has determined the depreciation rate allowed in class. These percentages are approximately equivalent to 150% declining balance switching to straight line depreciation. Table I lists ACRS recovery percentages for property placed in service after 1980.

TABLE I: ACRS Allowable Capital Recovery Percentages for Property Placed in Service on 1/1/1981 or After

Year	3-Year	5-Year	10-Year	15-Year Public Utility	15-Year Real Placed in Service 1st month of taxable year
1	25	15	8	5	12
2	38	22	14	10	10
3	37	21	12	9	9
4		21	10	8	8
5		21	10	7	7
6			10	7	6
7			9	6	6
8			9	6	6
9			9	6	6
10			9	6	5
11-15				6	5

Commerce Clearinghouse, Inc., 1983 U.S. Master Tax Guide

The depreciation allowance or recovery per year is obtained as follows:

$$D_j(k) = P_j(k) \times B$$

where $D_j(k)$ = Depreciation in year j for recovery property class k

$P_j(k)$ = ACRS percentage for year j in recovery property class k

B = Unadjusted basis of the recovery property (investment

TABLE II: Significant Changes in Tax Laws

Old	New
A. **Major Method** Straight Line Sum of the Year's Digits Declining Balance (200% new and 150% used)	A. **Major Method** *ACRS (150% Declining Balance Switching to Straight Line)
B. **Life** Useful Life as estimated	B. **Recovery Period** 3, 5, 10 or 15 years according to classification of asset
C. **IC** Full 10% for eligible assets with useful life greater than 7 years	C. **IC** Full 10% for assets with a recovery class greater than 3 years; also allows 6% for assets with a recovery class of 3 years
	* Tables are set up to allow only ½ the normal writeoff in the 1st year.

An example will illustrate the use of the ACRS side-by-side with the use of old depreciation methods. Consider a unit of automated machinery having a useful life of 9 years and costing $50,000 installed. Salvage value is expected to be $5,000. Operating costs of this machine will be $5,000 per year, but it will generate a cost savings of $20,000 per year. The firm uses a after-tax minimum attractive rate of return equal to 15% and has a tax rate of 46%. Table III details the cash flow profile over nine years using the ACRS. Note that the investment qualifies as a 5 year recovery property. Table IV presents the cash flow profile using sum of the years digits depreciation. Table III reflects the $2500 reduction in basis due to the $5000 ITC, as well as the expected $5000 income due to salvage value. In addition, Table III uses the new corporate tax rate of 46% while Table IV uses the old rate of 48%.

TABLE III: Cash Flow Profile Using the ACRS

End of Year	Before Tax Cash Flow	ACRS Rate	Allowable Deduction	Taxable Income	Tax (.46)	Investment Credit	After Tax Cash Flow
0	-50,000					5,000	-45,000
1	15,000	.15	7,125	7,875	3,622.50		11,377.50
2	15,000	.22	10,450	4,550	2,093		12,907
3	15,000	.21	9,975	5,025	2,311.50		12,688.50
4	15,000	.21	9,975	5,025	2,311.50		12,688.50
5	15,000	.21	2,975	5,025	2,311.50		12,688.50
6	15,000			15,000	6,900		8,100
7	15,000			15,000	6,900		8,100
8	15,000			15,000	6,900		8,100
9	20,000			20,000	9,200		10,800

$$\text{Net Present Value} = -45,000 + 11,377.50(P/F_{15,1}) + 12,907(P/F_{15,2}) +$$
$$12,688.50(P/A_{15,3})(P/F_{15,2}) + 8,100(P/A_{15,3})(P/F_{15,5})$$
$$+ 10,800(P/F_{15,9}) =$$
$$\$8,822.87$$

TABLE IV: Cash Flow Profile Using Sum of the Years Digits Depreciation

End of Year	Before Tax Cash Flow	SOYD Depre- ciation	Taxable Income	Tax (.48)	Investment Tax Credit	After Tax Cash Flow
0	-50,000				5,000	-45,000
1	15,000	9,000	6,000	2,880		12,120
2	15,000	8,000	7,000	3,360		11,640
3	15,000	7,000	8,000	3,840		11,160
4	15,000	6,000	9,000	4,320		10,680
5	15,000	5,000	10,000	4,800		10,200
6	15,000	4,000	11,000	5,280		9,720
7	15,000	3,000	12,000	5,760		9,240
8	15,000	2,000	13,000	6,240		8,760
9	20,000	1,000	19,000	9,120		10,880

$$NPV = -45,000 + 12,120(P/F_{15,1}) + 11,640(P/F_{15,2}) +$$
$$11,160(P/F_{15,3}) + 10,680(P/F_{15,4}) + 10,200(P/F_{15,5}) +$$
$$9,720(P/F_{15,6}) + 9,240(P/F_{15,7}) + 8,760(P/F_{15,8}) +$$
$$10,880(P/F_{15,9}) =$$
$$\$6,488.62$$

It is apparent that the faster depreciation allowed with ACRS frees more capital for investment. By allowing the writeoff to occur more rapidly, the Net Present Value increased due to the time value of money and thus more capital is available.

We should remember that in order to accomplish the objectives which the new tax laws are based upon, business and industry must invest this newly found capital in projects which have potential to improve productivity. Projects such as these include improved production machinery, better materials-handling equipment and techniques, and improved plant layout.

Proper investment in new technologies can stimulate productivity. An increase in productivity can stimulate the economy. The Federal Government, through the AONO, is attempting to provide a mechanism for creating a new "pot" of investment capital. It is now up to the management of the country to capitalize on this opportunity. This will require aggressive efforts to identify productivity improvement strategies. It will require improved productivity management programs. It will require effective implementation of a strategy for productivity management, perhaps like the "Grand Strategy" discussed in this newsletter in Issue #2, 1982.

"It will require attention to management basics followed by the essence of management... hard work".

ENDNOTES

1. Meanley, Carolyn, Productivity Perspectives, American Productivity Center (Houston), 1979. p.12.

2. Ibid.

3. Productivity and the Economy: A Chartbook, Department of Labor. Bulletin N. 2084, Washington, D.C.: Government Printing Office, 1981

4. Judson, Arnold S., "The Awkward Truth About Productivity." Harvard Business Review. V. 60,#5. pp. 94.

REFERENCES

Lohmann, Jack R. and E. W. Foster. "A Comparative Analysis of the Effect of ACRS on Replacement Economy Decisions." Engineering Economist. V. 27. #4. pp. 247-260.

Canada, John R. "How the New Tax Laws Increase Attractiveness of Investment Projects," Industrial Engineering. V. 14. #5.pp. 77-81

Riggs, James L. Tax Supplement to Accompany Engineering Economics. Second Edition. McGraw-Hill (New York), 1982.

MEASUREMENT

D. Scott Sink, Ph.D., P.E.

"In turbulent times, the first task of management is to make sure of the institution's capacity for survival, to make sure of its structural strength and soundness, of its capacity to survive a blow, to adapt to sudden change, and to avail itself of new opportunities." (Drucker, Managing In Turbulent Times, 1978).

The basic management processes of planning, organizing, leading, controlling, and adapting have been getting more complex. Strategies that used to succeed seem to be failing in this dynamic political, economic, and social environment. Employee values and attitudes, technology, cost of capital, markets, external regulation and interference, and many other business factors all seem to be so much more unpredictable and unmanageable. The bottom line is that in "turbulent times" a premium should be placed on effective and efficient management control systems.

As a manager, you know your job is to control and manage resources. You likely have some sort of "control system" for employee behaviors, materials, finished goods, quality, production flow, data and information, decisions, etc. The key question you better be asking yourself right now is, when was the last time you evaluated your control system. If you are like most managers, you developed the system incrementally or else you inherited it from someone else.

In a recent article in Industrial Engineering, "Establishing a Productivity Management Function," October 1982, pp. 42-51, I discuss six key measures that can be used to evaluate the performance of an organizational system. By organizational system I mean a firm, a division, a work group, a plant, etc. They reflect good old management basics. Whether you are a first line supervisor with a group of 10 employees, a general manager overseeing a plant of 200 to 500 employees, or the president managing the whole operation, what you do with these key measures of performance is critical to your ability to succeed at controlling and improving the performance of the specific system you are managing. If your goal is to grow, to develop, to compete favorably, to be the best, then you must systematically develop performance measurement systems. These are the basics, there is no easy way out. Anything less will likely result in mediocrity.

Productivity strictly defined, is a ratio relationship of quantities of output from an organizational system (i.e., a plant, a firm, a work group, a function, etc.) to quantities of input from that same organizational system for some period of time. So, in the numerator we have outputs; in the denominator we have inputs. It is a simple concept that becomes difficult to operationalize in practice. Measuring outputs quantitatively, matching outputs to inputs, price weighting and indexing outputs and inputs, and interpreting results are examples of key difficulties firms have in actually measuring productivity. Oddly enough, however, the biggest roadblock to productivity measurement is in getting management to agree on its very definition even at an abstract and conceptual level. Most managers do not have a solid understanding of the differences between effectiveness, efficiency, quality, productivity, quality of work life, profitability and innovation.

The basic assumption driving most of the interest in productivity and in particular productivity measurement is that it is a missing element in many American organization control systems. Further, it is hypothesized that pragmatic approaches can be taken to proactively address improving performance through more effective motivation. Participants are provided with an opportunity to begin to develop plans and skills for improved motivation programs and behaviors on the job.

The challenge then to productivity measurement lies in developing meaningful productivity ratios that identify, for management, what to change, when to change, where to change, and how much to change. How to change and who to change are, of course, still the responsibility of management. We should expect the productivity measurement system to tell us whether our change efforts have been "successful."

To date, there are two primary approaches that have been developed to measure productivity, as strictly defined. The first approach, a normative productivity measurement methodology utilized the Nominal Group Technique to develop consensus measures and/or ratios of productivity. The technique was designed and developed at The Ohio State University from 1975 to 1977. Since that time continued development and application has been widespread with major efforts being at Westinghouse, GE, The American Productivity Center, and at Oklahoma State University.

The second approach, a multi-factor, price weighted, indexed, and aggregated model utilizes accounting type data to drive formulations for productivity, price recovery, and profitability change ratios and indexes. This approach was pioneered by Hiram Davis in the middle 50's. Kendrick and Creamer refined Davis' work in the 60's and are primarily responsible for the productivity measurement model being promoted by the American Productivity Center. Bazil von Loggerenberg is another major contributor to this specific approach to productivity measurement.

Both approaches to productivity measurement adhere to the strict definition of the term productivity. The former is particularly useful for smaller units of analysis within the firm (i.e., work groups, departments, functions, etc.), while the latter approach is particularly appropriate for firm level diagnosis of productivity trends. Both of these approaches are being experimented with, developed, taught, and written about by the Oklahoma Productivity Center here at OSU.

"Without productivity objectives, a business does not have direction. Without productivity measurement, it does not have control."
(Peter Drucker)

For further information about either of these approaches or to other ways of measuring productivity, please feel free to contact Dr. Sink at (405) 624-6055. Future measurement columns will address these approaches in more detail.

IMPROVEMENT

"MOTIVATING FOR IMPROVED PERFORMANCE IN THE 80's"

Motivating employees for improved performance in the 80's will be a significant challenge for management. As might be expected, there is a strong relationship between employee performance (effectiveness, quality, efficiency, innovation) and productivity. The aggregate contribution of all individuals in an organization is likely the most significant factor in determining levels of organizational systems productivity. The task of motivating employees toward acceptable and perhaps even exemplary levels of performance continues to be one of the major challenges facing most supervisors, managers, and leaders in general.

It would appear that a critical mass of research, development, experience, and techniques has been generated on topics directly and indirectly related to motivation. At OSU, in the OPC, we have been assembling and analyzing this literature so as to begin to develop effective mechanisms for disseminating this knowledge to managers and supervisors in the Oklahoma region. A beginning was made over a year ago when Dr. Sink prepared and presented a paper to the Life Office Management Association (a life insurance "trade association") entitled "Motivating Employees for Improved Quality and Productivity." A revision of this paper was also presented at the Institute of Industrial Engineers 1982 Annual Conference in New Orleans last May. Both of these presentations were very well received. A total of 450 persons attended these presentations. In October, Dr. Sink also presented a one-hour briefing to 40 first line supervisors of the Western Electric Plant in Oklahoma City.

As a follow-up to these efforts, Dr. Sink has developed a two day short course on the topic. The short course focuses on presentation of the most advanced theories, techniques, and practices in relation to motivation. Although the course does focus on the most current concepts regarding motivation, it stresses attention to basics. The remainder of this column is devoted to a very brief presentation of a few of the concepts presented in that short course. By the way, the course will be offered in Stillwater on April 21-22, however, it can also be offered in-house should your company so desire. For more information contact either Dr. Sink at (405) 624-6055 or Dr. William Cooper at (405) 624-5146.

Motivation Basics

There are a few basic principles to motivating high levels of performance. They constitute the basics that should be mastered prior to attempting any "advanced" motivational strategies or techniques such as: incentive systems, quality circles, organizational behavior modification, or other OD type techniques. A few of these basics are:

Principle 1. Proper Selection and Placement - ensure you get the "right" people matched to the right jobs. Save yourself time, money, and lots of headaches.

Principle 2. Knowledge of what is expected - people actually behave in reasonable, predictable ways. Often, they do what they think they're expected to do and perform. Often managers aren't happy with results. Often it's because they did not clearly communicate what they wanted, when they wanted it, and the quality they expected.

Principle 3. Knowledge of Results (KOR) - probably the simplest and yet most violated or least adhered-to principle of behavior is the one involving feedback. Performance related knowledge of results can

often cause a 10-15% gain in performance. It's often there for the taking. Management simply needs to design a mechanism for tapping the potential.

Principle 4. Contingent reinforcement of performance, progress, achievement - employees will perform significantly better if they feel that they are succeeding. This does not constitute lowering standards. It does require more patience with personnel differences. Excellently managed companies demonstrate that there is no reason why we cannot reinforce the notion that we all are winners. Their systems reinforce degrees of winning rather than focusing on degrees of losing.

Motivating employees for increased performance in terms of effectiveness, efficiency, innovation, quality, etc., has become more complex and hence more difficult. Researchers and managers who are experimenting with new strategies and techniques for motivating are making great strides. Managers in the U.S. must move out of the era of thinking about motivation as only a hierarchy of needs.

"ROADBLOCKS TO PRODUCTIVITY"

by Jeff Swaim, Associate
Oklahoma Productivity Center

What are the major obstacles to improving productivity and which productivity-improvement activities have been most effective? These questions have been the focus of recent studies conducted by Dr. D. Scott Sink, the Institute of Industrial Engineers, and Arnold S. Judson, respectively. The results of these studies reveal answers that contradict much of what is generally believed about the causes of the decline of productivity in U.S. organizations.

Dr. Sink, Director of the Oklahoma Productivity Center, has conducted numerous Productivity Management Seminars, workshops, and shortcourses over the last few years. He has utilized the Nominal Group Technique to identify consensus from three hundred managers regarding roadblocks to productivity improvement in their respective organizations. Naturally, a multitude of obstacles have been suggested, but there are several which consistently receive a large percentage of the votes. Among these are:

Inadequate training of managers, management deficiencies
Poor planning (unclear goals and objectives)
Lack of communication, cooperation and coordination among groups
Lack of understanding and commitment by top management to productivity measurement and improvement
Management's resistance to change
Labor/Management relations

A recent survey conducted by the Institute of Industrial Engineering (IIE), their second annual productivity survey, received responses from 713 industrial engineers (Industrial Engineering, November 1982). The industrial engineers (IEs) were asked the extent (major, moderate, or minor) to which they felt the choices were obstacles to productivity improvement (Table I).

sense, "managing is loving," writes Dr. Hollomon— "the caring of one person for the people he serves...I believe the essence of the Japanese economic miracle is trust—the trust that exists between the worker and the manager."

Technology Review, May/June 1983.

TOOLS TO IMPROVE PRODUCTIVITY

Results of the June Productivity Survey show that PRODUCTIVITY readers are well ahead of the national average in productivity improvement and that they are successfully experimenting with a variety of advanced productivity improvement techniques.

Readers were asked: Are you using any of these tools to improve productivity? If you are, please rate:

1 = very effective   2 = effective   3 = neutral
4 = detrimental   5 = very detrimental

APPROACHES	RATING
Most Effective	
Extensive factory automation	1.73
Quite Effective	
Participative Management	1.84
Total Quality Control	1.86
Rapid Tool Setting	1.88
Statistical Quality Control	1.91
QWL Program	1.92
Flexible Workforce	1.93
CAD/CAM/CAE	1.94
Kanban (Just-in-time)	1.95
QC Circles	1.96
Somewhat Less Effective	
Bonus System for Workers	2.05
MRP I or II	2.10
Robotics	2.12
CEDAC	2.16
Least Effective	
Extensive Office Automation	2.20
Suggestion System	2.32
Profit Sharing	2.33

The same approaches are rated below according to the number of respondents using them:

APPROACHES	% OF FIRMS
Participative Management	68%
QC Circles	60
Suggestion System	59
Statistical Quality Control	44
Total Quality Control	36
CAD/CAM/CAE	32
Extensive Office Automation	31
Extensive Factory Automation	30
MRP I or II	28
Flexible Workforce	27
Bonus System for Workers	27
Robotics	27
Profit Sharing	25
QWL Program	19
Rapid Tool Setting	11
Kanban (Just-in-time)	7
CEDAC	3

Interestingly enough, popularity is not always closely correlated with perceived effectiveness. For example:

— QC Circles are used by 60% of respondents, but they are rated only moderately effective.
— Extensive factory automation is rated by far the most effective program, but it is used by less than one-third of respondents.
— Rapid tool setting which is rated quite effective, is used by only 11% of responding companies
— Suggestion Systems are very popular, but considered relatively ineffective.

Readers have very clear ideas on who is helping and who is hindering corporate productivity growth. When asked to rate the reactions of different groups in the organization to productivity improvement efforts, top management was rated most positive. Middle management, lower management, supervisors, workers and foremen were rated positive. Union leaders were clearly set off from the other groups as being judged least positive.

Readers also have a very keen perception as to where the keys to increasing productivity lie. They were almost unanimous in their perception that management attitudes are the single most important factor in national productivity improvement. The question asked: please rate these factors on their impact on American productivity growth:
1 = very important   2 = important   3 = unimportant

Very Important	Average Rating
Management Attitudes	1.03
Quite Important	
Training of the Workforce	1.28
Diligence of Workers	1.38
Research and Development	1.50
Investment in Plants	1.64
Structure of our Business Organizations	1.66
Marketing Strategies	1.76
Government Regulations	1.87
Unions	1.90
Lower Rate of Inflation	1.96
Tax Policy	1.98
Unimportant	
Japanese Productivity Techniques	2.12
Tariffs and Import Quotas	2.29
More Women in the Workforce	2.61

Survey Base: 149 respondents, a majority of which are manufacturing firms.

Excerpt from Productivity (July, 1983)

they interview and pass judgment on prospective employees.

For a more complete definition of how Skippy Little Rock emerged, please see Productivity V.3 #11, November, 1982. This fine newsletter is published by Productivity, Inc., Stanford, CT.

ENERGY UPDATE

Dr. Markus Fritz, energy consultant to the United Nations Educational, Scientific and Cultural Organization (UNESCO) is now a visiting associate professor for Oklahoma State University's Institute for Energy Analysis. Dr. Fritz is a native of Vienna, Austria and as taught at the University of Salzburg, Austria. Dr. Fritz's research efforts have been *inter alia* in the field of alternative energy sources, a subject which will become even more important to America in the future, when our traditional source of energy is depleted.

In a recent article entitled Societal Complexity of Energy Management, Dr. Fritz points out that the "energy problem" is commonly understood as a matter of technology and economics. However, this narrow view ignores social and political features which are both inherent within any proposed system and which potentially could disrupt any implementation.

Using the recent oil based individual transportation system as an example, Dr. Fritz demonstrates the far-reaching societal consequences when switching to a mass transportation system. Especially three criteria have to be considered:

    (1) Funding
    (2) Timing, and
    (3) Other economic and social implications

1. Funding

It has to be understood that no amortization of the investment is possible in case of an overall rail network covering the U.S. This asks for direct funding of the federal and state governments. However, a federal government facing an unemployment rate of more than 10% recently (without prospects of diminishing) and a budget deficit of (at least) $155 billion in 1983, will hardly be in the position or willing to fund the restructuring of the U.S. transportation system which would require billions of dollars.

2. Timing

Supply of the additional amount of energy required by most transportation systems in terms of electricity is not only a function of direct energy conversion, but also of organization and construction time - in advance. It is commonly understood that it takes at least one decade to get a planned coal mine or power plant finally operating. Considering the high amount of electricity an overall electrified transportation system would require, the time period with this amount available after the year 2000 would certainly be a conservative projection.

3. Other Economic and Social Implications

While the time factor may delay operation of an effective alternative transportation system, other economic and social implications, may even avoid its implementation in advance. Replacement of the recent transportation system means severe structural changes in industry and the economy.

That is, it would not only affect the truck manufacturing and transportation industries in general, (which would lead naturally to unemployment and bankruptcy in these sectors), but the consequences would be even more severe and evident on a national scale in the automobile industries. The possible consequences of the almost-bankruptcy of Chrysler Corporation may have been a clear 'warning.' The same is true with the oil companies: economic recession on their side (if in favor of an electrified transportation system) would raise the unemployment rate in the country, steeply.

A copy of the complete article may be obtained by contacting Dr. Fritz at (405) 624-5536 or the OPC (405) 624-6055, 301 Engineering North, Stillwater, OK 74078.

NEWS FROM OTHER CENTERS AROUND THE WORLD

The 1983 Regional Productivity and QWL Centers Consortium, sponsored by the Productivity Council of the Southwest will be held April 29, 1983 on the Queen Mary, Long Beach, California. In addition, the second annual conference on White Collar Productivity will be held April 27-28, also on the Queen Mary. Make plans now to attend one, or both of these events. For more information, and a registration form, contact The Productivity Council of the Southwest, 5151 State University Drive, STF 124, Los Angeles, CA 90032, (213) 224-2975.

The Second National Productivity Conference will be held in New York, at John Jay College of Criminal Justice on March 24-25, 1983. This conference is sponsored by the National Center for Public Productivity. For further information contact the National Center at 445 W. 59th Street, New York, NY 10019, (212) 489-5030.

The 4th International Productivity Congress will be held May 13-16, 1984 in Oslo, Norway. Preliminary application for the conference may be made by contacting the Norwegian Productivity Institute, P.O. Box 8401, Hammersborg, Oslo 1, Norway.

CALENDAR OF EVENTS

March 10, 1983 — Implementing CAD/CAM; sponsored by the O.S.U. CAD/CAM Consortium.

March 24-25, 1983 — Introduction to Computer Programming in BASIC; Dr. Philip Wolfe, Stillwater, Oklahoma.

April 21-22, 1983 — Improving the Effectiveness of Motivated Skills (short course); Dr. D. Scott Sink in Stillwater, Oklahoma.

April 25-27, 1983 — Managing Quality Costs; jointly sponsored by Oklahoma State and The American Society of Quality Control (Las Vegas).

April 28-29, 1983 — Microcomputer Applications for Engineers and Managers; Dr. Philip Wolfe, Dr. J. Bert Keats, Stillwater, Oklahoma.

May 12-13, 1983 — Microcomputer Fundamentals: Software and Hardware; Dr. Philip Wolfe, Stillwater, Oklahoma.

ABOUT THE VIRGINIA PRODUCTIVITY CENTER

The VPC is located in the Department of Industrial Engineering and Operations Research at VPI and State University. Founded in 1980 as the Productivity Evaluation Center, it has undergone a recent transformation to the VPC. Dr. D. Scott Sink became the Director of the Center in September 1984 and will provide new direction for the Center. The Center will have a broad productivity management perspective with primary focus on measurement, planning, automation, management of change, traditional industrial engineering, quality control, human factors, operations management, and fundamental management practice. The VPC is entirely self supporting from grants, contracts, and revenues from products and services offered. The goal of the VPC is to become a leader in productivity management theory and technique design and development.

For a current copy of the VPC's Programs, Products, Publications and Services Price List, please check the appropriate box on the subscription form and return.

(Return this Entire Panel in an envelope to subscribe or renew your subscription to Productivity Management or to receive more information about the VPC)

The VPC Productivity Management Newsletter is sent 4 times per year, at the rate of **$25.00 per year** for U.S., Canada, and Mexico ($35.00 for all others). To subscribe, complete the form below and return to VPC, 290 Whittemore Hall, VPI and State University, Blacksburg, VA 24061 (703) 961-4568.

NAME_____
        first   middle initial   last

COMPANY _____

ADDRESS _____

CITY _____ STATE _____ ZIP _____

TELEPHONE _____

METHOD OF PAYMENT: _____ Payment Enclosed

_____ Bill Company

☐ I do not wish to subscribe but do wish to remain on your mailing list for other mailings.

☐ Please send a copy of VPC Programs/Services list.

☐ Please send a copy of brochure describing the VPC in more detail.

PRODUCTIVITY MANAGEMENT

Virginia Productivity Center
Industrial Engineering and
    Operations Research
290 Whittemore Hall
VPI and State University
Blacksburg, VA 24061
(703) 961-4568

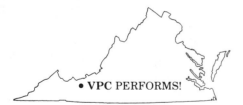

• **VPC** PERFORMS!

APPENDIX B

ONE-DAY MANAGEMENT BRIEFING AGENDA & CONTENTS

About the VPC
Productivity Management Publication
how to get on our distribution list

THREE-DAY SHORT COURSE
PRODUCTIVITY MANAGEMENT:
MEASUREMENT, EVALUATION,
CONTROL, & IMPROVEMENT
STRATEGIES AND TECHNIQUES
AGENDA AND CONTENTS

Book
Chapter

GENERAL INFORMATION

I. PRODUCTIVITY PERSPECTIVES AND BASICS   1

Discussion Paper: "Productivity Management:
  Measurement and Improvement Strategies,"
  Sink, D.S., 1980.

Optional Resource: "A Productivity Film" by the
  American Productivity
  Center (8 minutes)

Productivity Perspectives Exercise

A. Organizational Systems Performance/Measurement Control Basics   2

  Terms
  Control System Components and Design
  Practical Application Exercise

B. Productivity Basics   2

C. Productivity Process Modeling   2

Optional Resource: NBC White Paper, "If Japan Can,
  Why Can't We?" (Optional)

D. Selected Overhead and Slide Copies

IV. **DESIGNING, DEVELOPING, AND EXECUTING SUCCESSFUL**   14
    PRODUCTIVITY MANAGEMENT PROGRAMS

    A. "Building a Program for Productivity Management:
       A Strategy for IE's"

    B. "Designing, Developing, and Implementing Productivity Programs"

    C. "Designing Successful Productivity Management Systems"

    D. "Much Ado about Productivity: Where Do We Go From Here?"

V. **ABOUT THE VIRGINIA PRODUCTIVITY CENTER**

   SHORT-COURSE/SEMINAR EVALUATION

10- OR 15-WEEK GRADUATE CLASS (IEOR 5133) INDEN 5813 PRODUCTIVITY MEASUREMENT AND IMPROVEMENT

Professor: D. Scott Sink, Ph.D, P.E.
Office: 290A, 4568
Office Hours: By appt.
Class Time: 7–9:50 p.m.

Course Outline

WEEK/DATE	TOPIC	ASSIGNMENT	ACTIVITIES	DATE DUE
1 1/11	Organizational meeting Introduction to Productivity	HO's: Class note-book, Section I notebook O.R. list (#'s 8, 35, 72b, 72a, 5, 6, 7, 12, 15, 26)	Set-up Class discussion with slides	
2	Productivity perspectives	Productivity audit, design and development term project	Film: NBC White Paper, "If Japan Can, Why Can't We?"	
3	Productivity perspectives wrap-up	Case study #1	FB #1	3

4	Productivity basics —Definitions —Relationship with effectiveness, efficiency, quality, quality of work life, profitability, and innovation	Section II notebook, Chapter 2 draft (OR List: nos. 1, 10, 11, 14, 36, 41	Productivity process model Film: APC, "A Productivity Film"	5
5	Productivity basics		Productivity process models presentation FB #2	
6	Productivity basics wrap-up	Case study #2		6
7	Productivity measurement —Measurement basics —Multifactor model —Control systems	Section III notebook; Chapter 3 draft (OR List: #'s 1, 2, 4, 7, 10, 11, 13, 31, 33, 36, 41, 44, 60, 72f)	VPC/MFPMM (a) Base period to current period analysis; data sources	8
8	Productivity measurement —Multifactor model		VPC/MFPMM (b) Forecasted period simulation and sensitivity analysis	9
			FB #3&4	8
		SPRING BREAK—HAVE FUN!		
9	Productivity measurement —Multifactor model review; discuss simulation results —Partial-factor approaches	HO: Yale mgmt.— *Guide to Productivity*		

10	Productivity measurement —Partial-factor models —Normative approaches —Consultative approaches —Audits	*R&D Productivity:* OR #1 *Yale Mgmt. Guide*	Nominal group technique exercise	
11	Productivity measurement —Wrap-up —Other approaches critique	HO: Measurement articles and other resources bibliography	Discussion, other measurement approaches FB #5	11
	Productivity strategies	Section IV notebook (OR #'s 1, 3, 5, 6, 9, 16, 17, 18, 19, 20, 21, 22, 23, 27, 30, 34, 35, 36, 37, 38, 39, 40), Case Study #3	Film: NBC White Paper, "America Works When America Works"	12
13	Productivity improvement—selected techniques —Motivation basics —Employee involvement PATs, QCs, etc.	Section V notebook, Case Study #3	FB #6	13
14	Productivity improvement—selected techniques —Reinforcement Theory, incentive systems, gainsharing systems —Scanlon, Rucker, Improshare	Class resource files on each system	Team analysis of various systems	15

15	Productivity improvement—selected techniques		Team presentations, incentive systems	15
	Wrap-up: Designing, Developing, and Implementing successful productivity management programs	"Grand Strategy" article	Productivity audits due	
16	FINALS		FB #7 and 8	16

Note: All assignments are due the week assigned. Fifty points of subjective evaluation comes from the instructor's perception of each student's preparation for the week's classes. No partial credit. Case studies are due the week assigned. Student must be prepared to discuss in class. One-page memo/briefing/analysis is due the last day of class. The incentive system analysis is a team activity requiring a five-page report and an oral briefing. This is a graduate class: The student is responsible and accountable for learning the material. The more you read, the more you will learn.

APPENDIX C

PRODUCTIVITY MANAGEMENT

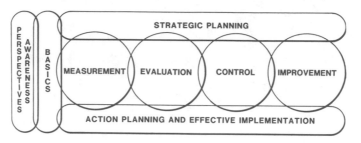

Fall 1983   **Volume II**   **Number Four**   **EDITOR:** Jeff Swaim

TABLE OF CONTENTS

FEATURE ARTICLE

MULTI-CRITERIA PERFORMANCE/ PRODUCTIVITY MEASUREMENT TECHNIQUE

by D. Scott Sink, Ph.D., P.E.
Director, Oklahoma Productivity Center
School of Industrial Engineering & Management
Oklahoma State University
Stillwater, OK 74078

ABSTRACT

An innovative, widely-applicable, and reasonably simple approach to measuring performance or productivity is being developed at the Oregon Productivity Center and at the Oklahoma Productivity Center. The technique can be nicely integrated with the Normative Productivity/Performance Measurement Methodology (Participative approach utilizing the Nominal Group Technique—see issue number 2 of *Productivity Management*) in order to facilitate more effective use of the tremendous number of measures and ratios of productivity or performance that can be attained from groups in organizations. The technique will be briefly presented in this paper.

VPI and State University
Blacksburg, VA 24061

BACKGROUND

Studies beginning at The Ohio State University in 1975 developed a participative, yet highly structured methodology for identifying consensus productivity/ performance measures for a given organizational system. The value of a participative approach lies in the creation of an "ownership" for the resulting measures. Successful implementation of ensuring measurement, evaluation, and control systems is more assured with effective participative approaches.

However, difficulties in operationalizing measurement systems that have origins in a participative process hindered early efforts. The question of how to evaluate performance against a list of measures that is often highly heterogeneous became a critical one to continued development. William Stewart addressed the issue of how to aggregate and hence evaluate performance against many measures or criteria in his dissertation effort at Ohio State. His approach was to develop a prioritized set of productivity/performance measures utilizing the NGT. (This approach evolved from the NSF sponsored Ohio State Studies of 1975-1977). He then developed a "utility" curve for each of the priority (top eight to ten) measures. A ranking and rating process was executed so as to weight the relative importance of each productivity/performance measure. The utility curve was utilized to transform actual performance against each specific measure or criteria to a common 0 to 1.0 performance score. This performance score, 0 to 1.0, was then multiplied by the relative weight for each measure to obtain a performance value. The various performance values for each of the top priority measures are then added together to obtain a productivity/performance index. Stewart based the procedure upon the works of Morris (1975, 1977), The Ohio State Productivity Research Group (1977), and Keeney and Raiffa (1976).

Since those early developments in 1976-1978 at Ohio State, several other efforts have been made in this general area. In 1980, Stewart applied this approach to the common carrier industry. In 1981, William Viana, while a graduate associate in The Oklahoma Productivity Center at Oklahoma State University applied a hybrid design of this procedure in a fairly large, diversified manufacturing firm (gate valves, ball valves, etc.) in Brazil. More recently, Riggs and Felix have developed and published an analogous approach called the "objectives matrix". (1983)

MCPMT PROCEDURE

Assume you have just generated a consensus and prioritized list of productivity/performance measures for a given organizational system utilizing the NGT (See *Productivity Management* issue number 2 and Sink, 1982). You have a list of heterogeneous measures (i.e., apples, oranges, peaches, etc.) You are interested in aggregating or evaluating performance against these criteria in an integrated fashion.

A common performance scale or utility scale needs to be developed that converts all the uncommon measures into some common denominator. The performance scale commonly is rather arbitrarily allowed to range from either 0 to 1.0, 0 to 10.0, or 0 to 100.

> 10.0 ┬ "Excellence"
>
> 5.0 ┼ "Acceptable"
>
> 0 ┴ "Lowest Possible"

Common Performance Scale for each Productivity/ Performance Measure

Level 0 represents the lowest level of performance possible for a given measure. Level 5 represents a minimally acceptable performance level (MAPL). And, level 10 represents the perception of best performance or excellence. Levels 0, 5, and 10 should be clearly defined and accepted benchmarks.

Each productivity/performance measure or criterion has at least one "natural" scale that it can be or is measured with. Often this "natural" scale is simply an industry consensus or norm. For example, measuring liquids in gallons or liters, measuring coal in tons, measuring profitability performance in terms of ROI, etc.

no complaints   100 complaints
                                          per month

Customer Satisfaction

The objective in the MCPMT is to develop a valid set of natural scales used to measure performance against a given criteria and to match levels of performance on that scale to levels of performance on the common utility scale.

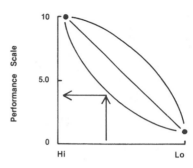

Customer Satisfaction
"Natural" Scale(s) (x-axis)

A utility curve, (such as curves a, b, or c) which can, and often will be, subjective, is developed and used to transform performance on one scale to a common scale. There exist techniques for developing these curves in a valid manner.

The M in the MCPMT stands for multi, which signifies that there are many criteria or measures of productivity/performance against which we are attempting to perform. Therefore, there will often be many of the utility function graphs as depicted above. The question becomes one of how to aggregate performance scores. Since these scores are all from a common scale we might be tempted to just add them up. However, the question of relative impact of performance against various criteria on overall performance becomes critical.

The NGT will have given you a ranked list of productivity/performance measures. The next step is to rate this list. Figure 1 depicts this process. The first step, once the criteria have been ranked, is to arbitrarily assign 100 points to the top priority measures or criteria. Next, the relative importance of the second most important measure is assessed. In the example depicted in Figure 1, which is for a computer center, customer satisfaction is seen as being equally important to projects completed/constant value budget $'s. So, customer satisfaction is also assigned 100 points. This paired comparison relative assignment of points is done for each successive criterion. (i.e., the most important (1st) relative to the second most important (2nd); the 2nd to the 3rd; 6th to 7th, etc.) The total points allocated for all measures or criteria is summed. Relative weights are then determined by dividing the individual points assigned by the total points (i.e., 100/730, . . . 80/730). One then has a sense for the relative importance or contribution of each measure or criterion to overall performance/productivity. There are some critical nuances to this procedure that are not described herein, however the approach is basically as straightforward as it seems. These weights can be determined unilaterally by the manager of the group or by an analyst or participatively by the same persons who identified the criteria and their rankings.

#	Criterion	Rank/Priority	Rating	Weight
1	Reports/Projects completed and accepted — Constant value Budget $	1	100	$\frac{100}{730}=.137$
2	Customer Satisfaction	2	100	$\frac{100}{730}=.137$
3	Quality of Decision Support From Systems Developed	3	100	$\frac{100}{730}=.137$
4	Meeting User Flexibility Requirements	4	90	$\frac{90}{730}=.123$
5	Existence of and use of work scheduling/project management	5	90	$\frac{90}{730}=.123$
6	Projects completed on time — Total projects completed	6	85	$\frac{85}{730}=.116$
7	# of requests for rework/ redoing a project	7	85	$\frac{85}{730}=.116$
8	Existence of and Quality of strategic planning for facilities, equipment, staffing, management processes, and operational systems	8	80	$\frac{80}{730}=.111$
			730	1.000

Figure 1. Ranking and Rating Procedure

The next step in the MCPMT is to integrate the performance (utility) graphs (scales and curves) with the criteria weightings. This will allow the development of one performance/productivity indicator which will indicate the overall performance of the organizational system. Figure 2 conceptually depicts what is happening in this step. Actual performance as measured against the scales represented on the x-axis is transformed into a performance score (0 to 10) on the y-axis. Those performance scores are then multiplied by the criteria weighting factors to obtain weighted scores. Note that these weighted scores all have common units while the x-axis reflects a variety of units. Note also from the computer center example that only one of the eight measures is a pure productivity measure (ratio) and that is criterion #1.

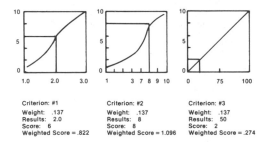

Criterion: #1
Weight: .137
Results: 2.0
Score: 6
Weighted Score = .822

Criterion: #2
Weight: .137
Results: 8
Score: 8
Weighted Score = 1.096

Criterion: #3
Weight: .137
Results: 50
Score: 2
Weighted Score = .274

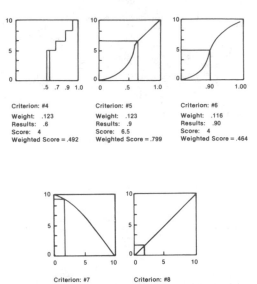

Criterion: #4
Weight: .123
Results: .6
Score: 4
Weighted Score = .492

Criterion: #5
Weight: .123
Results: .9
Score: 6.5
Weighted Score = .799

Criterion: #6
Weight: .116
Results: .90
Score: 4
Weighted Score = .464

Criterion: #7
Weight: .116
Results: 1
Score: 9
Weighted Score = 1.044

Criterion: #8
Weight: .111
Results: 1
Score: 2
Weighted Score = .222

Sum of
Weighted scores = 5.213

Figure 2. Multi-Criteria Performance/Productivity Measurement Technique

The final computational step in this procedure is to add together all the weighted scores. The value for the computer example is 5.213 out of a maximum score of 10.0. The individual performance scores in addition to the total weighted score or overall performance indicator can be tracked over time and utilized to develop evaluation and control systems.

A general matrix format for this technique is presented in Figure 3. (Riggs and Felix, 1983) Note that column eight of the matrix represents the y-axis and columns 1-7; rows 3-13 represent the x-axis of the utility curves. Note also that it is possible to have sub-criteria for a given measure. In other words, there could be more than one way to operationalize a given performance/productivity measure. In that case, the weighting for the given measure or criteria is simply divided up among the sub-criteria. For instance, if customer satisfaction were operationalized with two independent measures rather than one, then the .137 weighting would need to be divided among the two sub-criteria. This is done by repeating the ranking, rating, and weighting procedure within the criteria itself.

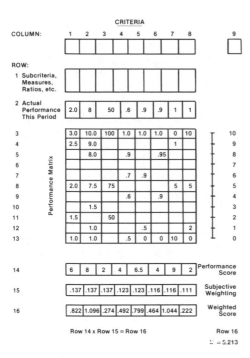

Row 14 x Row 15 = Row 16   Row 16

$\Sigma = 5.213$

Figure 3. Productivity/Performance Measurement Matrix General Format

CONCLUSION

This approach and technique for performance and productivity measurement and evaluation for organizational systems has tremendous potential. The roots of this technique lie in multi-attribute decision analysis which is at least twenty-five years mature. The ideas have been there; American managers and researchers have simply failed to innovate with the basic ideas, theories, concepts and techniques. This is changing. The Japanese have taught us that things don't have to be complex to work.

Current developments with this technique are in the area of validation, methodology refinements and documentation, and creation of a computerized decision support package for the IBM-PC. The IBM-PC software for this technique will be available from our Center in approximately six months. Those interested in learning more about this technique are urged to call our Center and talk to either myself or one of our research associates.

(This paper is excerpted from an upcoming book entitled Productivity Management: Planning, Measurement, Evaluation, Control and Improvement by Dr. Sink to be published by John Wiley in Fall, 1984.)

SPECIAL FEATURE

THE CONSISTENCY OF WORK PERFORMANCE
William W. Rambo and Anna Chomiak

As a general rule, studies dealing with job performance are carried out within a research design that involves only a narrow segment of time. Practical considerations coupled with the need to maintain controlled experimental conditions often limit researchers to data collection periods that are relatively short, and as a consequence they are forced to assume that their results remain valid over time just as long as certain basic conditions associated with the work task and the worker do not change. Under such stable conditions it is generally believed that production rates observed during one interval of time allow one to predict future production levels with reasonably high accuracy. In actual fact, however, we know relatively little about the consistency of work performance, and therefore we have little knowledge about the extent to which we can extend our observations of production behaviors over time.

The research that has been done in this area has tended to place emphasis on the role of motivational factors in sustaining week to week consistency in work performance. It is thought that when functioning under an effective incentive system, i.e., one that generates high mean levels of work motivation, workers will display high consistency in their performance. Under conditions of high motivation, transitory changes in mood, working conditions and fatigue are countered by work incentives that keep output at a steady pace. Presumably, as the effectiveness of a work incentive drops, output rates begin to vary in conjunction with social, emotional and motivational influences that change from day to day.

Consistency of output rates is also thought to reflect the complexity of the work task and the general stability of job conditions. There is some evidence, although it is by no means definitive, indicating that high consistency of work performance is most likely to be found on jobs that involve routine and stable duties. Also, high levels of output consistency are more probable in situations in which job duties require that workers draw on well developed skills that have become established parts of their performance repertoire. Hence, these three factors: incentives, stability of working conditions and the complexity of job duties, have been the main focus of the research that has been reported in the research literature. However, even under optimal conditions that would be expected to give rise to high consistency in week to week output rates, fluctuations in job performance may still take place. Perhaps inconsistency is a fundamental property of human performance, and one should expect to find unpredictable swings in employee performance just as a function of the natural changes in mood, motivation and physiology that occur within the individual as time passes. To the extent that this is true, then it is by no means certain that a distribution of output rates obtained from a work group will correlate closely with rates from the same group two, three or more weeks in the future.

A series of studies have just been completed by William Rambo, Anna Chomiak and James Price that analyzed output consistency of a group of production workers who were observed over a period of three and a half years. Carried out in the garment industry, the workers observed were a group of sewing machine operators and a group of workers involved in the folding and packaging of the completed garments. In both instances the job duties were routine and highly specialized, and all workers had at least a year and a half experience on the job before the study started. All available evidence indicated that throughout the study, job duties and group production rates remained unchanged. Employees worked under a piece rate incentive plan which gave evidence of sustaining high levels of motivation to work. For example, it was not uncommon for an employee to work through rest pauses and to shorten lunch breaks in order to increase earnings. Finally, there was no evidence indicating that restrictive production norms were in force. Rather, if any work norms were operating, they tended to be in the form of competition between employees to achieve high earnings.

The results of these studies indicate that for relatively short time intervals (3-4 months), there appears to be a high correlation between the output rates obtained from these production workers. However, as the time separating two work periods increased, the correlation between them tended to decrease to moderate levels. Figure 1 depicts the general trends observed for both the sewing machine operators and the nonsewing positions. Here the functions represent the mean correlation coefficients obtained between production periods that were separated by different intervals of time; maximum separation being 178 weeks or almost 3.5 years.

Surprisingly, neither function approached zero, as was anticipated at the onset of the study. Rather, both functions appeared to move toward some asymptotic value that represented moderately high levels of output consistency. There were, to be sure, marked differences between the two types of jobs; sewing machine operations, being more complex in terms of their psychomotor demands, displayed less output consistency than did the simpler nonsewing jobs. Nevertheless, for both jobs it appeared that one could do a moderately good job of predicting output between intervals separated by as much as 3.5 years. Subsequent analyses also revealed that most of this consistency reflected the fact that workers tended to maintain the same absolute rates of production for long periods of time. Therefore, for these jobs that involved a narrow range of skills and stable working conditions, it appeared that employees tend to maintain reasonably steady rates of production over the long term. Of course if one were to observe a more complicated set of job duties or jobs for which incentive pay is ineffective or job duties that are subject to change, the consistency of work performance that was observed in the present studies would probably not be found. The present studies examined productivity in situations that were unusual in the sense that working conditions were thought to approach the stable controlled circumstances that one tends to associate with a laboratory experiment. To the extent that this is in fact true, one might consider these results to be a reference point representing the upper levels of production consistency, that may be found when observing work activities over long periods of time. Under less stable working conditions or in instances where financial incentives are not as effective, considerable risk may occur in predicting long term worker performance from current output records. Nevertheless, it is comforting to find that in some employment settings human work performance does maintain relatively high levels of week to week consistency. Without such assurance as that provided by these studies, one would have to question whether the results of behavioral studies have significance much beyond the time during which experimental observations are taken.

FIGURE 1.   General Trends Representing the Regression of r on Time Interval for Nonsewing & Sewing Operators.

POINTS OF INTEREST

MANAGING AS LOVING

When we ponder how to increase the productivity of American industry, we think too much about capital investment and not enough about management, says J. Herbert Hollomon, Japan Steel Industry Professor of Engineering at M.I.T.

"A good manager has three primary functions," writes Dr. Hollomon in AT&T's Business Service Quarterly: "To encourage people to work in a way that allows them to make a substantial personal contribution, to recognize success, and to give new encouragement after failure."

The manager's basic job is to "make it possible for others in the organization to function and to assure them of continuity in their work." In that

TABLE I: Obstacles to Productivity Improvement

	Major Obstacle (%)	Moderate Obstacle (%)	Minor Obstacle (%)
Inability of labor and management to work towards common productivity goals.	47	32	21
Management failing to understand how productivity can be improved.	41	39	20
High interest rates squeezing investment of sufficient capital to improve productivity.	41	38	21
Management failing to authorize sufficient manpower to direct productivity improvement.	40	40	20

In regard to improvement activities, the IEs were asked to indicate which activities had been undertaken at their facilities within the last five years. They were also asked to evaluate the effectiveness (high, moderate, or low) of the activities and whether or not these activities were continuing (Table II).

TABLE II: Productivity Improvement Activities

	Undertaken Activity Yes (%)	Level of Effectiveness High (%)	Mod. (%)	Low (%)	Is Activity Continuing? Yes (%)
Capital investment for new or automated machinery (not including robotics)	81	48	44	8	93
Introduction or improvement of inventory control methods	73	30	52	18	98
Formal employee involvement in productivity improvement planning and evaluation:					
Suggestion programs	63	11	46	43	87
Quality circles	45	14	53	33	85
Improvement of quality of product through worker training	62	23	59	18	94
Evaluating performance and establishing specific PI targets	54	20	56	24	93

Arnold S. Judson is chairman of Gray-Judson, a Boston-based consulting firm. The focus of his consulting work over the past two decades has been in the area of productivity of organizations. His recent study involved 236 top-level executives representing a cross-section of 195 U.S. industrial companies (Judson, 1982). The results of his study are summarized in Table III.

TABLE III: Causes of Success or Disappointment in Improving Productivity

Reasons for Success	Percent Selected	Reasons for Disappointment	Percent Selected
Capital investment in plant, equipment, and process	72	A piecemeal, unplanned approach to improving productivity	66
Top Management commitment and involvement	61	Inadequate coordination among departments or functional areas	42

Good financial controls and information systems	45	Insufficient investment in management and supervisory training and development	41
Good employee relations	38	Lukewarm commitment and involvement by top management	40
Good communications	35	Insufficient awareness by engineering of the manufacturing implications of product and process designs	39
Competent middle managers in all departments	34	Weaknesses in industrial and manufacturing engineering	39

According to Mr. Judson, the "awkward truth about productivity" appears to be that managers have:

1. defined their productivity improvement efforts too narrowly,
2. addressed the symptoms rather than the causes of low productivity,
3. directed efforts toward quick fixes with short time horizons,
4. neglected to base efforts on explicit plans that are consistent with and support overall business strategy, and
5. implemented efforts without the wholehearted involvement or commitment of top management.

Business "experts" may blame industry's poor productivity on government regulations, tax disincentives, the decline of the work ethic, or uncooperative unions. However, managers (as these findings indicate) place the burden of responsibility elsewhere—most often, on themselves.

REFERENCES

"IE's Gauge State of the U.S. Work Ethic, Effectiveness of Productivity Activities," Industrial Engineering, November 1982.

Judson, Arnold S., "The Awkward Truth About Productivity," Harvard Business Review, September-October 1982.

POINTS OF INTEREST

The November 1982 issue of Productivity details an exciting case study of a Skippy Peanut Butter plant in Little Rock, Arkansas. This plant is operating under a highly unusual management system and succeeding. There are no supervisors, job descriptions or inspectors at Skippy Little Rock. Yet, the plant consistently turns out the highest quality products at lowest costs, of any plant that its parent company, Best Foods, operates. The plant does have a general manager, a human resource manager and a quality assurance manager, who serve primarily as advisors to the plant's 100 employees. "These employees literally are their own supervisors." The salaries are based upon the employees knowledge and skill, objectively tested. Employees volunteer for their work assignments, and are trained to perform every job in the plant.

Other interesting aspects of the Skippy Little Rock "phenomenon" are: (1) the lack of "prima donnas" whose egos need to be fed, (2) the career path laid out for each employee when they are hired, (3) the involvement of employees in the hiring process;

REFERENCES OF INTEREST

Sink, D. S., "State-of-the-Art Approaches to the Problem of Unlocking Employee Potential", Industrial Engineering, August 1983.
Managers need to take the time to learn the latest management techniques for motivating improved performance. This article explains how and why the latest management approaches work.

Sink, D. Scott, "Much Ado About Productivity: Where Do We Go From Here?" Industrial Engineering, October 1983.
This article presents a very disciplined view of productivity and focuses on how productivity relates to the broader concept of performance. It examines the various components of performance and discusses the role productivity can play in the measurement, control, and evaluation of organizational systems.

Argote, L., P. S. Goodman, and D. Schkade, "The Human Side of Robotics: How Workers React to a Robot", Sloan Management Review, Spring 1983.
This article examines workers' reactions to the introduction of a robot in a factory and presents a set of strategies for introducing robots in the factory.

Shimada, Haruo, "Japan's Success Story: Looking Behind the Legend", Technology Review, May/June 1983.
This article points out that the ingredients of Japan's economic success are not uniquely Japanese. They derive from principles of labor-management relations that are eminently transferable.

Baird, Lloyd, and Kathy Kram, "Career Dynamics: Managing the Superior/Subordinate Relationship", Organizational Dynamics, Spring 1983.
This article presents a checklist for analyzing how the superior-subordinate relationship operates as an exchange and how the resources of the parties mesh or fail to mesh. The authors show how this relationship can be productive and satisfying both for the parties concerned and for the organization.

Latham, Gary P. and Kenneth N. Wexley, Increasing Productivity Through Performance Appraisal, Addison-Wesley, Reading, Mass., 1981.
This book describes an effective approach to measuring an individual's performance that not only provides a solid basis for promotion and compensation decisions, but stimulates employee productivity as well.

Grove, Andrew S., High Output Management, Random House, New York, 1983.
The President of Intel, one of America's premier high technology companies, shows how managers can increase their productivity dramatically by applying the management techniques perfected at Intel.

Mohr, William L. and Harriet Mohr, Quality Circles: Changing Images of People at Work, Addison-Wesley, Reading, Mass., 1983.

The authors explain in detail how to assess your organization's readiness for a quality circles program. They show how to identify the objectives, problem-solving techniques, problems and pitfalls to sustain the momentum toward fulfillment of your organization's goals.

Naisbitt, John, Megatrends: Ten New Directions Transforming Our Lives, Warner Brooks, 1982.
This content analysis of America in transition describes how the patterns that are now developing at the grass-roots level are, in fact, the broad outlines that will define the society of the future.

Working Smarter, the Editors of Fortune, Viking Press, 1982.
Through examples of successful productivity programs at many major corporations, including IBM, Proctor & Gamble, and AT&T, this book shows how new growth can be achieved "by imparting a sense of teamwork and giving employees more say about how they do their jobs."

OPC ACTIVITIES

OPC RECEIVES EIGHTEEN MONTH CONTRACT TO ASSIST IN THE DEVELOPMENT OF A LABOR/MANAGEMENT PARTICIPATION PROGRAM
by Leva Swim, OPC Associate

The OPC has recently completed contract negotiations with a steel company in the Tulsa area. The OPC, under the direction of Dr. Sink, will be developing and implementing a Performance Action Team (PAT) Process within this unionized organization. The PAT Program design follows a format that allows for much up-front preparation to include several briefings with various levels of plant personnel to discuss the characteristics of the process, some leadership effectiveness training, and indoctrination of top level individuals to the process.

Briefings were held to explain the aspects of the PAT Process and its potential to enhance organizational performance and improve labor/management communication, cooperation, and coordination. These briefings were held first with the Top Management of the organization, and then with the USW Local Executive Board. A follow-up briefing, with both Top Management and Union Executive Board in attendance, was held to provide an opportunity for joint, open discussion of the proposed PAT Program and the perceptions of its ability to succeed or fail within the shared organizational environment. Two additional briefings were conducted by the OPC to inform certain "key" people who will not be participating directly in the PAT Program initially, but who can play a role in the success or failure of the PAT Project, and who will be participating in the Program eventually. This briefing procedure was designed to "tap into" all levels of the organization in an attempt to communicate, in a pervasive manner, what the PAT Program involves (i.e., to get everyone "on-board" to what changes the PAT Program would introduce within the standard operating procedure of the organization). An additional desired outcome of the briefing procedure was to identify potential "pockets of resistance" that might threaten the success of the participative management program.

With "all systems go," the next phase of the program design, leadership effectiveness training, was introduced. Dr. Sink and Dr. Kiser, an Oklahoma State University professor of Sociology, have conducted two sessions of "Situational Leadership," a program designed by Dr. Paul Hersey. The focus of this phase of the participative management program is to sharpen the skills of session participants with respect to the appropriate management styles to use when dealing with subordinates in a situation specific context. Attendees of both Situational Leadership Sessions included key people from management and labor and were selected by the Vice President-Human Resources of the Company. Both of the Situational Leadership Sessions were very well received by the attendees.

The third phase of the up-front preparation has been a joint labor/management strategic planning retreat, held off-site. This strategic planning session involved Top Management and the Union Executive Board in a 2-day process that utilized a structured technique to obtain group consensus on 2 to 5 year goals, objectives, and action plans, strategic to the organization. The session outcomes will be used in the development of a formal strategic plan for the company. An additional outcome of this retreat was the indoctrination of the top organizational members to the procedures inherent to the PAT Process itself (i.e., the structured group consensus seeking technique). The intent of this indoctrination is to foster further commitment to the PAT Program through the development of a broadened understanding of its robust potential.

A final phase of the "up-front" preparation will be the involvement of middle managers in a "shortened" version of the PAT Process that will allow them to experience its techniques, similar to the exposure received by Top Management and the Union Executive Board through the Strategic Planning Retreat. Upon completion of this part of the program, one of the middle management participants will be selected to begin an "actual" PAT, comprised of his immediate workgroup. This PAT will focus upon "ways to improve performance within their department", and will set the stage for the initiation of additional departmental PAT's. Our goal, within the eighteen-month time constraint, is to complete at least two departmental PAT's. Concurrent with this goal is the desired objective of providing the necessary training for the organization to be able to carry on with the PAT Program throughout the plant once the OPC discontinues its direct involvement.

We are extremely excited about this contract and look forward to assisting this company in its quest to improve overall performance, productivity, and quality of work life. We have already made progress in the improvement of labor/management cooperation, coordination, and communication—only 4 grievances have been filed since the initiation of the PAT Program, compared to 100 to 140 within the same time frame of the previous year. Labor and Management appear committed to working together to improve company performance and we in the OPC are happy they have selected us as a third party to help them accomplish this goal.

For more information on our Performance/Productivity Action Team Process, a truly American Employee Involvement Program, please write or call the OPC.

OPC RECEIVES CONTRACT TO DEVELOP PRODUCTIVITY MEASUREMENT TAXONOMY
by Sandy DeVries, OPC Associate

The Oklahoma Productivity Center has recently received a contract from the Air Force to develop a Productivity Measurement Taxonomy. The OPC received an RFP (request for a proposal) based upon its reputation in the field of productivity measurement. The Taxonomy is actually a small component of the Department of Defense's Industrial Modernization Incentives Program (IMIP), a tri-service program designed to encourage productivity improvement investment on the part of Defense Industry subcontractors and vendors. The IMIP program evolved from what are known as the Carlucci Initiatives (23 recommendations and eight issues for decision). These initiatives were developed in 1980 and 1981 and were promulgated throughout the Department of Defense by a DEPSECDEF Memorandum of April 30, 1981. Many of the recommendations focus specifically on productivity issues. In particular, one deals with encouraging capital investment to enhance productivity. The intent is to share measured productivity gains with subcontractors and vendors. The key word is **measured**. It appears that the DOD is moving in the direction of attempting to develop a Military Standard for Productivity Measurement such as the ones for Work Measurement and Quality Control. The OPC's early efforts in this project have focused on data base searches, literature reviews, site visits to organizations with exemplary productivity measurement efforts, and the categorization of theories and techniques. We will keep you updated on this project's development.

PAT PROCESS FEEDBACK

The OPC recently completed a Performance Action Team (PAT) project with the managerial staff of a service-oriented organization on the Oklahoma State University campus (see the summer 1983 issue of Productivity Management.) The following excerpts are from a letter that Dr. Sink received from the organization's director in regard to the PAT process.

- "I first want to thank you very much for the time you and your staff so generously gave to the PAT project; the patience each of you exhibited; and the professional manner in which you approached the task."

- "A key question in the evaluation of any endeavor must be, if I had it to do over again would I do it? The answer is yes."

- "I particularly appreciated your flexibility and willingness to help blend your process with our need for planning and budget preparation. That was particularly beneficial to our managerial staff."

The OPC's involvement with the organization has ceased but the organization is continuing the PAT process on its own and is progressing nicely.

NEWS
FROM OTHER CENTERS
AROUND THE WORLD

CALL FOR MANUSCRIPTS AND REVIEWERS

The *Public Productivity Review* is published quarterly (in March, June, September, and December) by the National Center for Public Productivity. Initiated in 1975, the *PPR* focuses on the need to create a more detailed understanding of public sector productivity. The *Review's* objectives are:

- to encourage a wider application of already proven techniques.

- to draw together citations to the literature.

- to present integrated analyses of productivity theories and concepts, including those that tie measurement to labor relations, capital investment to human capital, efficiency to effective ness, and organizational goals to system goals.

- to establish a persistent, recurrent impetus for productivity improvement at all levels of government and across a range of public sector functions, including police, fire, hospitals, education, and housing.

- to provide a forum for practitioner-academician exchange.

The *Public Productivity Review* welcomes manuscripts from practitioners and academicians alike. Proposals for special issues or symposia are also encouraged. Members of the *Review's* Editorial Board decide on the acceptability of a manuscript within four to six weeks, and publication normally occurs a few months thereafter. Manuscripts are evaluated for substance, methods, salience, utility to *PPR* readers, and their position viz-a-viz related literature.

Among the public sector fields covered by the *Public Productivity Review* are measurement, labor relations, training and development, management, budget and finance, police, fire, sanitation, maintenance, housing, health, welfare, and federal, state, and local jurisdictions. Prospective contributors may order a sample copy of the *Review* for $3.00.

There is no stipulation as to the length of articles submitted to the *Public Productivity Review;* however pieces from fifteen to twenty pages (3000-6000 words) are preferred. Shorter articles (fewer than fifteen pages) on a particular topic may be placed in the *Review's* "Current Research" section. Authors should submit five copies of their article, only one of which can be returned. It is not necessary to submit the original, and it is always advisable for authors to retain a copy themselves. The *Public Productivity Review* can accept no responsibility for unsolicited manuscripts.

Guidelines for Contributors

Articles should be typewritten on standard-size paper, double-spaced, with generous margins. They should be accompanied by a 150-200 word abstract, and a short biographical sketch of the author or authors.

Footnotes should be prepared separately at the end of the manuscript. Footnotes should follow the short-title style after the initial full entry. The latin abbreviations *loc. cit., op. cit.,* and *idem.,* are not used; *ibid.,* however, is retained. Sample footnotes consisting of an original entry with its short title are listed below:

Book	Dennis Pirages, and Paul R. Ehrlich, *Ark II: Social Response to Environmental Imperatives* (New York: Viking, 1974), 65.
	Pirages and Ehrlich, *Ark II,* 78.
Periodical	James D. Thompson, "Society's Frontiers for Organizing Activities," *Public Administration Review,* XXXIII (July/August 1973), 327-45.
	Thompson, "Society's Frontiers for Organizing Activities," 336.
Dissertation	Kenneth N. Fortier, "Implementation of a Computer-Aided Dispatch System for the San Diego Police Department" (MA thesis, San Diego State University, 1976), 20.
	Fortier, "Implementation of a Computer-Aided Dispatch System," 45.

The *PPR* is especially interested in review essays of three or more related titles. A list of titles available for review will be sent upon request. Reviews of other titles are also encouraged. Prospective book reviewers are encouraged to submit their vitae for consideration.

All inquiries and manuscripts should be addressed to—

Prof. Marc Holzer
Public Productivity Review
445 West 59 Street
New York, New York 10019

DIRECTORY OF MAJOR PRODUCTIVITY AND QUALITY OF WORKING LIFE CENTERS

The National Productivity Network (NPN) is a consortium of nongovernmental, nonprofit organizations that devote their efforts to productivity and quality of work life research, development, and dissemination/extension. The NPN has been in existence since 1978, but there is a lot of active work going on now to publicize and promote the importance of the network. The organizations within the NPN work on a direct, one-to-one basis to help business and government clients identify and implement management and technical systems that can improve productivity and quality of work life. In this issue, we present a list of major U.S. centers and productivity related publications.

Major U.S. Productivity Centers

Dr. Keith McKee
Manufacturing Productivity Center
Illinois Institute of Technology
10 W. 35th Street
Chicago, IL 60616
(312) 567-4808

Dr. Thomas Tuttle
Maryland Center for Productivity
    and Quality of Working Life
University of Maryland
College Park, MD 20742
(301) 454-6688

Mr. David Clifton
Georgia Productivity Center
Georgia Tech. Eng. Exp. Station
Atlanta, GA 30332
(404) 894-3404

Dr. James Riggs
Oregon Productivity Center
100 Merryfield Hall
Oregon State University
Corvallis, OR 94331
(503) 754-3249

Dr. William Smith/Tom Stephenson
Productivity Research
    and Extension Program
P.O. Box 5192
North Carolina State University
Raleigh, NC 27650
(919) 733-2370

Mr. Marc Holzer
National Center for Public Productivity
John Jay College
445 W. 59th Street
New York, NY 10019
(212) 489-5030

Dr. Barry Macy
Texas Center for Productivity
    and Quality of Working Life
College of Business
Texas Tech University
Lubbock, TX 79409
(806) 742-2154

Dr. Gary Hansen
Utah State Center for Productivity
    and Quality of Working Life
Utah State University
Logan, UT 84322
(801) 750-1033

Dr. LeRoy Marlow, Director
PENNTAP
Penn State University
501 J. Orris Keller Pldg.
University Park, PA 16802
(814) 865-0427

Dr. Scott Sink
Virginia Productivity Center
Dept. of Industrial Engineering
    and Operations Research
Virginia Polytechnic Institute
    and State University
Blacksburg, VA 24061

Mr. John Hermann
Productivity Center of the Southwest
1015 Gayley Avenue, Suite 1000
Los Angeles, CA 90024
(213) 208-3596

Dr. Carl Estes
Oklahoma Productivity Center
School of Industrial Engineering
    and Management
322 Engineering North
Oklahoma State University
Stillwater, OK 74078
(405) 624-6055

Mr. Carl Thor
American Productivity Center
123 North Post Oak Lane
Houston, TX 77024
(713) 681-4020

Dr. Jerome M. Rosow
Work in America Institute
700 White Plains Road
Scarsdale, NY 10583
(212) 823-5144

Productivity Related Publications

MANUFACTURING PRODUCTIVITY FRONTIERS:
This monthly journal is published by the Manufacturing Productivity Center at the Illinois Institute of
Technology at 10 West 35th St., Chicago, IL 60616.
Annual subscriptions are $100. This is probably the
best of the monthlies as it contains about 50 commercial free pages of up-to-date information on
topics such as: construction, government, manufacturing, book reviews, seminar schedules, etc.
(contact: Keith McKee, 312-567-4808).

THE MARYLAND WORKPLACE is the bi-monthly
newsletter of the Maryland Center for Productivity
and Quality of Working Life, University of Maryland,
College Park, MD 20742. It is a short newsletter
focusing on labor, government and management
news on productivity issues. (contact: Tom Tuttle,
301-454-6688).

NATIONAL PRODUCTIVITY REPORT is published
twice monthly and is four pages long. It is a provoca-
tive newsletter, typically covering one topic in some
depth. $48 per year. Information and samples are
available from NPR, 1110 Greenwood Rd., Wheaton,
IL 60187. (contact: Bill Schleicher, 312-668-5146).

PRODUCTIVITY IN ACTION is a monthly newsletter
published by the Georgia Productivity Center,
Georgia Tech Eng. Experiment Station, Atlanta, GA
30332. A very informative, well done newsletter.
(contact: Rudy Yobs, 404-894-3404).

PRODUCTIVITY is one of the newest monthlies and
the most expensive at $126. It is 12 pages long and
describes itself as "management's tool for improving productivity through worker satisfaction and innovation." Information and samples from: Productivity Inc., P.O. Box 3831, Stamford, CT 06905. (contact: Norman Bodek, 203-322-8388).

PRIMER is a four page newsletter of the Oregon Productivity Center, 100 Merryfield Hall, Oregon State University, Corvallis, OR 97331. It is a very well done and very informative newsletter. (contact: Jim Riggs, 503-754-3249).

THE PRODUCTIVITY LETTER is a bi-monthly publication of the American Productivity Center, 123 North Post Oak Lane, Houston, TX 77024. The objectives are to develop a useful and comprehensive newsletter and to expand national awareness of the need for productivity improvement. Available to Founders, Sponsors and Members as well as to other interested parties. (contact: Carole Roper, 713-681-4020).

INDUSTRIAL ENGINEERING is the monthly periodical of THE INSTITUTE OF INDUSTRIAL ENGINEERS. It is impossible to talk about productivity and not include *INDUSTRIAL ENGINEERING. IIE* has provided state of the art insights into productivity management for well over 32 years. *INDUSTRIAL ENGINEERING* is recognized internationally as being a premier source of information about productivity management.

NATIONAL PRODUCTIVITY REVIEW: THE JOURNAL OF PRODUCTIVITY MANAGEMENT is a journal in the style of *Harvard Business Review, California Management Review,* and *Sloan Management Review.* It is a quarterly journal focusing specifically on productivity-related issues such as: effective utilization of human resources, effective utilization of capital, effective uses of materials and energy resources, and organizational technical systems and processes that relate to productivity. The intended audience is executives and managers as well as consultants, faculty, and students. For subscription information, write to: 33 W. 60th St., N.Y., N.Y. 10023. Annual subscriptions are $96. This journal may well emerge as being a significant source of quality information on the topic of productivity.

PREP NEWS is the newsletter for the Productivity Research and Extension program at North Carolina State University. It is an informative newsletter focusing upon communicating activities at NC State and in North Carolina in general. The newsletter is very well done. (contact: Bill Smith, 919-733-2370).

ABOUT THE VIRGINIA PRODUCTIVITY CENTER

The VPC is located in the Department of Industrial Engineering and Operations Research at VPI and State University. Founded in 1980 as the Productivity Evaluation Center, it has undergone a recent transformation to the VPC. Dr. D. Scott Sink became the Director of the Center in September 1984 and will provide new direction for the Center. The Center will have a broad productivity management perspective with primary focus on measurement, planning, automation, management of change, traditional industrial engineering, quality control, human factors, operations management, and fundamental management practice. The VPC is entirely self supporting from grants, contracts, and revenues from products and services offered. The goal of the VPC is to become a leader in productivity management theory and technique design and development.

For a current copy of the VPC's Programs, Products, Publications and Services Price List, please check the appropriate box on the subscription form and return.

(Return this Entire Panel in an envelope to subscribe or renew your subscription to Productivity Management or to receive more information about the VPC)

The VPC Productivity Management Newsletter is sent 4 times per year, at the rate of **$25.00 per year** for U.S., Canada, and Mexico ($35.00 for all others). To subscribe, complete the form below and return to VPC, 290 Whittemore Hall, VPI and State University, Blacksburg, VA 24061 (703) 961-4568.

NAME_____
        first   middle initial   last

COMPANY _____

ADDRESS_____

CITY _____ STATE _____ ZIP _____

TELEPHONE _____

METHOD OF PAYMENT: _____ Payment Enclosed

_____ Bill Company

☐ I do not wish to subscribe but do wish to remain on your mailing list for other mailings.

☐ Please send a copy of VPC Programs/Services list.

☐ Please send a copy of brochure describing the VPC in more detail.

PRODUCTIVITY MANAGEMENT

Virginia Productivity Center
Industrial Engineering and
   Operations Research
290 Whittemore Hall
VPI and State University
Blacksburg, VA 24061
(703) 961-4568

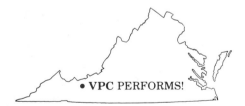

• **VPC** PERFORMS!

INDEX

STRATEGY

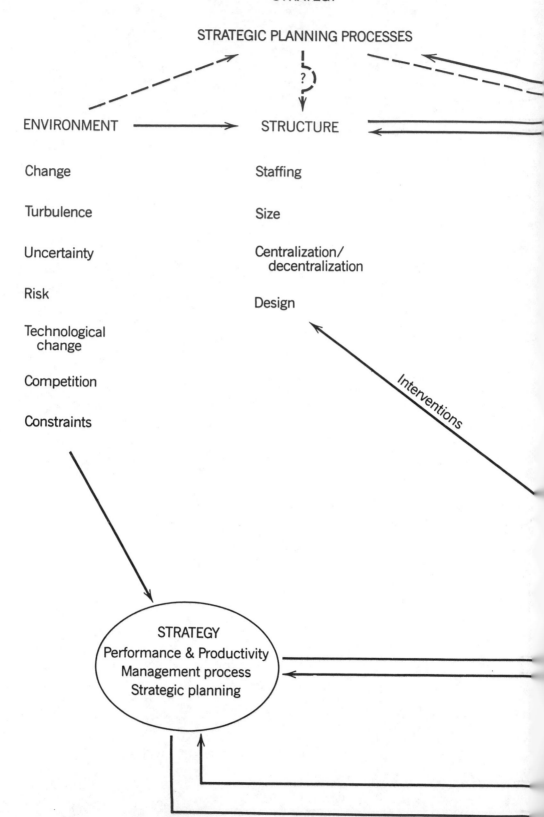

STRATEGIC PLANNING PROCESSES

ENVIRONMENT → STRUCTURE

Change

Turbulence

Uncertainty

Risk

Technological change

Competition

Constraints

Staffing

Size

Centralization/ decentralization

Design

Interventions

STRATEGY
Performance & Productivity
Management process
Strategic planning

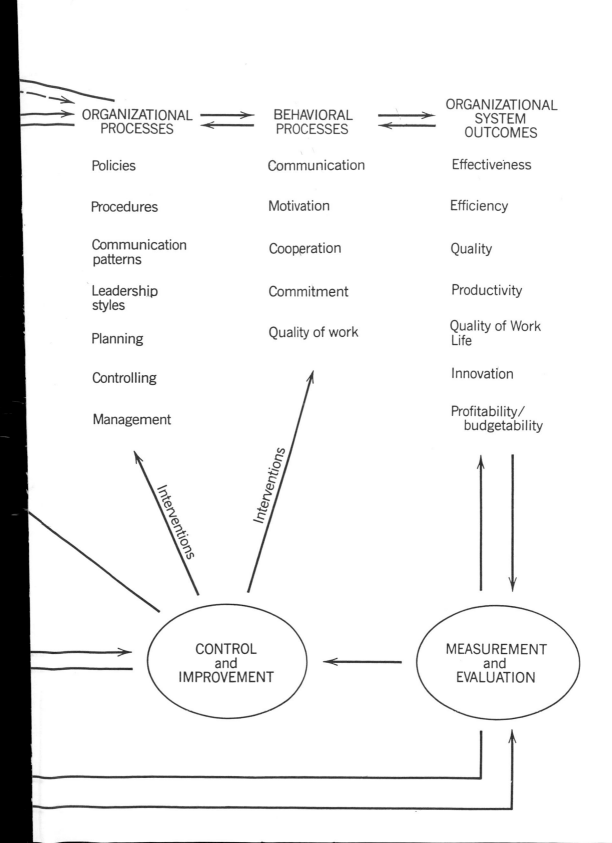

DATE DUE